Use Cases of Discrete Event Simulation

Steffen Bangsow (Ed.)

Use Cases of Discrete Event Simulation

Appliance and Research

Springer

Editor
Steffen Bangsow
Freiligrathstraße 23
Zwickau
Germany

ISBN 978-3-642-28776-3 e-ISBN 978-3-642-28777-0
DOI 10.1007/978-3-642-28777-0
Springer Heidelberg New York Dordrecht London

Library of Congress Control Number: 2012934760

Printed on acid-free paper

Springer is part of Springer Science+Business Media (www.springer.com)

Preface

Over the last decades discrete event simulation has conquered many different application areas. This trend is, on the one hand, driven by an ever wider use of this technology in different fields of science and on the other hand by an incredibly creative use of available software programs through dedicated experts.

This book contains articles from scientists and experts from 10 countries. They illuminate the width of application of this technology and the quality of problems solved using simulation. Practical applications of discrete event simulation dominate in the present book.

The practical application of discrete event simulation is always tied to software products and development environments. The increase in software quality and increased mastery in handling the software allows modeling increasingly complex tasks. This is also impressively reflected in the use cases introduced here.

This project began with an inquiry by Mr. Hloska (thanks for the impetus) and a following discussion in a number of web forums. The response was just amazing. Within a short time, interested parties had signed up to fill at least two books.

This was followed by a period of despair. A large portion of the potential authors had to withdraw their offer of cooperation. The largest part of the discrete event simulation projects is subject of confidentiality and the majority of companies are afraid to loose their competitive advantage when reporting on simulation projects. This makes a real exchange of experience among simulation experts extraordinarily difficult apart from the software manufacturers sales presentations.

I would like to thank all authors who contributed to this book.

I also want to especially thank those authors who have agreed to contribute an article, but did not receive approval for publication from their superiors.

Steffen Bangsow

Contents

1 Investigating the Effectiveness of Variance Reduction Techniques in Manufacturing, Call Center and Cross-Docking Discrete Event Simulation Models

Adrian Adewunmi* and Uwe Aickelin**

Variance reduction techniques have been shown by others in the past to be a useful tool to reduce variance in Simulation studies. However, their application and success in the past has been mainly domain specific, with relatively little guidelines as to their general applicability, in particular for novices in this area. To facilitate their use, this study aims to investigate the robustness of individual techniques across a set of scenarios from different domains. Experimental results show that Control Variates is the only technique which achieves a reduction in variance across all domains. Furthermore, applied individually, Antithetic Variates and Control Variates perform particularly well in the Cross-docking scenarios, which was previously unknown.

1.1 Introduction

There are several analytic methods within the field of operational research; simulation is more recognized in contrast to others such as mathematical modeling and game theory. In simulation, an analyst creates a model of a real - life system that describes some process involving individual units such as persons or products. The constituents of such a model attempt to reproduce, with some varying degree of accuracy, the actual operations of the real workings of the process under consideration. It is likely that such a real - life system will have time - varying inputs

Adrian Adewunmi · Uwe Aickelin
Intelligent Modelling & Analysis Research Group (IMA)
School of Computer Science
The University of Nottingham
Jubilee Campus
Wollaton Road
Nottingham NG8 1BB
UK
e-mail: adrian.a.adewunmi@googlemail.com,
uwe.aickelin@nottingham.ac.uk
* Corresponding author.
** Co – author.

and time - varying outputs which may be influenced by random events (Law 2007). For all random events it is important to represent the distribution of randomness accurately within input data of the simulation model. Since random samples from input probability distributions are used to model random events in simulation model through time, basic simulation output data are also characterized by randomness (Banks et al. 2000). Such randomness is known to affect the degree of accuracy of results derived from simulation output data analysis. Consequently, there is a need to reduce the variance associated within simulation output value, using the same or less simulation effort, in order to improve a desired precision (Lavenberg and Welch 1978).

There are various alternatives for dealing with the problem of improving the accuracy of simulation experimental results. It is possible to increase the number of replications as a solution approach, but the required number of replications to achieve a desired precision is unknown in advance (Hoad et al. 2009) , (Adewunmi et al. 2008). Another solution is to exploit the source of the inherent randomness which characterizes simulation models in order to achieve the goal of improved simulation results. This can be done through the use of variance reduction techniques.

> “A variance reduction technique is a statistical technique for improving the precision of a simulation out-put performance measure without using more simulation, or, alternatively achieve a desired precision with less simulation effort" (Kleijnen 1974).

It is know that the use of variance reduction techniques has potential benefits. However, the class of systems within which it is guaranteed to succeed and the particular technique that can achieve desirable magnitudes of variance reduction is ongoing research. In addition, applicability and success in the application of variance reduction techniques has been domain specific, without guidelines on their general use.

> “Variance reduction techniques cannot guarantee variance reduction in each simulation application, and even when it has been known to work, knowledge on the class of systems which it is provable to always work has remained rather limited" (Law and Kelton 2000).

The aim of this chapter is to answer the research question; which individual application of variance reduction techniques will succeed is achieving a reduction in variance for the different discrete event simulation scenarios under consideration. The scope of this chapter covers the use of variance reduction techniques as individual techniques on a set of scenarios from different application domains. The individual variance reduction techniques are:

i. Antithetic Variates
ii. Control Variates and
iii. Common Random Numbers.

In addition, the following three real world application domains are under consideration: (i) Manufacturing System (ii) Distribution System and (iii) Call Centre System. The rest of the book chapter is laid out as follows; the next section gives a background into the various concepts that underpin this study. This is followed by a case study section which describes the variance reduction techniques experimentation according to application domain. Further on is a discussion on the results from experimentation.

1.2 Reduction of Variance in Discrete Event Simulation

The development of simulation models requires a specific knowledge that is usually acquired over time and through experience. Since most simulation output results are essentially random variables, it may be difficult to determine whether an observation is as a result of system interrelationships or the randomness inherent in simulation models. Furthermore, simulation as a process can consume a lot of time, despite advances in computer technology. An example of a time consuming task is one which is statistically based i.e. output data analysis. However, it is known that advances in computer simulation have allowed the modeling of more complicated systems. Moreover, even when simpler systems are simulated, it can be difficult to judge the precision of simulation results. In general, output analysis is the examination of data generated by simulation experimentation, and its purpose is to predict the performance of a system or to compare the performance of two or more alternative system design (Law 2007).

However, simulation models differ from one another insofar as they have different values or types of system parameters, input variables, and behavioral relationships. These varying parameters, variables, and relationships are called "factors" and the output performance measure is called "response" in statistical design terminology (April et al. 2003). The decision as to which parameters are selected as fixed aspects of the simulation model and which are selected as experimental factors depends on the goals of the study rather than on the inherent form of the model. Also, during simulation studies there are usually a wide range of different responses or performance measure, which can be of interest. As a result, output performance measures for the three different simulation models considered within this study have been carefully selected after considering literature which reports on the most common performance metric for judging the performance of each simulation model (i.e. Manufacturing simulation, Call Centre simulation, and Cross-docking simulation). In addition, selection of output performance measures have been carried out in order to achieve a research goal of reducing simulation output variance through manual experimentation (Adewunmi 2010).

For simulation models, where the performance of such models is measured by its precision, i.e. mean, standard deviation, confidence interval and half width, for the selected output performance measure, it is sometimes difficult to achieve a target precision at an acceptable computational cost because of variance. This variance is usually that which is associated with the performance measure under consideration. For example, (Adewunmi et al. 2008), investigated the use of the Sequential Sampling Method (Law and Kelton 2000) to achieve a target variance reduction for a selected simulation output performance measure. Results from experimentation indicate that this technique for reducing variance requires a huge number of simulation runs to achieve any success for this particular simulation model. In a wider context, the variance associated with a simulation or its output performance measure may be due to the inherent randomness of the complex system under study. This variance can make it difficult to get precise estimates on the actual performance of the system. Consequently, there is a need to reduce the variance associated with the simulation output value, using the same or less simulation runs, in order to achieve a desired precision (Wilson 1984). The scope of this investigation covers the use of individual variance reduction techniques on different simulation models. This will be carried out under the assumption that all the simulation models for this study are not identical. The main difference between these models is the assumed level of inherent randomness. Where such randomness has been introduced by the following:

a. The use of probability distributions for modeling entity attributes such as inter arrival rate and machine failure. Conversely, within other models, some entity attributes have been modeled using schedules. The assumption is; the use of schedules does not generate as much randomness as with the use of probability distribution.
b. In addition, to the structural configuration of the simulation models under consideration i.e. the use of manual operatives, automated dispensing machines or a combination of both manual operatives and automated dispensing machines.

As a result, the manufacturing simulation model is characterized by an inter arrival rate and processing time which are modeled using probability distribution, the call centre simulation model's inter arrival rate and processing time are based on fixed schedules. The cross-docking simulation model is also characterized by the use of probability distribution to model the inter arrival rate and processing time of entities. The theoretical assumption is that by setting up these simulation models in this manner, there will be a variation in the level of model randomness. This should demonstrate the efficiency of the selected variance reduction techniques in achieving a reduction of variance for different simulation models, which are characterized by varying levels of randomness. In addition, as this is not a full scale simulation study, but a means of collecting output data for the variance reduction experiments, this investigation will not be following all the steps in a typical simulation study (Law 2007).

1.2.1 Variance Reduction Techniques

Within this section, the discussion has been restricted to a selected subset of variance reduction techniques which have proven to be the most practical in use within the discrete event simulation domain (Lavenberg and Welch 1978), (Cheng 1986). Furthermore, these techniques have been chosen because of the manner each one performs variance reduction i.e. through random number manipulation or the use of prior knowledge. The three selected variance reduction techniques fall into two broad categories; the first class manipulates random numbers for each replication of the simulation experiment, thereby inducing either a positive or a negative correlation between the mean responses across replications. Two methods of this category of variance reduction techniques are presented. The first method, Common Random Numbers, only applies when comparing two or more systems. The second method, using Antithetic Variates, applies when estimating the response of a variable of interest (Cole et al. 2001).

The second class of variance reduction techniques incorporates a modeler's prior knowledge of the system when estimating the mean response, which can result in a possible reduction in variance. By incorporating prior knowledge about a system into the estimation of the mean, the modeler's aim is to improve the reliability of the estimate. For this technique, it is assumed that there is some prior statistical knowledge of the system. A method that falls into this category is Control Variates (Nelson and Staum 2006). The following literature with extensive bibliographies is recommended to readers interested in going further into the subject i.e. (Nelson 1987), (Kleijnen 1988) and (Law 2007). In next section is a discussion on the three variance reduction techniques that appear to have the most promise of successful application to discrete event simulation modeling is presented.

1.2.1.1 Common Random Numbers (CRN)

Usually the use of CRN only applies when comparing two or more alternative scenarios of a single systems, it is probably the most commonly used variance reduction technique. Its popularity originates from its simplicity of implementation and general intuitive appeal. The technique of CRN is based on the premise that when two or more alternative systems are compared, it should be done under similar conditions (Bratley et al. 1986). The objective is to attribute any observed differences in performance measures to differences in the alternative systems, not to random fluctuations in the underlying experimental conditions. Statistical analysis based on common random numbers is founded on this single premise. Although a correlation is being introducing between paired responses, the difference, across pairs of replications is independent. This independence is achieved by employing a different starting seed for each of the pairs of replications. Unfortunately, there is no way to evaluate the increase or decrease in variance resulting from the use of CRN, other than to repeat the simulation runs without the use of the technique (Law and Kelton 2000).

There are specific instances where the use of CRN has been guaranteed. Gal et.al. present some theoretical and practical aspects of this technique, and discuss its efficiency as applied to production planning and inventory problems (Gal et al. 1984). In addition, Glasserman and Yao state that

> "common random numbers is known to be effective for many kinds of models, but its use is considered optimal for only a limited number of model classes".

They conclude that the application of CRN on discrete event simulation models is guaranteed to yield a variance reduction (Glasserman and Yao 1992). To demonstrate the concept of CRN, let X_a denote the response for alternative A and X_b denote the response for alternative B, while considering a single system. Let D, denote the difference between the two alternatives, i.e. $D = X_a - X_b$. The following equation gives the random variable $D's$ variance.

$$Var(D) = Var(X_a X_b) + Var(X_a) - 2Cov(X_a, X_b) \tag{1.1}$$

1.2.1.2 Antithetic Variates (AV)

In comparison to CRN, the AV technique reduces variance by artificially inducing a correlation between replications of the simulation model. Unlike CRN, the AV technique applies when seeking to improve the performance of a single system's performance. This approach to variance reduction makes *n* independent pairs of correlated replications, where the paired replications are for the same system. The idea is to create each pair of replications such that a less than expected observation in the first replication is offset by a greater than expected observation in the second, and vice versa (Andreasson 1972), (Fishman and Huang 1983). Assuming that this value is closer to the expected response than the value that would result from the same number of completed independent replications, the average of the two observations is taken and the result used to derive the confidence interval.

A similar feature that AV shares with CRN is it can also be difficult to ascertain that it will work, and its feasibility and efficacy are perhaps even more model dependent than CRN. Another similarity it shares with CRN is the need for a pilot study to assess its usefulness in reducing variance for each specific simulation model (Cheng 1981). In some situations, the use of AV has been known to yield variance reduction, and as mentioned earlier it can be model specific. In his paper, Mitchell considers the use of AV to reduce the variance of estimates obtained in the simulation of a queuing system. The results reported in this paper, show that a reduction in variance of estimates was achieved (Mitchell 1973). The idea of AV is more formally presented. Let random variable *X*, denote the response from the first replication and X' denote the replication from the second replication, within a pair. The random variable *Y* denotes the average of these two variables, i.e. $Y = (X + X')/2$. The expected value of *Y* and the variance of *Y* are given as follows:

$$E(Y) = \frac{[E(X) + E(X')]}{2} = E(X) = E(X') \tag{1.2}$$

and

$$Var(Y) = \frac{[Var(X) + Var(X') + 2Cov(X, X')]}{4} \tag{1.3}$$

1.2.1.3 Control Variates (CV)

This technique is based on the use of secondary variables, called CV. This technique involves incorporating prior knowledge about a specific output performance parameter within a simulation model. It does not however require advance knowledge about a parameters theoretical relationship within the model as would other variance reduction techniques such as Indirect Estimation (IE). As compared with CRN and AV, CV attempts to exploit the advantage of the correlation between certain input and output variables to obtain a variance reduction. Of course depending on the specific type of CV that is being applied, the required correlation may arise naturally during the course of a simulation experiment, or might arise by using CRN in an auxiliary simulation experiment (Law 2007).

In order to apply the CV technique, it has to be assumed that a theoretical relationship exists between the control variate *X*, and the variable of interest *Y*. This approach does not require that a modeler knows the exact mathematical relationship between the control variates and the variable of interest; all the knowledge needed is to only know that the values are related. This relationship can be estimated by using the data recorded for instance from a pilot simulation study. Information from the estimated relationship is used to adjust the observed values of *Y* (Sadowski et al. 1995). Let *X* be the random variable that is said to partially control the random variable *Y*, and hence, it is called a control variate for *Y*. Usually it is assumed that there is a linear relationship between the variable of interest and the control variate. The observed values of the variable of interest *Y* can then be corrected, by using the observed values of the control variates *X*, as follows:

$$Y_i(n) = Y(n) - a(X(n) - E(X)(n)) \tag{1.4}$$

And

$$a = \frac{Cov(Y(n), X(n))}{Var(X)} \tag{1.5}$$

Where a is the amount by which an upward or downward adjustment of the variable of interest Y is carried out, $E(X)$ is the mean of X, and n is the number of replications.

There are, however, some classes of discrete event simulation models for which the application of control variates has proven to be successful. In a recent article on the use of variance reduction techniques for manufacturing simulation by Eraslan and Dengiz, CV and Stratified Sampling were applied for the purpose of improving selected performance measures, results from this paper suggest that CV yields the lowest variance for selected performance measures (Eraslan and Dengiz 2009). The main advantage of using CV as a technique for variance reduction is that they are relatively easy to use. More importantly, CV can essentially be generated anywhere within the simulation run, so they add basically nothing to the simulation's cost; thus they will prove worthwhile even if they do not reduce the variance greatly (Kelton et al. 2007).

1.3 Case Studies

This section proceeds to present 3 case studies:

- The application of individual variance reduction techniques in a manufacturing system,
- The application of individual variance reduction techniques in a call centre system,
- The application of individual variance reduction techniques in a cross-docking distribution centre.

1.3.1 Manufacturing System

1.3.1.1 Description of a Manufacturing System / Simulation Model

Typically, the simulation of manufacturing systems is performed using a commercial software, rather than through a purpose built application. The manufacturing simulation model has been developed using the Arena^{TM} simulation software. It is common that one of the activities during a simulation study is the statistical analysis of output performance measures. Since random samples from input probability distributions are used to model events in a manufacturing simulation model through time, basic simulation output data (e.g., average times in system of parts) or an estimated performance measure computed from them (e.g., average time in system from the entire simulation run) are also characterized by randomness (Buzacott and Yao 1986). Another source of manufacturing simulation model randomness which deserves a mention is unscheduled random downtime and machine failure which is also modeled using probability distributions. It is known that inherent model randomness can distort a true and fair view of the simulation model output results. Consequently, it is important to model system randomness correctly and also design and analyze simulation experiments in a proper manner (Law 2007).

There are a number of ways of modeling random unscheduled downtimes, interested readers are directed to Chapter 13, section 3, Discrete Event System Simulation, Banks et.al. (Banks et al. 2000). The purpose of using variance reduction techniques is to deal with the inherent randomness in the manufacturing simulation model. This is through the reduction of variance associated with any selected measure of model performance. This reduction will be gained using the same number of replications that was used to achieve the initial simulation results. Improved simulation output results obtained from the application of variance reduction techniques has been known to increase the credibility of the simulation model.

An investigation into the application of variance reduction techniques on a small manufacturing simulation model is herein presented. The simulation model under consideration has been adapted from chapter 7, Simulation with Arena, Kelton et.al. (Kelton et al 2007), purely for research purposes. Experimentation is based on the assumption that the output performance measures are of a terminating, multi scenario, single system discrete event simulation model. The simple manufacturing system consists of parts arrival, four manufacturing cells, and parts departure. The system produces three part types, each routed through a different process plan in the system. This means that the parts do not visit individual Cells randomly, but through a predefined routing sequence. Parts enter the manufacturing system from the left hand side, and move only in a clockwise direction, through the system. There are four manufacturing cells; Cells 1, 2, and 4 each have a single machine, however, Cell 3 has two machines. The two machines at Cell 3 are not identical in performance capability; one of these machines is newer than the other and can perform 20% more efficiently than the other. Machine failure in Cells 1, 2, 3, and 4 in the manufacturing simulation model was represented using an exponential distribution with mean times in hours. Exponential distribution is a popular choice when modeling such activities in the absence of real data. A layout of the small manufacturing system under consideration is displayed in figure 1.1.

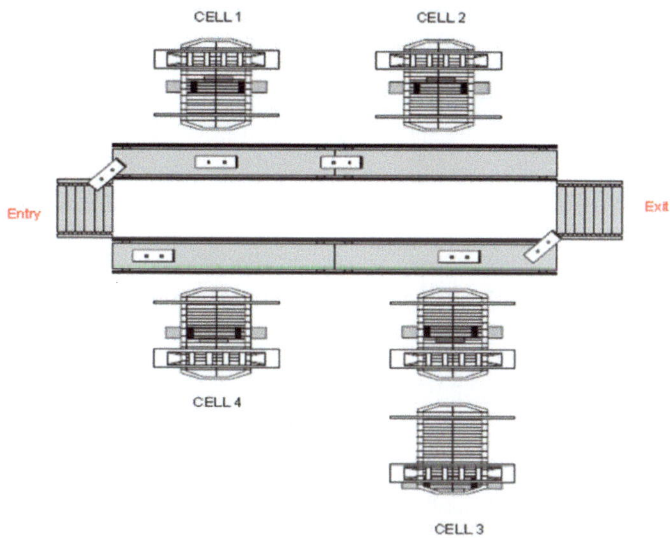

Fig. 1.1 Small Manufacturing System Layout adapted from (Kelton et al. 2007) Chapter 7.

Herein is a description of the simulation model under consideration. All process times are triangularly distributed, while the inter arrival times between successive part arrivals are exponentially distributed. These are the probability distributions which were already implemented in the simulation model, and there was no reason not to continue using them. The Arena ™ simulation model incorporates an animation feature that captures the flow of parts to and fro the cells, until they are finally disposed or exist out of the system. The inter arrival times between successive parts arrival are exponentially distributed with a mean of 13 minutes, while the first part arrives at time 0.

Here is a brief description of the Arena™ control logic which underlines the animation feature. Parts arrival are generated in the create parts module. The next step is the association of a routing sequence to arriving parts. This sequence will determine the servicing route of the parts to the various machine cells. Once a part arrives at a manufacturing cell (at a station), the arriving part will queue for a machine, and is then processed by a machine. This sequence is repeated at each of the manufacturing cells the part has to be processed. The process module for Cell 3 is slightly different from the other three Cells. This is to accommodate the two different machines, a new machine and an old machine, which process parts at different rates. Figure 1.2 shows the animation equivalent and control logic of the small manufacturing system simulation model.

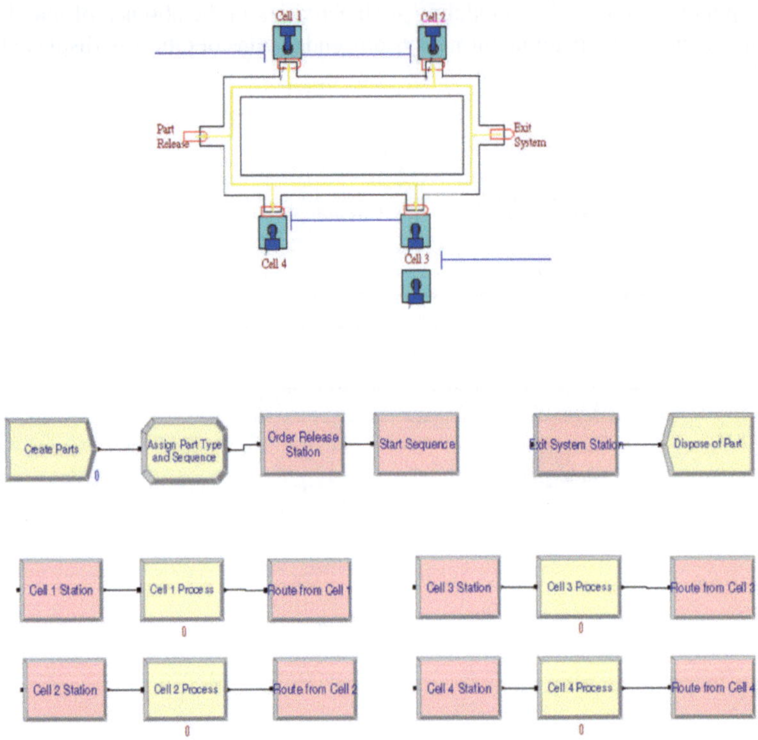

Fig. 1.2 Manufacturing system simulation animation and control logic adapted from (Kelton et al. 2007) Chapter 7

1.3.1.2 Variance Reduction Experiments

This section of the chapter is divided into two parts; the first describes the design of the variance reduction experiments and the second details the results of the application of individual variance reduction techniques.

1.3.1.2.1 Experimental Design

In designing the variance reduction experiment, data on time persistent performance measures was utilized for experimentation as opposed to both time and cost data. This is due mainly to the availability of time based data as opposed to cost based data during the performance of the case study. Although both types of data would have given a greater insight into the performance of the variance reduction techniques, using different classes of time based data should be sufficient for this level of experimentation. Here is a list of the three performance measures utilized:

- Entity Total Average Time (Base): This is the average of the total time each entity will travel over the total length of the conveyor through the manufacturing system.
- Resource Utilization (Base): This variable records the instantaneous utilization of a resource during a specific period.
- Average Total WIP (Base): This metric records the average quantity of total work in process for each entity type.

The experimental conditions are as follows:

- Number of Replications: 10
- Warm up Period: 0
- Replication Length: 30 Days
- Terminating Condition: None

The performance measures have been labeled (Base), to highlight their distinction from those that have had variance reduction techniques applied and those that have not. As this is a pilot study where the goal is to establish the effectiveness of the variance reduction techniques under consideration, in this instance 10 simulation replications is deemed sufficient for collecting enough data for this purpose. An extensive bibliography on an appropriate number of replications for simulation experimentation and such like issues can be found in Robinson (Robinson 1994) and Hoad et.al (Hoad et al. 2009).In addition, for a full discussion on design issues such as warm up, replication length and simulation model termination condition for this study, readers are encouraged to see (Adewunmi 2010).

In addition, performance measures have been classed according to variance reduction techniques, i.e. Average Total WIP (Base), Average Total WIP (CRN), and Average Total WIP (AV). This means for each performance measure, the appropriate variance reduction that has been applied to it is stated, i.e. CRN and that which has not been treated to a variance reduction technique is labeled (Base). Under consideration is a two scenario, single manufacturing discrete event simulation model. The scenario which has performance measures labeled (Base) is characterized by random number seeds dedicated to sources of simulation model randomness as selected by the simulation software Arena ™. The other scenario which has performance measures labeled common random number (CRN) has its identified sources of randomness, allocated dedicated random seeds by the user. So these two scenarios have unsynchronized and synchronized use of random numbers respectively (Law and Kelton 2000).

At this stage of experimental design, an additional performance measure Entity Wait Time is being introduced. This performance measure will be used for the CV experiment, with a view to applying it to adjusting upward or downwards the performance measure Entity Total Average Time (Base). Initial simulation results show a linear relationship between both variables, which will be exploited for variance reduction.

Here is the hypothesis that aim's to answer the research question:

- There is no difference in the standard deviations of the performance measure.

The hypothesis that tests the true standard deviation of the first scenario μ_1 against the true standard deviation of the second scenario μ_2,... scenario μ_k is:

$$H_0 : \mu_1 = \mu_2 = \ldots = \mu_k \tag{1.6}$$

Or

$$H_1 : \mu_i \neq \mu_k \; for\ at\ least\ one\ pair\ of (i,k) \tag{1.7}$$

Assuming we have samples of size n_i from the $i-th$ population, $i = 1, 2, \ldots, k$, and the usual standard deviation estimates from each sample:

$$\mu_1, \mu_2 = \ldots = \mu_k \tag{1.8}$$

Test Statistic: Bartlett's Test

The Bartlett's Test (Snedecor and Cochran 1989) has been selected as a test for equality of variance between samples, as it is assumed that our data is normally distributed. Furthermore, this is one of the most common statistical techniques for this purpose. However, an alternative test like the Levene's test (Levene 1960) could have been used. In this instance, it will not be appropriate because Levene's test is less sensitive than the Bartlett test to departures from normality.

Significance Level: A value of $\alpha = 0.05$

Next is a summary of results from the application of individual variance reduction techniques on a manufacturing simulation model.

1.3.1.2.2 Results Summary

In this section, a summary of results on the performance of each variance reduction technique on each output performance measure is presented. In addition, a more in-depth description of results from the application of individual variance reduction techniques is presented in (Adewunmi 2010).

- At a 95% confidence interval (CI), homogeneity of variance was assessed by Bartlett's test. The P-value (0.000) is less than the significance level (0.05), therefore "reject the null hypothesis". The difference in variance between Average Total WIP (Base, CRN, AV, and CV) is "statistically significant". On the basis of the performance of the variance reduction techniques, CV technique achieved the largest reduction in variance for the simulation output performance measure, **Average Total WIP**.

- At a 95% confidence interval (CI), homogeneity of variance was assessed by Bartlett's test. The P-value (0.003) is less than the significance level (0.05), therefore "reject the null hypothesis". The difference in variance between Entity Total Average Time (Base, CRN, AV, and CV) is "statistically significant". On the basis of the performance of the variance reduction techniques, AV technique achieved the largest reduction in variance for the simulation output performance measure, **Entity Total Average Time**.
- At a 95% confidence interval (CI), homogeneity of variance was assessed by Bartlett's test. The P-value (0.006) is less than the significance level (0.05), therefore "reject the null hypothesis". The difference in variance between Resource Utilization (Base, CRN, AV, and CV) is "statistically significant". On the basis of the performance of the variance reduction techniques, CRN technique achieved the largest reduction in variance for the simulation output performance measure, **Resource Utilization**.

1.3.2 Call Centre System

1.3.2.1 Description of a Call Centre System / Simulation Model

With the progression towards skill based routing of inbound customer calls due to advances in technology, Erlanger calculations for call centre performance analysis has become outdated since it assumes that agents have a single skill and there is no call priority (Doomun and Jungum 2008). On the other hand, the application of simulation ensures the modeling of human agent skills and abilities, best staffing decisions and provides an analyst with a virtual call centre that can be continually refined to answer questions about operational issues and even long term strategic decisions (L'Ecuyer and Buist 2006).

A close examination of a typical call centre reveals a complex interaction between several "resources" and "entities". Entities can take the form of customers calling into the call centre and resources are the human agents that receive calls and provide some service. These incoming calls, usually classified by call types, then find their way through the call centre according to a routing plan designed to handle specific incoming call type. While passing through the call centre, incoming calls occupy trunk lines, wait in one or several queues, abandon queues, and are redirected through interactive voice response systems until they reach their destination, the human agent. Otherwise, calls are passed from the interactive voice response system to an automatic call distributor (Doomun and Jungum 2008).

An automatic call distributor is a specialized switch designed to route each call to an individual human agent; if no qualified agent is available, then the call is placed in a queue. See figure 1.3 for an illustration of the sequence of activities in typical call centre, which has just been described in this section. Since each human agent possesses a unique skill in handling incoming calls, it is the customers' request that will determine whether the agent handles the call or transfers it to another agent. Once the call is handled, it then leaves the call centre system. During all of these call handling transactions, one critical resource being consumed is time. For example time spent handling a call and the time a call spends in the system. These are important metrics to consider during the evaluation of the performance of a call centre.

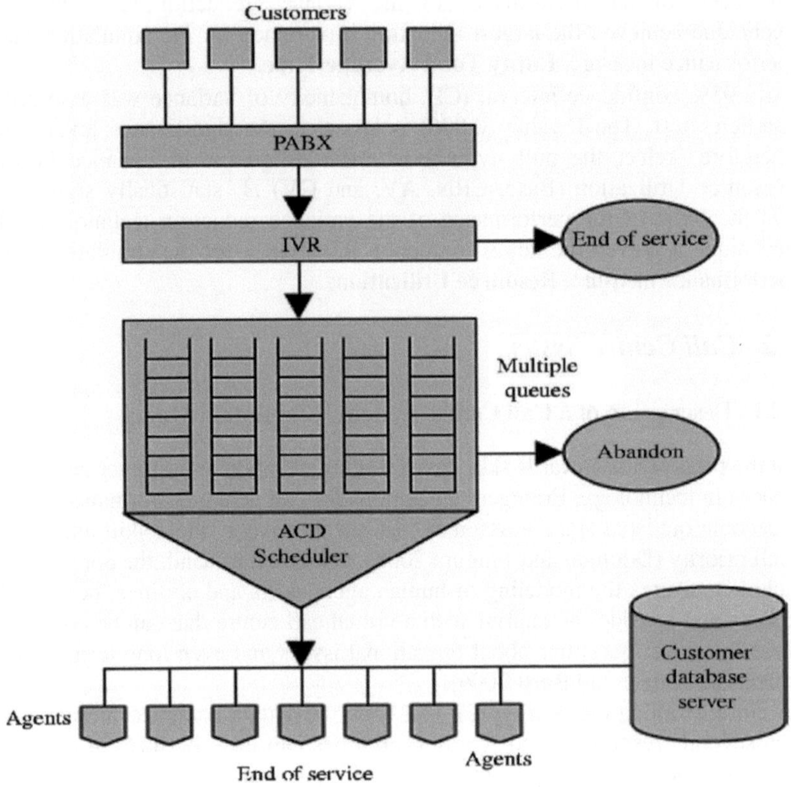

Fig. 1.3 A Simple Call Centre adapted from (Doomun and Jungum 2008).

Herein is a description of the simulation model under consideration. The simple call centre system under consideration has been adapted from the Chapter 5, Simulation with Arena, (Kelton et al 2007). This call centre system, although theoretical in nature, contains the essential working components of a typical real life call centre, i.e. technical support, sales and customer order status checking. Arrival of incoming calls is generated using an arrival schedule. The purpose for using an arrival schedule instead of modeling this event using a probability distribution and a mean in minutes is to cause the system to stop creating new arrivals at a designated time into the simulation experiment. An answered caller has three options: transfer to technical support, sales information, or order status inquiry.

The estimated time for this activity is uniformly distributed; all times are in minutes.

In simulation terms, the "entities" for this simple call centre model are product type 1, 2 and 3. The available "resources" are the 26 trunk lines which are of a fixed capacity, and the sales and technical support staff. The skill of the sales and technical staff is modeled using schedules which show the duration during which for a fixed period, a resource is available, its capacity and skill level. The simulation model records the number of customer calls that are not able to get a trunk line and are thus rejected from entering the system similar to balking in queuing system. However, it does not consider "reneging", where customers who get a trunk line initially, later hang up the phone before being served. Figure 1.4, shows an Arena ™ simulation animation of the simple call centre simulation model.

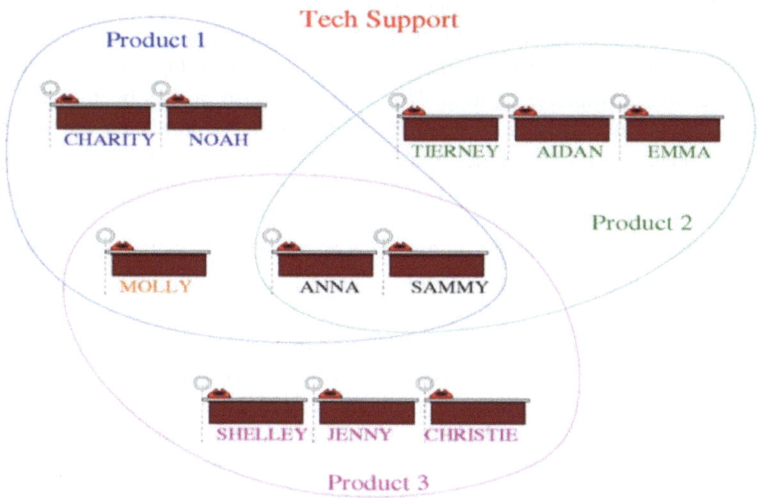

Fig. 1.4 Call Centre Simulation Animation adapted from (Kelton et al. 2007) Chapter 5

1.3.2.2 Variance Reduction Experiments

This section of the chapter is divided into two parts; the first describes the design of the variance reduction experiments and the second details the results of the application of individual variance reduction techniques.

Experimental Design

For the design of the call centre variance reduction experiments, the three output performance measures which have been chosen are both time and cost persistent in nature. Here is a list of these performance measures:

- Total Average Call Time (Base): This output performance measure records the total average time an incoming call spends in the call centre simulation system.

- Total Resource Utilization (Base): This metric records the total scheduled usage of human resources in the operation of the call centre over a specified period in time.
- Total Resource Cost (Base): This is the total cost incurred for using a resource i.e. a human agent.

The experimental conditions are as follows:

- Number of Replications: 10
- Warm up Period: 0
- Replication Length: 660 minutes (27.5 days)
- Terminating Condition: At the end of 660 minutes and no queuing incoming

The call centre simulation model is based on the assumption that there are no entities at the start of each day of operation and the system will have emptied itself of entities at the end of the daily cycle. For the purpose of variance reduction experimentation, it is a terminating simulation model, although a call centre is naturally a non terminating system. No period of warm up has been added to the experimental set up. This is because experimentation is purely on the basis of a pilot run and the main simulation experiment, when it is performed, will handle issues like initial bias and its effect on the performance of variance reduction techniques. The performance measures have been labeled (Base), to highlight their distinction between those that have had variance reduction techniques applied and those that have not. These experiments assume that the sampled data is normally distributed.

In addition, the performance measures have been classed according to variance reduction techniques, i.e. Total Average Call Time (Base), Total Average Call Time (CRN), and Total Average Call Time (AV).Under consideration as in the previous manufacturing simulation study is a two scenario, single call centre simulation model. The scenario which has performance measures labeled (Base) is characterized by random number seeds dedicated to sources of simulation model randomness as selected by the simulation software Arena ™. The other scenario which has performance measures labeled CRN has its identified sources of randomness, allocated dedicated random seeds by the user. So these two scenarios have unsynchronized and synchronized use of random numbers (Law and Kelton 2000).

The research question hypothesis remains the same as that in the manufacturing system; however an additional performance measure Total Entity Wait Time is introduced at this stage. This performance measure will be used for the CV experiment, with a view to adjusting the variance value of the performance measure Total Average Call Time (Base).

Results Summary

In this section, a summary of results on the performance of each variance reduction technique on each output performance measure is presented. In addition, a more in-depth description of results from the application of individual variance reduction techniques is presented in (Adewunmi 2010).

- At a 95% confidence interval (CI), homogeneity of variance was assessed by Bartlett's test. The P-value (0.000) is less than the significance level (0.05), therefore "reject the null hypothesis". The difference in variance between Total Aver-age Call Time (Base, CRN, AV, and CV) is "statistically significant". On the basis of the performance of the variance reduction techniques, CV technique achieved the largest reduction in variance for the simulation output performance measure, **Total Average Call Time**.
- At a 95% confidence interval (CI), homogeneity of variance was assessed by Bartlett's test. The P-value (0.995) is greater than the significance level (0.05), therefore "do not reject the null hypothesis". The difference in variance between Total Resource Utilization (Base, CRN, AV, and CV) is "statistically insignificant". On the basis of the performance of the variance reduction techniques, there was no reduction in variance for the simulation output performance measure, **Total Resource Utilisation**.
- At a 95% confidence interval (CI), homogeneity of variance was assessed by Bartlett's test. The P-value (0.002) is less than the significance level (0.05), therefore "reject the null hypothesis". The difference in variance between Total Re-course Cost (Base, CRN, AV, and CV) is "statistically significant". On the basis of the performance of the variance reduction techniques, AV technique achieved the largest reduction in variance for the simulation output performance measure, **Total Resource Cost**.

1.3.3 Cross-Docking System

1.3.3.1 Description of Cross-Docking System / Simulation Model

Many systems in areas such as manufacturing, warehousing and distribution can sometimes be too complex to model analytically; in particular, Just in Time (JIT) warehousing systems such as cross-docking can present such difficulty (Buzacott and Yao 1986). This is because cross-docking distribution systems operate processes which exhibit an inherent random behavior which can potentially affect its overall expected performance. A suitable technique for modeling and analyzing complex systems such as cross-docking systems is discrete event simulation (Magableh et al. 2005). Normally, such a facility would consist of a break up area where inbound freight is received and sorted as well as a build up area which handles the task of picking customer orders for onward dispatch via out bound dock doors. The usual activities of the cross-docking distribution centre begin with the receipt of customer orders, batched by outbound destinations, at specified periods during the day. As customer orders are being received, inbound freight arranged as pallet load is being delivered through inbound doors designated according to destination.

Customer orders batched by destination can differ in volume and variety; also they are released into the order picking system at the discretion of an operator in order to even out the work load on the order picking system. Once pallet load is sorted by a floor operative i.e. during the break up process, individual items in packs of six to twelve units can be placed in totes (A plastic container which is used for holding items on the conveyor belt). Normally, totes will begin their journey on a conveyor belt, for onward routing to the order picking area. Just before the order picking area is a set of roof high shelves where stock for replenishing the order picking area is kept. A conveyor belt runs through the order picking area and its route and speed are fixed. Figure 1.5, below provides a representation of the cross-docking distribution centre.

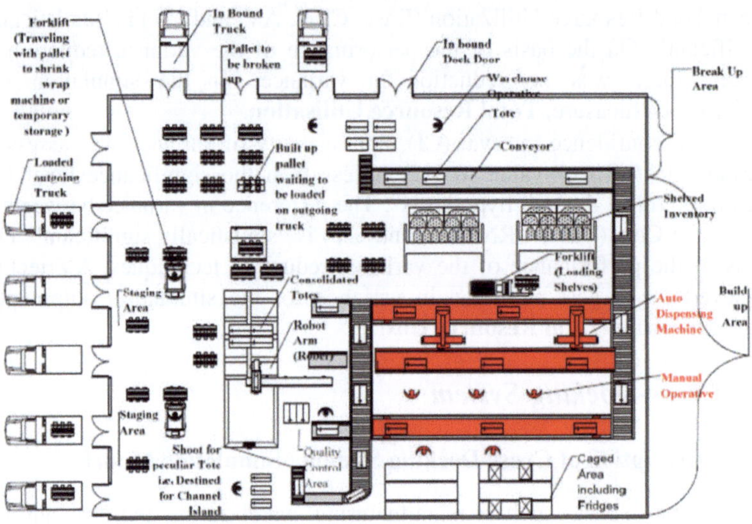

Fig. 1.5 A Typical Cross-docking Distribution Centre (Adewunmi 2010).

Within the order picking area, there are two types of order picking methods; automated dispensing machines and manual order picking operatives. These order picking resources are usually available in shifts, constrained by capacity and scheduled into order picking jobs. There is also the possibility that manual order picking operators possess different skill levels and there is a potential for automated order picking machines to breakdown. In such a situation, it becomes important for the achievement of a smooth cross-docking operation, to pay particular attention to the order picking process within the cross-docking distribution system. The order picking process essentially needs to be fulfilled with minimal interruptions and with the least amount of resource cost (Lin and Lu 1999). Below figure 1.6 provides a representation of the order picking function with a cross-docking distribution centre.

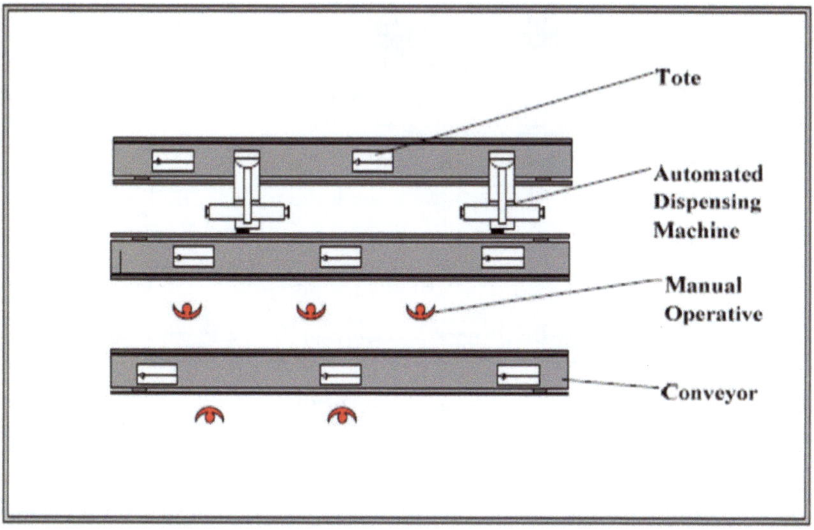

Fig. 1.6 An Order Picking Process within a Cross-docking Distribution Centre (Adewunmi 2010)

A description of the order picking simulation model, which will be the scope of the cross-docking simulation study is presented. The scope of this particular study is restricted to the order picking function as a result of an initial investigation conducted at a physical cross-docking distribution centre. It was discovered that amongst the different activities performed in a distribution centre, the order picking function was judged as the most significant by management. The customer order (entity) inter arrival rate is modeled using an exponential probability distribution, and the manual as well as the automated order picking process are modeled using triangular probability distribution. Customer orders are released from the left hand side of the simulation model. At the top of the model are two automated dispensing machines and at the bottom of the simulation model are two sets of manual order picking operatives, with different levels of proficiency in picking customer orders. Figure 1.7, displays a simulation animation of the order picking process cross-docking distribution centre.

1.3.3.2 Variance Reduction Experiments

This section of the chapter is divided into two parts; the first describes the design of the variance reduction experiments and the second details the results of the application of individual variance reduction techniques.

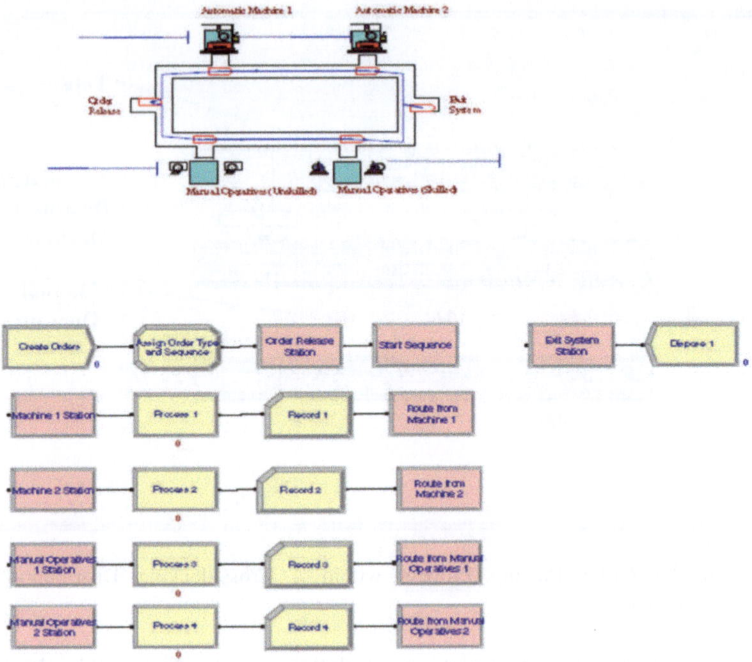

Fig. 1.7 Simulation animation of a Cross-docking order picking process (Adewunmi 2010).

Experimental Design

For the design of the cross-docking distribution system variance reduction experiments, the following performance measures were chosen:

- Total Entity Time (Base): This variable records the total time an entity spends in the simulation system.
- Total Resource Utilization (Base): The purpose of collecting data on resource utilization is to have statistics on the level of usage of the resources during a specified period.
- Total Resource Cost (Base): This is a cost based statistic that records the monetary amount expended on the use of re-sources for a specific period.

The experimental conditions are as follows:

- Number of Replications: 10
- Warm up Period: 0
- Replication Length: 30 Days
- Terminating Condition: None

The performance measures have been classed according to variance reduction technique, i.e. Total Resource Utilization (Base), Total Resource Utilization

(CRN), and Total Resource Utilization (AV). Under consideration is a two scenario, single cross-docking discrete event simulation model. The scenario which has performance measures labeled (Base) is characterized by random number seeds dedicated to sources of simulation model randomness as selected by the simulation software Arena ™. The other scenario which has performance measures labeled CRN has its identified sources of randomness, allocated dedicated random seeds by the user. So these two scenarios have unsynchronized and synchronized use of random numbers (Law and Kelton 2000).

The research question hypothesis remains the same as that in the manufacturing system; however an additional performance measure Total Entity Wait Time is introduced at this stage. This performance measure will be used for the CV experiment, with a view to applying it to adjusting the performance measure Total Entity Time. For those interested, detailed results from the application of individual variance reduction techniques are presented in (Adewunmi 2010).

Results Summary

In this section, a summary of results on the performance of each variance reduction technique on each output performance measure is presented. In addition, a more in-depth description of results from the application of individual variance reduction techniques is presented in (Adewunmi 2010).

- At a 95% confidence interval (CI), homogeneity of variance was assessed by Bartlett's test. The P-value (0.000) is less than the significance level (0.05), therefore "reject the null hypothesis". The difference in variance between Total Entity Time (Base, CRN, AV, and CV) is "statistically significant". On the basis of the performance of the variance reduction techniques, CV technique achieved the largest reduction in variance for the simulation output performance measure, **Total Entity Time**.
- At a 95% confidence interval (CI), homogeneity of variance was assessed by Bartlett's test. The P-value (0.000) is less than the significance level (0.05), therefore "reject the null hypothesis". The difference in variance between Total Re-source Cost (Base, CRN, AV, and CV) is "statistically significant". On the basis of the performance of the variance reduction techniques, AV technique achieved the largest reduction in variance for the simulation output performance measure, **Total Resource Cost**.
- At a 95% confidence interval (CI), homogeneity of variance was assessed by Bartlett's test. The P-value (0.003) is less than the significance level (0.05), therefore "reject the null hypothesis". The difference in variance between Total Resource Utilization (Base, CRN, AV, and CV) is "statistically significant". On the basis of the performance of the variance reduction techniques, AV technique achieved the largest reduction in variance for the simulation output performance measure, **Total Resource Utilization**.

1.4 Discussion

The purpose of this study is to investigate the application of variance reduction techniques (CRN, AV and CV) on scenarios from three different application domains. In addition, to finding out which class of systems the variance reduction techniques will prove to most likely succeed. It also seeks to provide general guidance to beginners on the universal applicability of variance reduction techniques. A review of results from the variance reduction experiments indicate that the amount of variance reduction by the techniques applied can vary substantially from one output performance measure to the other, as well as one simulation model to the other. Among the individual techniques, CV stands out as the best technique. This is followed by AV and CRN. CV was the only technique that achieved a reduction in variance for at least one performance measure of interest, in all three application domains. This can be attributable to the fact that the strength of this technique is its ability to generate a reduction in variance by inducing a correlation between random variates. In addition, control variates have the added advantage of being able to be used on more than one variate, resulting in a greater potential for variance reduction. However, implementing AV and CRN required less time, and was less complex than CV for all three domain application domains. This maybe because with CV, where there is a need to establish some theoretical relationship between the control variate and the variable of interest.

The variance reduction experiments were designed with the manufacturing simulation model being characterized by an inter arrival rate and processing time which were modeled using probability distribution. The cross-docking simulation model was also characterized by the use of probability distribution to model the inter arrival rate and processing time of entities. Conversely, the call centre simulation model inter arrival rate and processing time were based on fixed schedules. The assumption is that by setting up these simulation models in this manner, there will be a variation in the level of model randomness i.e. the use of schedules does not generate as much model randomness as with the use of probability distribution. For example, results demonstrate that for the call centre simulation model, the performance measure "Total Resource Utilization" did not achieve a reduction in variance with the application of CRN, AV and CV, on this occasion. However, for this same model, the performance measures "Total Average Call Time" and "Total Resource Cost" did achieve a reduction in variance. This expected outcome demonstrates the relationship between the inherent simulation model's randomness and the efficiency of CRN, AV and CV, which has to be considered when applying variance reduction techniques in simulation models.

This study has shown that the Glasserman and Yao (Glasserman and Yao 1992) statement regarding the general applicability of CRN is true, for the scenarios and application domains under consideration. As a consequence, this makes CRN a more popular choice of technique in theory. However, results from this study demonstrate CRN to be useful but not the most effective technique for reducing variance. In addition CV under the experimental conditions reported within this study did outperform CRN. While it is not claimed that CV is more superior a technique as compared with CRN, in this instance, it has been demonstrated that

CV achieved more instances of variance reduction as compared with CRN and AV. In addition, under current experimental conditions, a new specific class of systems, in particular the Cross-docking distribution system has been identified, for which the application of CV and AV is beneficial for variance reduction.

1.5 Conclusion

Usually during a simulation study, there are a variety of decisions to be made at the pre and post experimentation stages. Such decisions include input analysis, design of experiments and output analysis. Our interest is in output analysis with particular focus on the selection of variance reduction techniques as well as their applicability. The process of selection was investigated through the application of CRN, AV and CV in a variety of scenarios. In addition, this study seeks to establish which of the application domains considered, will the application of CRN, AV and CV be successful, where such success had not been previously reported. Amongst the individual variance reduction techniques (CRN, AV and CV), CV was found to be most effective for all the application domains considered within this study. Furthermore, AV and CV, individually, were effective in variance reduction for the cross-docking simulation model. Typically, a lot of consideration is given to number of replications, replication length, terminating condition, warm up period during the design of a typical simulation experiment. It would be logical to imagine that there will be a linear relationship between these factors and the performance of variance reduction techniques. However, the extent of this relationship is unknown unless a full simulation study is performed before the application of variance reduction techniques. The experimental conditions applied to this study were sufficient to demonstrate reduction. However, upcoming research will investigate the nature and effect of considering the application of variance reduction techniques during the design of experiments for full scale simulation study.

In future, research investigation will be focused on exploring the idea of combining different variance reduction techniques, with the hope that their individual beneficial effort will add up to a greater magnitude of variance reduction for the estimator of interest. These combinations could have a positive effect when several alternative configurations are being considered. To obtain more variance reduction, one may want to combine variance reduction techniques simultaneously in the same simulation experiment and use more complicated discrete event simulation models. The potential gain which may accrue from the combination of these techniques is also worth investigating because it will increase the already existing knowledge base on such a subject.

Authors Biography, Contact

Dr Adrian Adewunmi was a Post Graduate Researcher in the Intelligent Modelling & Analysis (IMA) Research Group, School of Computer Science, University of Nottingham. A summary of his current interest is Modeling and Simulation, Artificial Intelligence and Data Analysis.´

Professor Uwe Aickelin is an EPSRC Advanced Research Fellow and Professor of Computer Science at The University of Nottingham. He is also the Director of Research in the School of Computer Science and leads one of its four research groups: Intelligent Modeling & Analysis (IMA). A summary of his current research interests is Modeling and Simulation, Artificial Intelligence and Data Analysis.

Contact

adrian.a.adewunmi@googlemail.com
uwe.aickelin@nottingham.ac.uk
Intelligent Modelling & Analysis Research Group (IMA)
School of Computer Science
The University of Nottingham
Jubilee Campus
Wollaton Road
Nottingham NG8 1BB
UK

Bibliography

Adewunmi, A.: Selection of Simulation Variance reduction techniques through a Fuzzy Expert System. PhD Thesis, University of Nottingham (2010)

Adewunmi, A., Aickelin, U., Byrne, M.: An investigation of sequential sampling method for crossdocking simulation output variance reduction. In: Proceedings of the 2008 Operational Research Society 4th Simulation Workshop (SW 2008), Birmingham (2008)

Andradottir, S., Heyman, D.P., Ott, T.J.: Variance reduction through smoothing and control variates for markov chain simulations. ACM Transactions on Modeling and Computer Simulation 3(3), 167–189 (1993)

Andreasson, I.J.: Antithetic methods in queueing simulations. Technical Report, Royal Institute of Technology, Stockholm (1972)

April, J., Glover, F., Kelly, J.P., Laguna, M.: Simulation-Based optimisation: practical introduction to simulation optimisation. In: WSC 2003: Proceedings of the 35th Conference on Winter Simulation, New Orleans, Louisiana (2003)

Avramidis, A.N., Bauer Jr., K.W., Wilson, J.R.: Simulation of stochastic activity networks using path control variates. Journal of Naval Research 38, 183–201 (1991)

Banks, J., Carson II, J.S., Nelson, B.L., Nicol, D.M.: Discrete Event System Simulation, 3rd edn. Prentice - Hall, New Jersey (2000)

Bratley, P., Fox, B.L., Schrage, L.E.: A guide to simulation, 2nd edn. Springer, New York (1986)

Burt, J.M., Gaver, D.P., Perlas, M.: Simple stochastic networks: Some problems and procedures. Naval Research Logistics Quarterly 17, 439–459 (1970)

Buzacott, J.A., Yao, D.D.: Flexible manufacturing systems: A review of analytical models. Management Science 32(7), 890–905 (1986)

Cheng, R.C.H.: The use of antithetic control variates in computer simulations. In: WSC 1981: Proceedings of the 13th Conference on Winter Simulation. IEEE, Atlanta (1981)

Cheng, R.C.H.: Variance reduction methods. In: WSC 1986: Proceedings of the 18th Conference on Winter simulation. ACM, Washington D.C. (1986)

Cole, G.P., Johnson, A.W., Miller, J.O.: Feasibility study of variance reduction in the logistics composite model. In: WSC 2007: Proceedings of the 39th Conference on Winter Simulation. IEEE Press, Washington D.C. (2001)

Doomun, R., Jungum, N.V.: Business process modelling, simulation and reengineering: call centres. Business Process Management Journal 14(6), 838–848 (2008)

Eraslan, E., Dengiz, B.: The efficiency of variance reduction in manufacturing and service systems: The comparison of the control variates stratified sampling. Mathematical Problems in Engineering, 12 (2009)

Fishman, G.S., Huang, B.D.: Antithetic variates revisited. Communications of the ACM 26(11), 964–971 (1983)

Gal, S., Rubinstein, Y., Ziv, A.: On the optimality and efficiency of common random numbers. Mathematics and Computers in Simulation 26, 502–512 (1984)

Glasserman, P., Yao, D.D.: Some guidelines and guarantees for common random numbers. Management Science 38(6), 884–908 (1992)

Gordon, G.: System Simulation, 2nd edn. Prentice - Hill, New Jersey (1978)

Hoad, K., Robinson, S., Davies, R.: Automating discrete event simulation output analysis automatic estimation of number of replications,warm-up period and run length. In: Lee, L.H., Kuhl, M.E., Fowler, J.W., Robinson, S. (eds.) INFORMS Simulation Society Research Workshop, INFORMS Simulation Society, Warwick, Coventry (2009)

Kelton, D.W., Sadowski, R.P., Sturrock, D.T.: Simulation with Arena, 4th edn. McGraw-Hill, New York (2007)

Kleijnen, J.P.C.: Statistical Techniques in Simulation, Part 1. Marcel Dekker, New York (1974)

Kleijnen, J.P.C.: Antithetic variates, common random numbers and optimal computer time allocation in simulations. Management Science 21(10), 1176–1185 (1975)

Kleijnen, J.P.C.: Statistical tools for simulation practitioners. Marcel Dekker, Inc., New York (1986)

Kleijnen, J.P.C.: Experimental design for sensitivity analysis optimization, and validation of simulation models. In: Handbook of Simulation. Wiley, New York (1988)

Kwon, C., Tew, J.D.: Strategies for combining antithetic variates and control variates in designed simulation experiments. Management Science 40, 1021–1034 (1994)

Lavenberg, S.S., Welch, P.D.: Variance reduction techniques. In: WSC 1978: Proceedings of the 10th Conference on Winter Simulation. IEEE Press, Miami Beach (1978)

Law, A.M.: Simulation Modeling and Analysis, 4th edn. McGraw-Hill, New York (2007)

Law, A.M.: Statistical analysis of simulation output data: the practical state of the art. In: WSC 2007: Proceedings of the 39th Conference on Winter Simulation. IEEE Press, Washington, DC (2007)

Law, A.M., Kelton, D.W.: Simulation Modeling and Analysis, 3rd edn. McGraw Hill, New York (2000)

L'Ecuyer, P.: Effciency improvement and variance reduction. In: WSC 1994: Proceedings of the 26th Conference on Winter Simulation, Society for Computer Simulation International, Orlando, Florida (1994)

L'Ecuyer, P., Buist, E.: Variance reduction in the simulation of call centers. In: WSC 2006: Proceedings of the 38th Conference on Winter Simulation, Winter Simulation Conference, Monterey, California (2006)

Levene, H.: Robust Tests for Equality of Variances. In: Contributions to Probability and Statistics. Stanford University Press, Palo Alto (1960)

Lin, C., Lu, I.: The procedure of determining the order picking strategies in distribution center. The International Journal of Production Economics 60-61(1), 301–307 (1999)
Magableh, G.M., Ghazi, M., Rossetti, M.D., Mason, S.: Modelling and analysis of a generic cross-docking facility. In: WSC 2005: Proceedings of the 37th Conference on Winter Simulation, Winter Simulation Conference, Orlando, Florida (2005)
Mitchell, B.: Variance reduction by antithetic variates in gi/g/1 queuing simulations. Operations Research 21, 988–997 (1973)
Nelson, B.L.: A perspective on variance reduction in dynamic simulation experiments. Communications in Statistics- Simulation and Computation 16(2), 385–426 (1987)
Nelson, B.L.: Control variates remedies. Operations Research 38, 974–992 (1990)
Nelson, B.L., Schmeiser, B.W.: Decomposition of some well-known variance reduction techniques. Journal of Statistical Computation and Simulation 23(3), 183–209 (1986)
Nelson, B.L., Staum, J.: Control variates for screening, selection, and estimation of the best. ACM Transactions on Modeling and Computer Simulation 16(1), 52–75 (2006)
Robinson, S.: Successful Simulation: a Practical Approach to Simulation Projects. McGraw-Hill, Maidenhead (1994)
Sadowski, R.P., Pegden, C.D., Shannon, R.E.: Introduction to Simulation Using SIMAN, 2nd edn. McGraw-Hill, New York (1995)
Schruben, L.W., Margolin, B.H.: Pseudorandom number assignment in statistically designed simulation and distribution sampling experiments. Journal of the American Statistical Association 73(363), 504–520 (1978)
Shannon, R.E.: Systems Simulation. Prentice - Hill, New Jersey (1975)
Snedecor, G.W., Cochran, W.G.: Statistical Methods, 8th edn. University Press, Iowa (1989)
Tew, J.D., Wilson, J.R.: Estimating simulation metamodels using combined correlation based variance reduction techniques. IIE Transactions 26, 2–26 (1994)
Wilson, J.R.: Variance reduction techniques for digital simualtion. American Journal on Mathematics in Science 4(3-4), 277–312 (1984)
Yang, W., Liou, W.: Combining antithetic variates and control variates in simulation experiments. ACM Transactions on Modeling and Computer Simulation 6(4), 243–260 (1996)
Yang, W., Nelson, B.L.: Using common random numbers and control variates in multiple-comparison procedures. Operations Research 39(4), 583–591 (1991)

2 Planning of Earthwork Processes Using Discrete Event Simulation

Johannes Wimmer, Tim Horenburg, Willibald A. Günthner, Yang Ji, and André Borrmann

The planning of earthworks represents a complex task. The use of different machine configurations as well as alternative scenarios in the site layout (e.g. transport routes and temporal storage areas) must be evaluated and dimensioned consistently. Wrong decisions can lead to delays or an uneconomic solution and hence increase the costs and project duration. In practice, this planning process is based on the experience and knowledge of the persons in charge; however, decision support tools are not used in the planning of excavation and transportation equipment despite their central importance. Therefore an approach has been developed to support the planning of construction processes in earthworks by applying discrete event simulation. For this purpose, methods for calculating the performance of earthmoving equipment were extended based on statistical components, adapted for simulation, and implemented in a module library. Furthermore, the simulation tool has been coupled with a mathematical optimization procedure to reduce the cost of transport in earthworks by minimizing haul times.

2.1 Actual Situation in Earthwork Planning

Planners of earthworks are facing various influences and changing conditions that could lead to continual adjustments that inevitably impair the construction process during execution. The scheduling is therefore a dynamic process that is very difficult to control due to the fast pace of construction progress. Therefore an efficient and well coordinated schedule is the basis for an economic operation. In this

---

Johannes Wimmer · Tim Horenburg · Willibald A. Günthner · Yang Ji ·
André Borrmann
Dipl.-Ing. Johannes Wimmer
Technische Universität München,
fml - Lehrstuhl für Fördertechnik Materialfluss Logistik
Boltzmannstr. 15
D-85748 Garching bei München
Germany
e-mail: wimmer@fml.mw.tum.de

context, a number of individual processes have to be coordinated temporally and in terms of capacity. One way of modeling the dynamic processes and constraints on construction sites is by discrete event simulation (DES). But the complex on-site conditions complicate the modeling of construction processes.

Uncertainties such as changing weather conditions have a direct impact on the performance of earthworks, although they have not yet been investigated in detail. So far, these conditions were only taken into account by global reduction factors or average performances. Due to disturbances or unexpected delays it may not be possible to meet the construction schedule that was originally planned. To avoid cost-intensive, nonproductive times, the schedule is often changed spontaneously based on the current situation without considering the whole construction process. This adaption which is flexible in practice has to be modeled in the simulation. Another difficulty in modeling construction processes in earthworks is the collection of all necessary input data, as these are often missing or difficult to access. On the one hand, details of the soil layers in the construction site are only estimated. On the other hand, data of the location of buildings, transportation routes, and site equipment are usually stored in printed and manually enhanced 2D plans or in other formats which are difficult to access. Therefore all relevant data for the simulation must be explicitly transferred for each construction site.

2.2 Analysis of Requirements for DES in Earthworks

The primary objective of the simulation in earthworks is to ensure that all construction activities can be smoothly realized. To model uncertainties in scheduling which result from various influences and reflect changing conditions, a method of evaluating various scenarios before construction and comparing relevant parameters is provided. Besides the economic aspects, the clear visualization of construction processes in the simulation environment is an essential point. For the large number of participants the 3D animation of the construction process provides a clear representation of the actual plans, so that errors due to misunderstandings can be avoided.

In earthworks the use of simulation is mainly applied in two phases. Firstly it can be used in tender preparation, in which the construction process must be designed in a short period and respective costs must be calculated. Secondly the use of DES is suitable in work preparation, where different scenarios must be compared in order to generate reliable, highly detailed plans. Therefore it is useful to create a specific simulation model for a specified construction project which can be used consistently for an approximate calculation in tender preparation and for detailed planning in the works scheduling. Hereby the requirements shown in Figure 2.1 should be met.

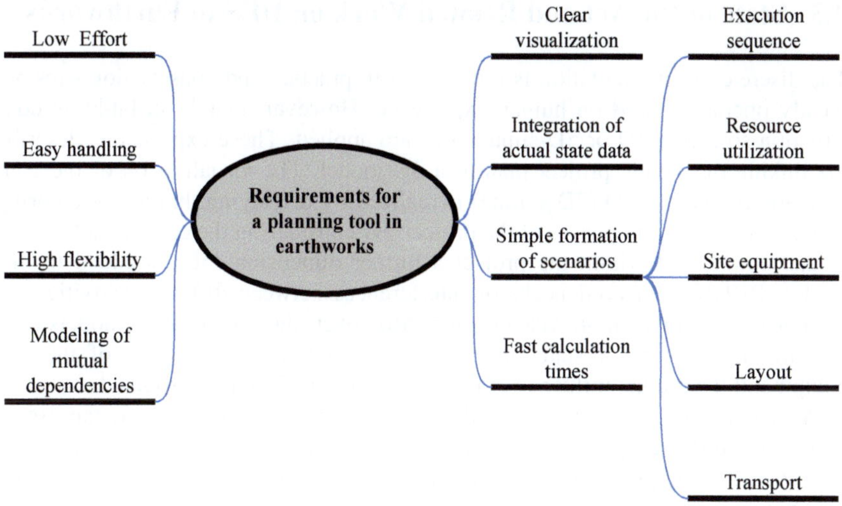

Fig. 2.1 Requirements for a planning tool in earthworks (Source: TUM-fml)

During execution of construction sites diverse variations arise, which must be modeled independently and flexibly in order to respond to the on-site constraints. Therefore, several scenarios must be compared for a secured scheduling of earthworks. Important parameters for the scenario design in earthmoving are the location of construction roads as well as the position of interim storage, disposal areas, and material sources. In addition, the execution order of the earthworks and the allocation of excavation-areas (cut) to dump-areas (fill) should be modifiable in the simulation. Another parameter for the formation of scenarios is the use of resources. Each resource is allocated flexibly to the individual activities and the type and number of resources are selected independently.

A detailed modeling of the interdependencies between the earthwork processes must also be possible with little effort, since the procedures vary with each construction site. To enable simulation runs for the current situation, the actual state of the site is to be integrated into the planning tool. Furthermore changes in the construction sequence have to be adapted quickly, because unexpected soil layers or equipment failures can cause changes which have to be solved within a few hours in order to prevent excessive downtime costs. Therefore, an earthworks decision support tool must be operated by the responsible supervisor and must deliver results within a short period, providing a clear visualization which can be easily interpreted by the user on the construction site.

Several approaches to meet these requirements and scheduling tasks in earthworks already exist and are explained briefly in the next chapter.

2.3 State of the Art and Related Work on DES in Earthworks

The discrete event simulation is rarely used in practice and construction sites are mostly operated based on human experience. However, mainly in building construction so-called 4D or 5D-simulations are applied. These expressions describe the linking of a static project plan to a 3D model. The visualization of the construction process in 4D (3D + time) is realized by displaying, hiding, or coloring components at certain phases of the project. By considering the cost of each activity and component within the project a further dimension (5D) is added to this model [RIB10]. But bottlenecks or interferences between different activities are not detected by this simple visualization. Moreover, these simulation models only consider the change of state of the building and in certain cases of the site equipment; the mutual influence of different activities is not investigated.

A method for event-oriented modeling of construction processes is the use of Petri nets. In these, cyclical works, for example, can be modeled simply [Fra99]. However, the complexity of modeling increases strongly with the number of relevant processes and their dependencies. An example of the application of attributed Petri nets in construction is the work of Chahrour, who has analyzed the link between CAD and simulation of earthwork processes based on Petri nets [Cha07].

Furthermore, several DES systems have been designed based on activity cycle diagrams. Two representative systems of this group are CYCLONE and STROBOSCOPE. CYCLONE is a well established, widespread system which is easy to learn and suited for the effective modeling of many basic construction projects. STROBOSCOPE, on the other hand, is a programmable and expandable simulation system for modeling complex construction projects which requires a longer training time and expert programming skills [MI99].

In factory planning, use of DES systems is widespread for modeling manufacturing processes (e.g. *Plant Simulation* or *Enterprise Dynamics*). These complex simulation systems are providing both programmable and prefabricated simulation modules that are designed for application in intralogistics. Owing to the modular structure, a large part of the effort of modeling and implementing can be transferred to a project-independent phase. A new specific simulation model is then created via the connection of multiple modules and graphical configuration, decreasing the cost of each individual simulation project. Another advantage of these module-based systems is the ability to clearly visualize the simulated processes.

Therefore, these simulation systems are increasingly used in the construction sector, although their standard modules are designed for production and logistic processes. Initial approaches for the simulation of construction processes have been implemented in research [KBSB07; Web07]. The goal is now to harness the advantages of module-based modeling in the scheduling of earthworks.

2.4 Modeling and Implementation of a Module Library for the Simulation of Earthworks

Flexibility is a significant characteristic for the application of simulation methods in earthworks due to various objectives as well as parametric attributes and constraints. Hence a module library for earthworks and civil engineering was implemented in *Siemens Plant Simulation*. The library includes models of construction site-related processes, internal management, and specific objects of construction site equipment as well as functionalities for import and export of required data. The following sections address the respective modeling and implementation.

2.4.1 Framework for Earthwork Simulations

As introduced above, the complexity and effort required to prepare and implement simulation experiments may not be too large. The embedding of simulation in present planning processes and the integration of available information from different sources is therefore obligatory.

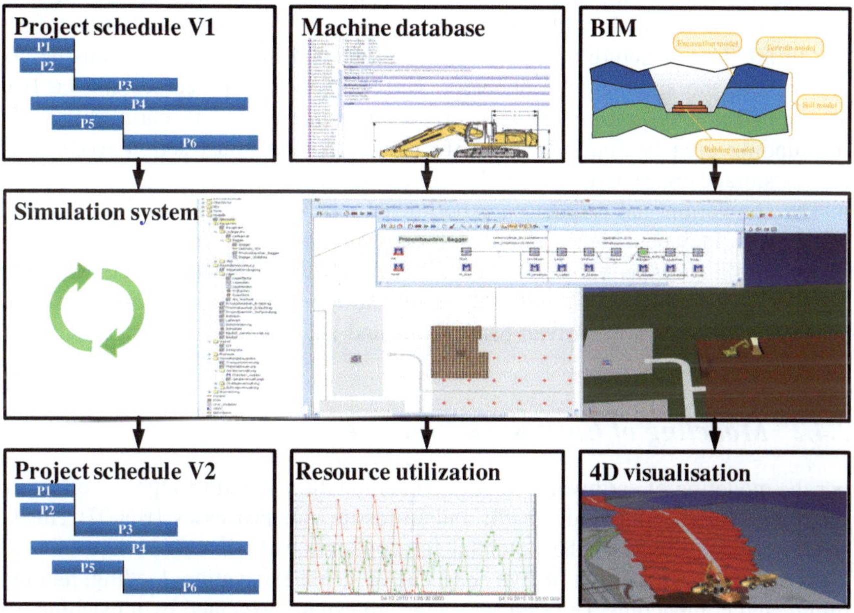

Fig. 2.2 Interfaces of the simulation system (source: TUM-fml)

The concept in Figure 2.2 shows respective input and output data for process simulation in earthworks. An existing project plan is imported from conventional project management tools such as MS Project, providing the basis for the simulation progress – start/end times, makespan of processes, relevant resources, and specific operating times. Within the simulation framework a project plan is converted to individual processes, which are executable without further information. Hard or non-predictive processes can be detailed, so that the activities and their corresponding progress are elaborated specifically.

A database provides all relevant data concerning the deployment of machines. Therefore an interface to the machine database Equipment Information System (EIS) was developed [GKF+08]. Integrated search functionalities allow easy handling and selection of compatible equipment and machinery. The specific properties of the selected machines can be imported directly into the simulation environment. In addition, corresponding 3D models of construction equipment are used for visualization.

Further important data for the simulation are the volume and position of all cut- and fill-areas in earthworks. Usually the volume calculations are based on Gauß-Elling operations [DD10]. From two cross-sections an average surface is generated and multiplied by the distance between these two sections. This technique is not very accurate, especially for curved surfaces, complex geometries, and large distances between the surfaces. Therefore, and also to ease mass determination, a tool was developed which combines the models of subsoil, surface, and structure in one integrated model. Hereby the volume and mass of cut- and fill-areas are determined. The volume of each area is subdivided into homogenous cuboids (voxels), which hold information on the position, volume, and type of soil [JLOB08]. This structure of all cut- and fill-voxels is adjusted to an xml-file and imported to the simulation framework. Hence earthwork processes can be both determined and visualized in much higher quality. Besides all earth masses the surrounding area is imported as a 3D model to plan construction roads, road access, storage areas, and so on in the overall 3D context.

The simulation and respective experiments result in an improved project plan which also includes highly detailed processes. Moreover information such as resource utilization is subject to the experiments and can be further evaluated. Another essential result is the 4D visualization, which enables a meaningful discussion between all involved participants and the detection of mistakes in early planning phases.

2.4.2 Modeling of Earthwork Processes

For the modeling of earthwork processes in simulation, existing approaches for the calculation of the individual earth and infrastructure processes [Bau07; Hüs92; Gir03] are analyzed and the potential for application in simulation is evaluated. Earthworks usually consist of five consecutive steps: excavation, loading, hauling, filling, and compacting. The first two steps are usually executed by a machine, for example, an excavator. The calculation of their single processes as well as those of the filling and compacting processes is well-established, is customizable through various parameters, and can therefore be used for the simulation of construction processes. These four steps were adapted to DES – as shown for an excavator in Figure 2.3 – and statistical components were added.

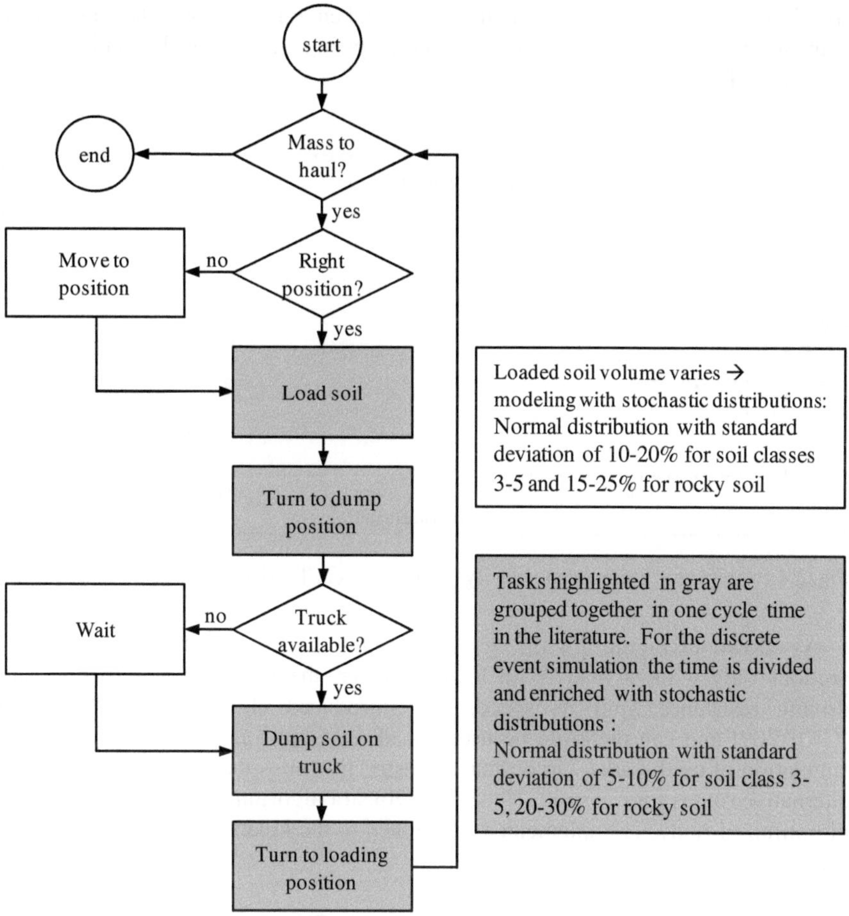

Fig. 2.3 Flow chart for the modeling of an excavator (source: TUM-fml)

The calculation of transport performance, however, has significant potential in optimization, especially in forecasting specific cycle times for transport vehicles. Existing approaches do not consider dynamic motions such as acceleration, deceleration, and lower velocity while turning. Depending on the corresponding road attributes this leads to large variations in transport times and therefore to relevant inaccuracy in the planning of transport capacities – even though transportation frequently accounts for most of the cost of earthworks.

The application of a kinematic simulation method can improve the calculation of transport times. This technique provides velocity profiles for individual vehicles depending on road attributes and current loading conditions. For every single time step the effective acceleration of any vehicle is calculated from the current velocity, vehicle characteristics, and road profile. When the driving force is lower than the driving resistances, overall speed decreases or increases otherwise. Static velocity limits can address both vehicles and road sections, so that speed limits or influences from traffic are taken into account. The kinematic simulation compares different vehicles and helps to select an ideal combination of machinery for earthworks.

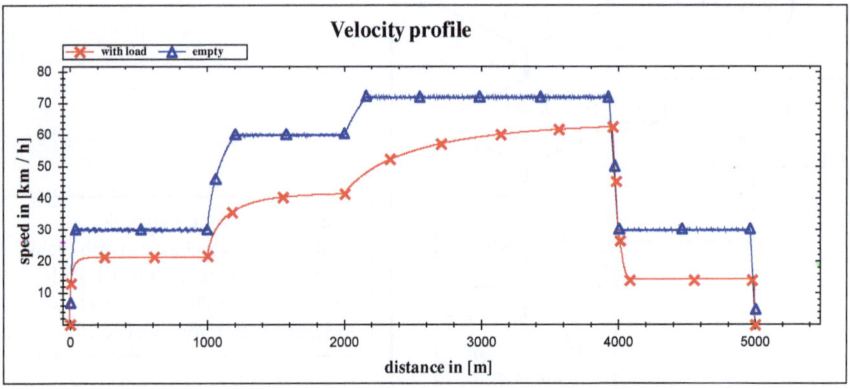

Fig. 2.4 Example for a simulated velocity profile (source: TUM-fml)

As shown in Figure 2.4 the vehicle reaches the velocity limits of the road sections only without load. With load, however, the vehicle's performance and driving resistance limit its speed. The introduced algorithm was evaluated [GKFW09] and can therefore be used for all relevant transport processes in the simulation of earthworks. On ordinary construction sites there are usually several alternative routes for transport. Hence the Dijkstra-algorithm for the determination of optimized routes is implemented and linked to the kinematic simulation and its algorithms.

2.4.3 Module Library for Simulation in Earthworks

The algorithms and calculations introduced above were implemented in various modules and merged into a library of intelligent objects in such a way that complex processes in earthworks could be analyzed by simulation experiments. Figure 2.5 shows the overall context of the developed modules. During the implementation of a specific problem in a simulation model the level of detail is determined – uncertain and critical processes require a higher level of detail and further investigations during simulation experiments.

The project plan and its specific processes are transformed into executable Gantt-modules. These are activated once their dedicated start time is reached and all predecessors are completed. If a process is not of particular interest, its level of detail remains low and the duration of the process depends only on the previously defined time. The activation and completion of the process can change the state of the construction site, which is used for the timescale and the visualization of the progress.

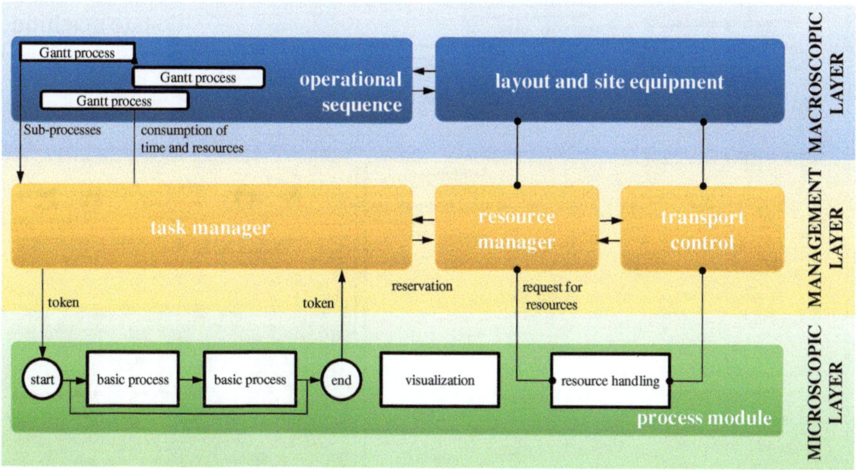

Fig. 2.5 Internal context of the simulation modules (source: TUM-fml)

In the case of further detailing of a Gantt-module, it is then transformed into sub-processes which are managed by a central task manager instead of being executed as a static duration. For this purpose the user has to input all required information. The individual sub-processes from all corresponding processes are then sorted in analogy to the constraint-based sequencing [KBSB07]. For every time step the task manager examines whether upcoming sub-processes comply with the following restrictions:

- All preceding (sub-)processes are completed successfully
- All required resources (personnel, equipment, area, material) are available

If all requirements are met, the corresponding sub-process starts. Resource managements have been developed to determine whether a resource is generally qualified for an operation. If several resources meet the criteria, various strategies determine the assignment. Therefore strategies such as minimal distance to workspace or balanced workload are implemented.

Once all required resources are available, they are reserved for the corresponding operation. A process module is created which holds all necessary information on a process-token. As shown in Figure 2.6, process modules consist of a state machine, resource handling, and visualization. The process token passes the different basic activities of the state machine – each activation and completion results in an updated state. Depending on the parameters of the process modules and the current conditions, single states can be passed several times or skipped entirely. Likewise the resource handling consists of basic activities which reserve and deallocate material, personnel, devices, and areas by sending requests to a global resource manager. The resource manager appoints the requested resource and assigns it to the respective sub-process. Once the resource arrives at its location of assignment, the token passes on to the next state and the successive basic activity continues the process.

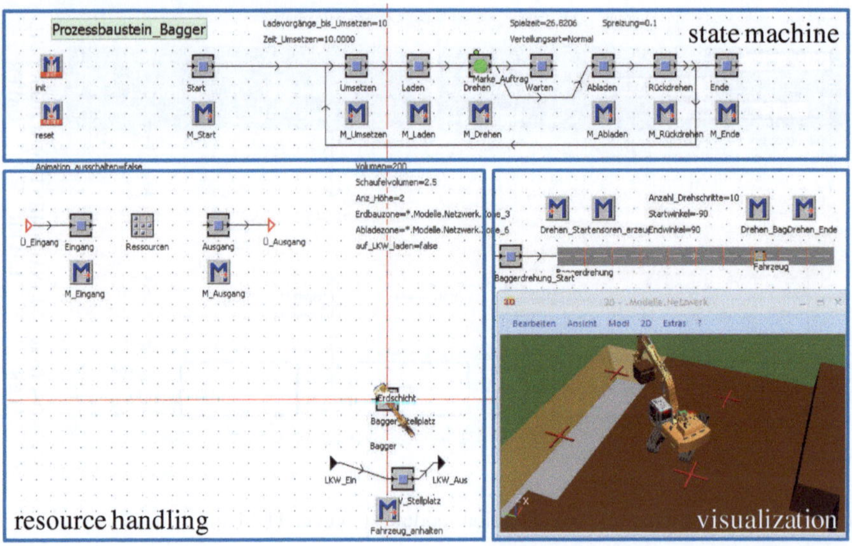

Fig. 2.6 Example of a process module (source: TUM-fml)

Furthermore movements such as rotation and translation of 2D and 3D objects are implemented to visualize the current progress of construction works.

2.5 Coupling DES in Earthworks with Mathematical Optimization Methods

In practice the average transport distance is a common metric for evaluating earthworks costs. The task in the planning of earthworks is now to assign the individual areas where soil is excavated (cut) to areas where the soil is dumped (fill) so that transport costs are minimized. To minimize the average transportation distance a linear optimization method has already been successfully applied [For-09]. In this approach graph-based methods (see Figure 2.7) are used which carry out the optimal allocation on the basis of the 3D position of each area.

Transport costs for earthworks depend primarily on the transport time, which is influenced not only by distance but also by the set-up of the roads and the resources used. However, these factors are not included in the graph-based optimization model. In the DES those influences on the transport time can be modeled through the application of the kinematic simulation.

For this purpose a coupling concept was developed that creates a bidirectional connection between the two techniques (linear optimization and simulation) in order for them to supplement each other. In this coupled system mathematically optimized transport assignments are now imported into the simulation system. There the simulation of transport processes is carried out on the basis of these data and the internally specified routes and resources. The results of the simulation are the duration of each earthwork process from a specific cut to a fill area. These times are re-imported into the optimization module in order to execute the mathematical optimization with these simulated earthwork durations instead of the transport distances.

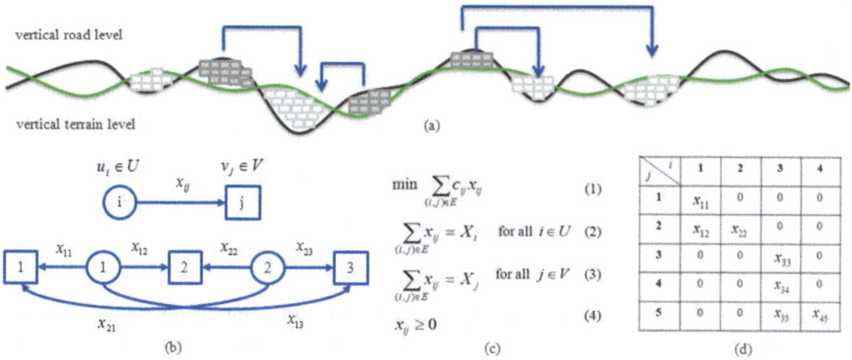

j \ i	1	2	3	4
1	x_{11}	0	0	0
2	x_{12}	x_{22}	0	0
3	0	0	x_{33}	0
4	0	0	x_{34}	0
5	0	0	x_{35}	x_{45}

Fig. 2.7 Graph-based approach for optimizing earth transport (Source: TUM-cms)

Figure 2.8 shows an analysis of simulation runs of hauling earth from a cut to a fill area using the same machinery (one excavator and three dumpers) but different road types and distances between the areas and thus also different cycle times. It can be seen that the earthwork duration per cubic meter increases linearly with the cycle time of the dumpers. However, at very short distances, the duration remains at a consistently high level and scatters accordingly strongly. This is explained by the fact that in this case the performance of the excavator and not the transport performance is decisive. Hence to reproduce this behavior in the optimization, the earthworks duration of all of possible cut-to-fill combinations should now be simulated. But this step is very computationally intensive, since several runs must be executed for each combination in order to receive an average duration despite the modeled stochastic effects. Thus, another method was chosen: In a first step the cycle times of the selected transport vehicles for all possible cut-to-fill combinations are determined with the kinematic simulation shown above. Then for some randomly selected cut-to-fill combinations the earthwork durations are simulated (see Fig. 2.8). In a last step the earthwork durations of all possible cut-to-fill combinations are determined by applying a sliding linear approximation to the randomly selected and simulated ones. In this way the cycle time of the hauling vehicles is taken as ordinate.

In this manner it is possible to minimize the average duration of earthworks on the basis of the determined cycle times in the simulation. This reduces the costs of earthworks, which increase almost linearly with the duration.

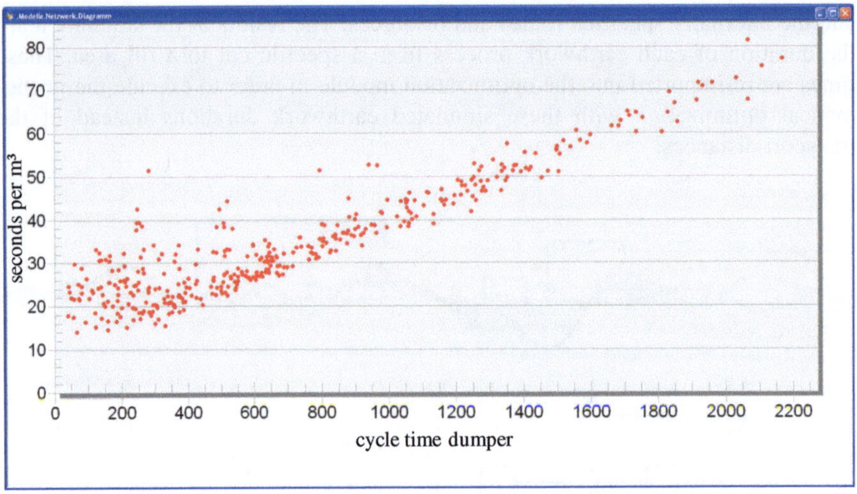

Fig. 2.8 Simulation analysis of randomly selected cut-to-fill combinations: the duration of earthworks operations is shown normalized to cubic meters and plotted against the cycle time of the hauling trucks

2.6 Evaluation and Case Study

The module library introduced above was evaluated on an actual construction project. Within this infrastructure project a new 14 km road was built with a total of 33 cut and fill areas. There was no balance of cut and fill volumes, so spare earth masses were driven to several landfills. Because of three bridge constructions the layout of transport routes and landfills was complicated enough to analyze various scenarios based on simulation experiments.

Input data were tender drawings including a 3D model of the original terrain, soil examinations, exact route coordinates, and a project plan for milestones and completion.

The combination into one integrated 3D model (see Section 2.4.1) generates cut- and fill-areas, which for reasons of accuracy are split into several groups. Figure 2.9 shows the generated voxel structure of a cut- and a fill-area as well as the 3D model of the terrain and the corresponding 2D plan within the simulation system. The large number of voxels required for an exact mass determination was a problem in the case study, since a bisection of the voxels' edges increases the number of voxels to eight times their original number and therefore results in long simulation runs. For this reason small voxels are used for mass determination and are merged into bigger voxels for use in the simulation. The latter then hold all information such as volume and soil type to keep the original accuracy.

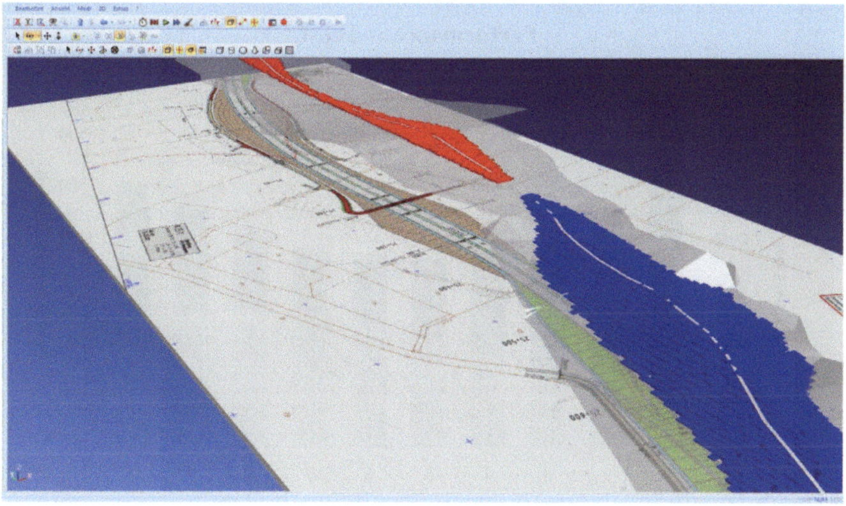

Fig. 2.9 Screenshot of the 3D visualization of one cut- and fill-area

As described in section 2.5, a linear optimization of earth transports provides an assignment of cut- and fill-areas to minimize average transport distances. Besides the tender information, further information is required from the company responsible on available machines, which machines are used on the construction site, and a project plan including constraints and durations. Besides the machine management, construction roads and further equipment are also defined by the construction company.

The roads of the case study usually lead along the future route. Because of the cohesive soil these roads are highly exposed to the weather, which results in unsteady cycle times or even production stoppages for the earthworks. The performance of vehicles on different soil conditions can be computed by the implemented kinematics simulation. Hence the model is based on realistic transport velocities, which is not possible within the linear optimization. The process simulation therefore evaluates the assignments resulting from the optimization.

In the case study earthworks operate with five parallel convoys consisting of an excavator (excavation and loading), a number of dumpers depending on the route (transport), a crawler (filling), and a compactor (compacting). Since the corresponding operations are part of the module library, the earthworks can be modeled from the implemented process modules. Further works such as embankments or drains are simply added to the overall time consumption of the basic operations considering stochastic influences.

Figure 2.10 shows the performance and total costs of several scenarios for the earthworks of one cut-to-fill assignment. The subjects of experiments were the number of dumpers which transport earth from excavation to fills and the corresponding costs of machinery. For less than seven vehicles transport performance is crucial, and overall performance increases with every extra vehicle. From there on the loading performance of the excavator dictates. Extra vehicles do not influence overall performance; costs, however, increase linearly with the number of dumpers.

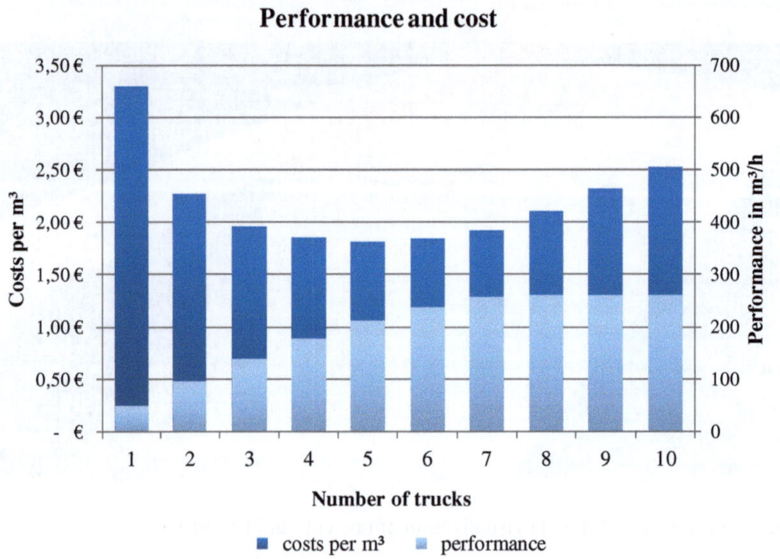

Fig. 2.10 Evaluation of different scenarios with regard to machinery (source: TUM-fml)

The concept introduced in Chapter 2.5 of coupling DES and mathematic optimization methods to minimize transport times was also applied to the case study. Therefore a scenario of one excavator and three dumpers was created within the simulation environment. The concept was evaluated and compared to different strategies for the cut-to-fill assignment as shown in Figure 2.11.

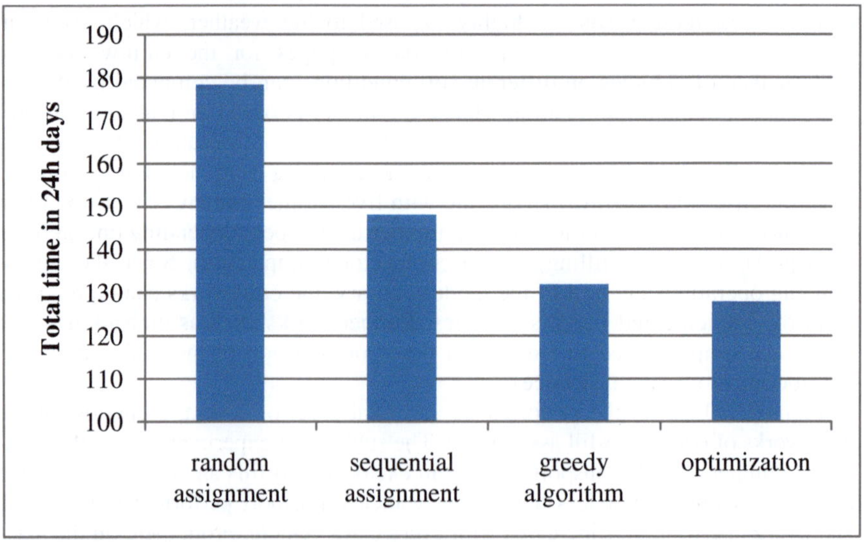

Fig. 2.11 Result of the different cut-to-fill assignments in the use case

The first experiment is based on random assignment. Due to a random generator transports and respective masses between fills and cuts are assigned. Obviously the random assignment of cut to fill areas on large linear construction sites does not result in an acceptable solution and causes various unneeded transports.

The second experiment complies with the principle "as easy as possible". Transports from cut- to fill-areas are allocated based on the geographical location along the route. Starting from the first cut-area earth is excavated and transported to the closest fill-area in sequence (see Figure 2.12). By applying such a simple strategy more than 30 days of work can be saved in this use case.

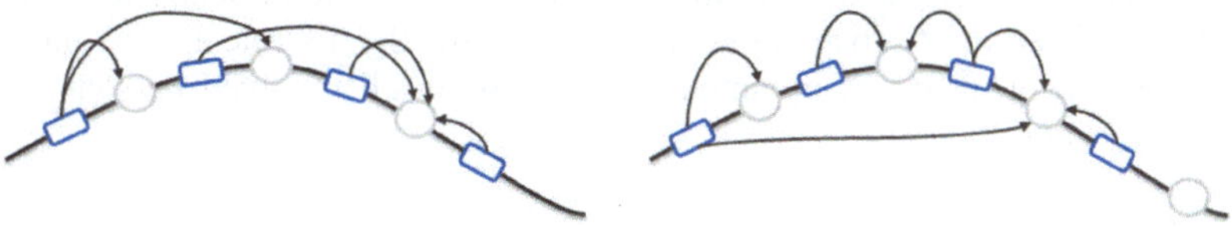

Fig. 2.12 Scenario sequential assignment (left) and greedy algorithm (right) (source: TUM-cms)

For the third experiment a heuristic approach (greedy algorithm) locates the closest fill-area for every cut-area. Figure 2.12 shows the difference between the approaches in experiments 2 and 3. The greedy algorithm chooses the shortest overall distance and assigns all masses possible between the corresponding cut- and fill-areas. Subsequently the earth is transported along the next largest distance, continuing until all necessary masses are relocated. Sixteen additional days can be saved in this case by using this heuristic approach.

The last experiment evaluates the concept of coupling simulation and linear optimization. Based on the same resources and the transport times determined in simulation the optimization reduces the number of days of work to 128. Twenty days can be saved compared to the traditional sequential assignment and another four can be saved compared to the greedy algorithm case. Even greater time savings are possible if the topography of the route includes larger gradients and differs strongly in its parameters.

The results therefore confirm the potential of the concept introduced as coupling of simulation and linear optimization. However it is important to mention that the applied model does not include data such as traffic conditions and weather influences, which may have a great impact on the progress of construction works. Hence the time of construction predicted in simulation is not completely realistic, but provides an essential contribution for construction planning.

2.7 Conclusion

Due to short planning periods and high costs, the discrete event simulation of earthwork processes is rarely used in practice. Hence a concept was created to significantly reduce the cost of simulation studies by using module-based modeling and reusing existing design data. Existing calculation methods have been adapted for the simulation, and the modeling of transportation has been refined with a kinematic simulation approach. The various processes of a construction site, the elements of the site equipment, and the resources can be combined independently by the use of the module structure shown in Figure 2.5. Thus, different scenarios with varying use of machines and different boundary conditions can be formed before the start of a construction project. Due to the selectable level of detail it is possible to examine all processes that are considered critical for the overall process. The simulation modules have standardized interfaces, so that any further activities can easily be implemented in the module library. Furthermore the DES is combined with an optimization algorithm which provides a supplementary high potential to rationalize earthworks. Additionally a combined 2D/3D visualization of processes is provided, so that the discrete event simulation can be used as a means of communication between all involved persons on the construction site.

Authors Biography, Contact

Institute for Materials Handling, Material Flow, Logistics

The institute fml perceives itself as an open research institution aiming to significantly contribute towards the scientific progress in the areas of material flow technology and logistics. An essential contribution to safeguarding Germany as a location of logistics is made by the acquired knowledge and its transfer towards practical applications, especially in small and medium-sized businesses. Insights gathered from fundamental research activities represent the basis for developing innovative solutions for current and practically relevant problems from research and industry. An integrated knowledge transfer and problem specific knowledge adaption belong to the institute's core responsibilities just like the education and training of the upcoming scientific generation by committed teaching activities. Along with aspects of technical logistics, the control and optimization of material flow processes by innovative identification technologies (RFID), the development of logistics planning by means of digital tools as well as the role of human beings in logistics represent the institute's essential research topics.

The institute fml is active both in the publicly financed domains of fundamental and applied research and in research cooperations with industrial partners. Research projects are usually carried out in interdisciplinary collaborations.

Contact

Dipl.-Ing. Johannes Wimmer
Technische Universität München,
fml - Lehrstuhl für Fördertechnik Materialfluss Logistik
Boltzmannstr. 15
D-85748 Garching bei München
Germany
Phone.: +49 (0)89 289-15914
Email: wimmer@fml.mw.tum.de

References

[Bau07] Bauer, H.: Baubetrieb. Springer, Heidelberg (2007)

[Cha07] Chahrour, R.: Integration von CAD und Simulation auf Basis von Produktmodellen im Erdbau. Kassel Univ. Press, Kassel (2007)

[DD10] Deutsches Institut für Normung; Deutscher Vergabe- und Vertragsausschuss für Bauleistungen: VOB; Beuth, Berlin, Deutsches Institut für Normung; Deutscher Vergabe- und Vertragsausschuss für Bauleistungen (2010)

[For-09] ForBAU: Zwischenbericht des Forschungsverbundes "Virtuelle Baustelle", Institute for Materials Handling, Materials Flow, Logistics. Technische Universität München, München (2009)

[Fra99] Franz, V.: Simulation von Bauprozessen mit Hilfe von Petri-Netzen. In: Fortschritte in der Simulationstechnik, Weimar (1999)

[Gir03] Girmscheid, G.: Leistungsermittlung für Baumaschinen und Bauprozesse. Springer, Berlin (2003)

[GKF+08] Günthner, W.A., Kessler, S., Frenz, T., Peters, B., Walther, K.: Einsatz einer Baumaschinendatenbank (EIS) bei der Bayerischen BauAkademie. In: Tiefbau, Jahrgang 52, vol. 12, pp. 736–738 (2008)

[GKFW09] Günthner, W.A., Kessler, S., Frenz, T., Wimmer, J.: Transportlogistikplanung im Erdbau. Technische Universität München, München (2009)

[Hüs92] Hüster, F.: Leistungsberechnung der Baumaschinen. Werner, Düsseldorf (1992)

[JLOB08] Ji, Y., Lukas, K., Obergriesser, M., Borrmann, A.: Entwicklung integrierter 3D-Trassenproduktmodelle für die Bauablaufsimulation. In: Tagungsband des 20. Forum Bauinformatik, Dresden (2008)

[KBSB07] König, M., Beißert, U., Steinhauer, D., Bargstädt, H.-J.: Constraint-Based Simulation of Outfitting Processes in Shipbuilding and Civil Engineering; In: 6th EUROSIM Congress on Modeling and Simulation, Ljubljana, Slovenia (2007)

[MI99] Martinez, J.C., Ioannou, P.G.: General-Purpose Systems for Effective Construction Simulation 125(4), 265–276 (1999)

[RIB10] RIB Software AG: transparent, `http://www.rib-software.com/de/ueber-rib/transparent-das-magazin.html` (accessed on August 12, 2010)

[Web07] Weber, J.: Simulation von Logistikprozessen auf Baustellen auf Basis von 3D-CAD Daten, Universität Dortmund, Dortmund (2007)

3 Simulation Applications in the Automotive Industry

Edward J. Williams and Onur M. Ülgen

Simulation analyses subdivide themselves conveniently into two major categories: discrete-event simulation and continuous simulation (Zeigler, Praehofer, and Kim 2000). Continuous simulation studies processes amenable to analysis using differential and difference equations, such as stability of ecological systems, chemical synthesis, oil refining, and aerodynamic design. Discrete-event simulation studies processes in which many of the most important variables are integer values, and hence not amenable to examination by continuous equations. Such processes almost invariably involve queuing, and the variables of high interest include current and maximum queue lengths, number of items in inventory, and number of items processed by the system. Many of the integer values are binary; for example, a machine is in working order or down, a worker is present or absent, a freight elevator is occupied or vacant. Processes with these characteristics are common in manufacturing, warehousing, transport, health care, retailing, and service industries.

3.1 Manufacturing Simulation

Historically, one of the first major application areas of discrete-event process simulation was within the manufacturing sector of the economy (Miller and Pegden 2000). Strategically minded managers, not to mention industrial and process engineers, quickly learned that simulation is a delightfully quick, inexpensive, and non-disruptive alternative to the potential purchase, installation, and integration of expensive machines or material-handling equipment "because it ought to improve productivity. Simulation permits "trial by proxy" of such proposals, often involving high capital expenditure and risk, before undertaking them on the manufacturing plant floor.

3.2 Automotive Industry Simulation

The automotive industry involves not only complex and variegated manufacturing contexts but also large and complex supply chains. At the apex of the supply chain

Edward J. Williams · Onur M. Ülgen
College of Business
B-14 Fairlane Center South
University of Michigan - Dearborn
Dearborn, Michigan 48126
USA
e-mail: ewilliams@pmcorp.com

lies the final assembly plant – no matter how many subsidiary plants, both those of the vehicle manufacturer and those of its suppliers, contribute to the manufacture of the vehicle, the manufacturing process must culminate with the integration of all the parts (engine, powertrain, body panels, interior trim, exterior trim….) into a vehicle. Underscoring the complexity of vehicle manufacturing and supply chain operations, automotive industry suppliers are routinely classified as Tier I (supply vehicle components to the final manufacturer), Tier II (supply components to a Tier I company), Tier III (recursively). Conceptually, the automotive company itself can be considered Tier Zero, although this term is seldom used. Accordingly, managers and engineers in the automotive industry, whether their employer is a vehicle manufacturer or a supplier thereto, have been eager and vigorous users of simulation for many years (Ülgen and Gunal 1998). As early as the 1970s, long before the advent of modern simulation software and animation tools, when GPSS [General Purpose Simulation System] (Gordon 1975) and GASP [General Activity Simulation Program] were relatively new special-purpose languages (GASP was FORTRAN-based), pioneers in automotive-industry simulation sought to accommodate increasingly frequent requests for simulation analyses. One of these early efforts, in use for many years, was GENTLE [GENeral Transfer Line Emulation] (Ülgen 1983).

3.2.1 Overview of Automobile Manufacturing

Historically and etymologically, the very word "automobile" reflects astonishment: a vehicle, unlike a wagon, cart, or buggy, which can move ["mobile"] by itself ["auto"], without need of an ox, donkey, horse, or mule. Automotive manufacture was an early pioneer of the assembly line, in which the work piece is brought to the worker, instead of the reverse. Over several generations, as vehicles (not just automobiles!) became more complex and diversified their manufacture naturally became subdivided into stages. At a very broad, overview level, these stages are (Hounshell 1995):

1. Press shop, in which sheet steel is stamped into recognizable vehicle components, such as roofs, doors, hoods, trunk lids, etc.
2. Weld shop, in which these components are joined; at this stage, the structural form or silhouette of a vehicle becomes readily visible and is called a "body in white [BIW]."
3. Paint shop, in which, under strict conditions of cleanliness, the BIW is pre-treated, sealed, and painted; the coat of paint is then baked dry, and perhaps waxed and polished.
4. Engine shop, in which the vehicle's motive components – engine and powertrain – are installed into the painted vehicle body.
5. Trim shop, in which components such as windshields, interior trim and seats, steering column, electronics, and tires are fitted to the vehicle; after final test in this shop, the vehicle is driven away under its own power.

All of these manufacturing processes entail numerous problems and concerns highly amenable to analysis via discrete-event process simulation. Stages four and five, in particular, involve integration of components which very often are supplied by Tier I (and higher) companies within the supply chain; hence reliability and tight integration of the supply chain assume great importance. Some of these concerns apply to all phases of vehicle manufacture, no matter the tier level. Highly important and visible concerns include:

1. Keeping work-in-process [WIP] inventory as low as possible consistent with no starved operations (operations idled because no work pieces reach them)
2. Achieving reasonably high (but not overly high) utilizations of both labor and capital equipment, both of which are very expensive
3. Avoiding bottlenecking and in-plant congestion of material-handling operations, especially those involving forklift trucks and automatic guided vehicles [AGVs]
4. Meeting throughput targets (often expressed as a jobs per hour [JPH] metric) without compromising quality

3.2.2 Simulation Studies Relative to Production Facility Lifecycles

A production facility's lifecycle comprises four phases. The first is the conceptual phase, during which the facility first exists "on the back of an envelope". During this phase, new methods of manufacturing, material handling, and testing are investigated both for their own practicality and for their ability to integrate well with traditional methods within the hypothesized work flow of the system under design. During this phase, the simulation modeler is likely to be called upon to build small models of the more innovative or experimental system components while working closely with mechanical and process design engineers. Next, during the design phase, the system design moves from "the back of the envelope" to engineering drawings (e.g., AutoCAD®) to permit the formulation of detailed layout plans and equipment specifications. At this phase, questions are raised concerning the relative floor locations of machines, conveyors, and buffers; the capacities of buffers, the speeds of conveyors, and the number of forklifts needed. Such questions typically become the concern of Tier I (and derivatively lower tiers) suppliers who will soon install the production line. During this phase, the simulation modeler will build larger, more inclusive models to assess the adequacy of the overall system. Third, during the launch phase, the system will actually operate below designed capacity to test its operation. Colloquially, the initiation of this phase is often called "Job 1", at which time the first production unit (e.g., instrument panel, engine, entire vehicle) is produced by the system. During this phase, the simulation modeler will often be called upon to assess operational policies competing for eventual adoption within the system. Examples of such policy investigations might be:

- If the same mechanic is responsible for repairing both machine A and machine B in case of malfunction, and machine B breaks down while the mechanic is repairing machine A, should the needed repair of machine B preempt the repair work at machine A?
- Should attendants at the tool crib prioritize requests by workers from part X of the line ahead of requests from part Y of the line, or take these requests on a first-come-first-served (FIFO, FCFS) basis?
- If the brazing oven is not full (its capacity was presumably decided during the previous design phase), how many parts should it contain and how long should its operator wait for additional parts before starting a brazing cycle?
- How large or small should batch sizes be (for example, how many dual-rear-wheel models should be grouped together to proceed through the system before single-rear-wheel models are again run through the system)?

During this phase, the simulation models will also be large, and will become more detailed, calling for additional modeling-logic power from the software tool(s) in use. During the fourth and last phase, the fully operational phase, the production facility will "ramp up" to its designed capacity. During this phase, simulation models often become, and should become, "living documents" used for ongoing studies of the system as market demands, product mix changes, new work rules; invention and introduction of new manufacturing, assembly, material handling, and quality control techniques; and other exogenous events impose themselves on system operation. The model run and analyzed during the launch phase will evolve, perhaps into several related and similar models, during this phase. This phase is significantly the longest (in total elapsed time) of the four phases – indeed, typically longer than the first three phases collectively. Due to this required model longevity, thorough, clear, and correct model documentation (both internal and external to the model) becomes not just important, but vital. The second author, during his career at an automotive manufacturer, was once asked to exhume and revise a model built eleven years previously.

Various categories of simulation applications assume high importance as the life cycle of a production facility proceeds through the four phases described above. Applications assessing equipment and layout of equipment (e.g., choice of buffer sizes, location of surge banks) are most commonly undertaken during the first three phases, particularly the design phase. Applications addressing the management of variation (e.g., examination of test-and-repair loops and scrap rates) first arise during the design phase, and maintain their usefulness throughout the fully operational phase. Much the same holds true for product mix sequencing applications, themselves conceptually also involved with the management of variation – exogenously imposed by the marketplace. Examination of detailed operational issues (e.g., scheduling of shifts and breaks and traffic priority management among material handling equipment) first arises during the design phase, and becomes steadily more important as the facility life cycle proceeds through launch to full operation. In particular, scheduling of shifts and breaks typically requires collaboration with union negotiators, usually occurring repeatedly during a facility life cycle which routinely extends across several periodic union contract negotiations.

3.2.3 Data Collection and Input Analysis Issues in Automotive Simulation

No simulation study, indeed, no computer program, is better than its input data. Typical input data for automotive-industry studies (and indeed for manufacturing-system studies in general) include, but are certainly not limited to:

1. Operation cycle times
2. Operation downtime data (both "how long before failure" and "how long to repair")
3. Reject rates, and what proportion of rejects are repaired and then used versus discarded (scrapped)
4. Material handling equipment capacities, speeds, accelerations, loading times, and unloading times
5. Material handling equipment downtime data
6. Resource allocation policy and travel times between various points of usage
7. Resource scheduling, including meal and rest breaks for workers
8. Absenteeism rates of workers
9. Skill sets of workers, i.e. interchangeability of workers among the various manual operations
10. Arrival rates of inputs at the upstream end of the process under study

Gathering these data involves many problems all too easily overlooked. For example, if the operation runs multiple shifts, data must be gathered and compared across shifts – workers may be fewer and/or less experienced on night shifts. Attempting to obtain accurate downtime data often engenders vigorous resistance from local supervisory personnel who fear (perhaps correctly) that the downtime data will make conspicuous their inattention to required machine maintenance policies. Downtime data are often markedly more difficult to collect and model than basic operational data; some of the considerations involved are discussed in (Williams 1994). As seemingly basic a statement as "Operator X is assigned to operate machine Y" should provoke the modeler to ask the manager or client questions such as:

1. Does operator X have any other duties? If so, how do those duties rank in priority relative to operating machine Y?
2. Is machine Y semi-automatic – e.g., does the operator load and unload the workpiece on the machine yet have freedom to do other tasks while the machine cycles, versus attending machine Y throughout its cycle?
3. If machine Y malfunctions (suffers a downtime), is operator X the person who will repair it, or does operator X call another worker such as a master mechanic to repair it? Does the answer to this question depend on the type of malfunction?
4. If, when machine Y has completed its cycle and the next machine downstream is blocked, can operator X unload machine Y and then do something else, or must operator X stay at machine Y until the blockage is lifted?

Correctly incorporating these data into the model also merits careful attention. The modeler of a vehicle manufacturing process must decide whether the model will be run on a terminating or a steady-state basis. Since most manufacturing operations run conceptually continuously – that is, the production line (unlike a bank or a restaurant) does not periodically "empty itself" and restart next morning – the analyst usually will, and should, run the model on a steady-state basis. Unless start-up conditions are of particular interest (almost always, long-run performance of the system is of primary interest), the modeler must then choose a suitably long warm-up time (whose output statistics will be discarded to avoid biasing the results with start-up conditions of an initially empty model). Various heuristics and formulas are available to choose a warm-up time long enough (but not excessively long) to accomplish this removal of initial bias (Law 2007).

Next, empirical data collected must be incorporated into the model. Whenever possible, a good-fitting probability density should be fitted to the empirical data, thereby smoothing the data, ensuring that behavior in the tails (especially the upper tail) is represented, and permitting investigative changes in the model later, such as a new procedure or machine which requires the same average time but reduces variability). Various techniques, such as Kolmogorov-Smirnov, Anderson-Darling, or Furthermore, careful attention to probabilistic models can prevent errors whose origin is overlooking correlations. Naively sampling either empirical distributions or fitted distributions can lead to errors such as this one:

At one operation, the vehicle is provided with its initial supply of motor oil. At the next operation, the vehicle is provided with its initial supply of transmission fluid. Naïve sampling of distributions for the two consecutive cycle times tacitly assumes independence of these two cycle times. Investigation of the input data via a scatterplot and calculation of the correlation coefficient reveals that these cycle times are positively correlated: larger vehicles need both more oil and more transmission fluid. (Williams et al. 2005).

Similar errors can occur when time-dependencies of data are overlooked: A manual operation may be done gradually faster over time because the worker is learning or more slowly over time because the worker is tiring. Operations done on the night shift may take longer on average than operations done on the day shift because the less desirable night shift is staffed with workers of lower experience.

As (Biller and Nelson 2002), experts on input data modeling, have alertly and trenchantly observed, "…you can *not* [emphasis added] simulate your way out of an inaccurate input model."´

3.2.4 Software Tools Used in Automotive Simulation

Successful and efficiently accomplished automotive simulation projects require appropriate software tools to support the client, production or manufacturing engineer, and the simulation analyst (Banks 1998). First and perhaps most conspicuously, the choice of simulation software package arises. Currently, numerous such packages compete in the marketplace. Selection of the most appropriate package requires thought and care – the more so if the first simulation project for which the software will be used will be one of many. Questions that need to be asked and answered prior to purchase or lease of software include (but are surely not limited to):

1. What compromise should be struck between ease of learning and use and highly detailed modeling power?
2. How conveniently will the software interface with desired input sources and output sinks (e.g., spreadsheets, relational databases)?
3. Are statistical distributions to be used (Poisson, exponential, lognormal, Johnson,....) incorporated in the software?
4. Is the random number generation algorithm used by the software vetted as algorithmically trustworthy?
5. Does the software incorporate built-in constructs that will be needed (e.g., conveyors, bridge cranes, manually operated material-handling vehicles, machines, mobile laborers, buffers...)? It may be insufficient to say "Yes, software package X can model machines." For example, can package X model semi-automatic machines (machines which require labor attention for parts of their cycle but run automatically during other parts of their cycle)? It may be insufficient to say "Yes, software package Y can model conveyors." For example, can it model situations in which a part gets on (or off) the conveyor even though the part is not at either end of the conveyor? Can it model situations in which two conveyors flow into a third conveyor?
6. Does the software contain built-in capability to model various queuing disciplines such as first-come-first-served, shortest job next, longest job next, most urgent job next, etc.?
7. For effective use, does the software presume that the simulation modeler is well acquainted with object-oriented programming concepts?
8. Does the software run on all computers and operating systems on which the model will need to run?
9. Does the software enable creation of an "executable" model which can be run for experimentation on a machine not having a full copy of the software installed?
10. Does the software produce useful standard reports, and can those reports be readily customized?
11. Does the software permit easy creation of an animation, and can the animation be either two- or three-dimensional?

In addition to the simulation software itself, two other items in the software toolkit merit attention. One is the need for a strong general statistical analysis software tool, which will surely be used for both examination of input data and for analysis of output results. Typical and often overlooked, statistical examinations of input data are:

1. Are time-based observations autocorrelated (for example, do long cycle times occur in clusters because of arriving product mix or worker fatigue)?

2. Do differences exist among shifts (for example, more scrap produced by less experienced workers assigned to the night shift)?
3. Are there outliers in the data which merit re-examination, and possible correction or deletion (for example, as a result of oral communication in a noisy factory, was "fifteen" misreported as "fifty"?

Later in the simulation project, this highly capable statistical software tool will be useful for running Student-t tests and analyses of variance (ANOVA) to comparatively assess the merits of various alternatives investigated by the simulation model.

After this statistical software tool is selected, it should be checked for distribution-fitting capabilities. "Distribution-fitting" refers to the task of assessing a collection of numbers representing a data set (e.g., a collection of manual cycle times or a collection of recorded machine repair times) and determining which standard canonical distribution (if any) and which parameter values for that distribution provide a good fit (as assessed by a chi-squared, a Kolmogorov-Smirnov, and/or an Anderson-Darling goodness-of-fit test). When a good-fitting canonical distribution can be found, its use (compared to using the available data to define an empirical distribution) increases both the ease of building the model and the mathematical power with which it can be analyzed. Some standard statistical software packages provide this capability for only one or a few distributions (most frequently, the normal distribution); hence, a distribution fitter may well be needed. A detailed explanation of the importance of a distribution fitter and its typical use, plus an example of one such software tool, appear in (Law and McComas 2003).´

3.3 Examples

In our first example (Lang, Williams, and Ülgen 2008), simulation was applied to reduce manufacturing lead times and inventory, increase productivity, and reduce floor space requirements within a company providing forged metal components to the automotive light vehicle, heavy lorry [truck], and industrial marketplace in North America. The company has six facilities in the Upper Midwest region of the United States which collectively employ over 800 workers. Of these six facilities, the one here studied in detail specializes in internally splined (having longitudinal gearlike ridges along their interior or exterior surfaces to transmit rotational motion along their axes (Parker 1994)) shafts for industrial markets. The facility also prepares steel for further processing by the other five facilities. Components supplied to the external marketplaces are generally forged metal components; i.e., compressively shaped by non-steady-state bulk deformation under high pressure and (sometimes) high temperature (El Wakil 1998). In this context, the components are "cold-forged" (forged at room temperature), which limits the amount of re-forming possible, but as compensation provides precise dimensional control and a surface finish of higher quality. In this study, the simulation results were summarized for management as a recommendation to buy 225 heat-treat pots (there were currently 204 heat-treat pots on hand). The disadvantage: this recommendation entailed a capital expenditure of $225,000 ($1,000 per pot). The advantages were:

1. One heat-treat dumping operator on each of the three shifts was no longer needed (annual savings $132,000).
2. Less material handling (dumping parts into and out of pots) entailed less risk of quality problems (dings and dents).
3. The work to be eliminated was difficult, strenuous, and susceptible to significant ergonomic concerns.

Hence, from a financial viewpoint, the alternative investigated with this simulation study has a payback period just under 1¾ years, plus "soft" but significant benefits. Management adopted these recommendations and a follow-up check nine months after conclusion of the study confirmed the benefits were indeed accruing at economic accuracy within 4%.

In our second example (Dunbar, Liu, and Williams 2009), simulation was used to evaluate, and assess various alternatives for, a portion of an assembly line and accompanying conveyor system currently under construction at a large automobile transmission manufacturing plant in the Great Lakes region of the north-central United States. Two important and beneficial practices appear here: (a) the project definition specified a careful examination of a subset of the collective manufacturing process instead of a superficial examination of all of it, and (b) the project entailed examination of a manufacturing system under construction (as opposed to currently in operation, with perhaps painfully obvious inefficiencies). Both aspects of this study, warmly recommended by numerous authors (e.g., [Buzacott and Shanthikumar 1993]) increase the benefits of simulation by enabling a simulation study to address strategic and tactical issues as well as shorter-term operational issues. In this study, six alternatives were compared, involving three prioritization strategies at conveyor join points and two hypothesized arrival rates, considered orthogonally. Interestingly, of the three prioritization strategies investigated, the one predicted to minimize work-in-progress (WIP) was also predicted to be the worst at minimizing maximum queue residence time (a "minimax" consideration – minimize the "badness" of worst-case behavior. Furthermore, a different strategy was dramatically the best at minimizing both maximum length of important queues and the closely related maximum queue residence time. Thus armed with detailed and useful predictions, management chose to implement the latter strategy, and subsequent measurement of performance and economic metrics of the revised system have matched the simulation study predictions within 5%.

In our third example (Williams and Orlando 1998), simulation was applied to the improvement of the upper intake manifold assembly process within the overall engine assembly process – yet another example of examining an intelligently restricted, problematic subset of an overall process rather than "trying to model everything in sight." Specifically, production managers wished to increase production per unit of time cost-effectively. Two key questions, whose answers were correctly suspected to be highly interrelated even before formal analytical study began, were:

1. How many pallets should be used for upper-intake-manifold transport along the recirculating spur line within assembly operations?
2. From which operating station on the main line should a broadcast signal be sent to the recirculating spur line?

There were sixteen possible alternatives to consider, resulting from the orthogonal combination of four possible broadcast-point locations and four possible pallet quantities. These alternatives were tested against four performance metrics, which, in decreasing order of importance, were:

1. System throughput in jobs per hour (JPH)
2. Average time in system ("makespan")
3. Queue lengths (average and maximum) for the three different queues of pallets in the pallet loops
4. "Queue disparity," the difference between the average length of the longest of these three queues and the average length of the shortest of these three queues

After extensive simulation analysis and evaluation of the simulation predictions using design-of-experiments statistical methods, three of these alternatives, involving two different broadcast points and pallet quotas of either 15 or 22, emerged as "finalists." Of the three "finalist" scenarios, management chose one which performed well on these performance metrics and also (as confirmed via additional runs of the simulation under hypothesized increased workload) offered the best protection against potential production ramp-up requirements (which did indeed occur 2½ years later). After implementation, system performance agreed with the simulation study predictions within 5% on all four performance metrics.

3.4 A Glimpse into the Future of Simulation in the Automotive Industry

The authors foresee the following trends pertinent to the use of simulation in the automotive industry:

1. Increased awareness of simulation capabilities up and down the supply chain, so that small niche automotive-part suppliers, for example, become nearly as likely to use simulation beneficially as the large OEMs [original equipment manufacturers] already are.
2. Increased routine and ongoing electronic data collection via control devices and cameras monitoring production lines.
3. Increased ease and prevalence of importing such automatically collected data into simulation models seamlessly, without elaborate data-processing techniques requiring manual intervention.
4. Further penetration of simulation tools offering three-dimensional animation into this market.
5. Migration of simulation usage from outside consultants and specialized internal staff personnel toward line personnel who receive ongoing training and support from the consultants and specialized internal staff.

6. Increased collaboration of two (or even more) separate corporate entities within the supply chain on a simulation study whose results will benefit both.

Authors Biography, Contact

ONUR M. ÜLGEN - PMC

For over 30 years, PMC has been a leading provider of manufacturing, engineering, supply chain, and operations productivity solutions. Our data-driven productivity solutions help customers shorten product life cycles, increase quality and throughput, reduce lead time, and improve their return on capacity and technology investments (ROI).

PMC also provides Technical Staffing solutions designed to offer cost-effective, one-stop-shop solutions.

Our solutions are primarily targeted for automotive, aerospace and defense, AEC (Architecture, Engineering, & Construction), healthcare, and industrial manufacturing.

Source: http://www.pmcorp.com

ONUR M. ÜLGEN is the president and founder of Production Modeling Corporation (PMC), a Dearborn, Michigan, based industrial engineering and software services company as well as a Professor of Industrial and Manufacturing Systems Engineering at the University of Michigan-Dearborn. He received his Ph.D. degree in Industrial Engineering from Texas Tech University in 1979. His present consulting and research interests include simulation and scheduling applications, applications of lean techniques in manufacturing and service industries, supply chain optimization, and product portfolio management. He has published or presented more that 100 papers in his consulting and research areas.

Under his leadership PMC has grown to be the largest independent productivity services company in North America in the use of industrial and operations engineering tools in an integrated fashion. PMC has successfully completed more than 3000 productivity improvement projects for different size companies including General Motors, Ford, DaimlerChrysler, Sara Lee, Johnson Controls, and Whirlpool. The scientific and professional societies of which he is a member include American Production and Inventory Control Society (APICS) and Institute of Industrial Engineers (IIE). He is also a founding member of the MSUG (Michigan Simulation User Group).

EDWARD J. WILLIAMS - University of Michigan-Dearborn

Since its founding in 1959 with a gift of 196 acres from Ford Motor Company, the University of Michigan-Dearborn has been distinguished by its commitment to providing excellent educational opportunities responsive to the needs of southeastern Michigan. Shaped by a history of interaction with business, government, and industry of the region, the University of Michigan-Dearborn has developed into a comprehensive university offering undergraduate and master's degrees in arts and sciences, education, engineering and computer science, and management.

One third of the campus, more than 70 acres, is maintained as one of the largest natural areas in metropolitan Detroit, serving as a research and educational resource for the campus and the region. The Henry Ford Estate, home to the automotive pioneer and his wife, Clara, for more than 30 years and a National Historic Landmark, is located on the University of Michigan-Dearborn campus.

For the 8,600 enrolled students and 381 full-time instructional faculty, the University of Michigan-Dearborn is a place where students learn and grow, explore new ideas, and acquire the knowledge and skills they need to achieve their personal and professional goals. As graduates of University of Michigan-Dearborn, students will have a broad knowledge of the many fields of human achievement, and will be prepared for their careers with imagination, reasoning, and creative problem-solving abilities.

The University of Michigan-Dearborn is fully accredited by The Higher Learning Commission and is a member of the North Central Association of Colleges and Schools.

Source: http://www.umd.umich.edu

EDWARD J. WILLIAMS holds bachelor's and master's degrees in mathematics (Michigan State University, 1967; University of Wisconsin, 1968). From 1969 to 1971, he did statistical programming and analysis of biomedical data at Walter Reed Army Hospital, Washington, D.C. He joined Ford Motor Company in 1972, where he worked until retirement in December 2001 as a computer software analyst supporting statistical and simulation software. After retirement from Ford, he joined PMC, Dearborn, Michigan, as a senior simulation analyst. Also, since 1980, he has taught classes at the University of Michigan, including both undergraduate and graduate simulation classes using GPSS/H™, SLAM II™, SIMAN™, ProModel™, SIMUL8™, or Arena®. He is a member of the Institute of Industrial Engineers [IIE], the Society for Computer Simulation International [SCS], and the Michigan Simulation Users Group [MSUG]. He serves on the editorial board of the International Journal of Industrial Engineering – Applications and Practice. During the last several years, he has given invited plenary addresses on simulation and statistics at conferences in Monterrey, México; İstanbul, Turkey; Genova, Italy; Rīga, Latvia; and Jyväskylä, Finland. He served as a co-editor of Proceedings of the International Workshop on Harbour, Maritime and Multimodal Logistics Modelling & Simulation 2003, a conference held in Rīga, Latvia. Likewise, he served the Summer Computer Simulation Conferences of 2004, 2005, and 2006 as Proceedings co-editor. He is the Simulation Applications track co-ordinator for the 2011 Winter Simulation Conference.

Contact

Edward Williams
College of Business
B-14 Fairlane Center South
University of Michigan - Dearborn
Dearborn, Michigan 48126
USA
ewilliams@pmcorp.com

References

Banks, J.: Software for Simulation. In: Banks, J. (ed.) Handbook of Simulation: Principles, Methodology, Advances, Applications, and Practice, pp. 813–835. John Wiley & Sons, Incorporated, New York (1998)

Biller, B., Nelson, B.L.: Answers to the Top Ten Input Modeling Questions. In: Yücesan, E., Chen, C.-H., Snowdon, J.L., Charnes, J.M. (eds.) Proceedings of the 2002 Winter Simulation Conference, vol. 1, pp. 35–40 (2002)

Buzacott, J.A., George Shanthikumar, J.: Stochastic Models of Manufacturing Systems. Prentice-Hall, Incorporated, Englewood Cliffs (1993)

Dunbar III, J.F., Liu, J.-W., Williams, E.J.: Simulation of Alternatives for Transmission Plant Assembly Line. In: Balci, O., Sierhuis, M., Hu, X., Yilmaz, L. (eds.) Proceedings of the 2009 Summer Computer Simulation Conference, pp. 17–23 (2009)

El Wakil, S.D.: Processes and Design for Manufacturing, 2nd edn. PWS Publishing Company, Boston (1998)

Gordon, G.: The Application of GPSS V to Discrete System Simulation. Prentice-Hall Incorporated, Englewood Cliffs (1975)

Hounshell, D.A.: Planning and Executing 'Automation' at Ford Motor Company, 1945-1965: The Cleveland Engine Plant and its Consequences. In: Shiomi, H., Wada, K. (eds.) Fordism Transformed: The Development of Production Methods in the Automobile Industry, pp. 49–86. Oxford University Press, Oxford (1995)

Lang, T., Williams, E.J., Ülgen, O.M.: Simulation Improves Manufacture and Material Handling of Forged Metal Components. In: Louca, L.S., Chrysanthou, Y., Oplatkov, Z., Al-Begain, K. (eds.) Proceedings of the 22nd European Conference on Modelling and Simulation, pp. 247–253 (2008)

Law, A.M.: Simulation Modeling & Analysis, 4th edn. The McGraw-Hill Companies, Incorporated, New York (2007)

Law, A.M., McComas, M.G.: How the ExpertFit Distribution-Fitting Software Can Make Your Simulation Models More Valid. In: Chick, S.E., Sánchez, P.J., Ferrin, D., Morrice, D.J. (eds.) Proceedings of the 2003 Winter Simulation Conference, vol. 1, pp. 169–174 (2003)

Miller, S., Pegden, D.: Introduction to Manufacturing Simulation. In: Joines, J.A., Barton, R.R., Kang, K., Fishwick, P.A. (eds.) Proceedings of the 2000 Winter Simulation Conference, vol. 1, pp. 63–66 (2000)

Parker, S.P. (ed.): McGraw-Hill Dictionary of Scientific and Technical Terms, 5th edn. McGraw-Hill, Incorporated, New York (1994)

Ülgen, O.M.: GENTLE: Generalized Transfer Line Emulation. In: Bekiroglu, H. (ed.) Proceedings of the Conference on Simulation in Inventory and Production Control, pp. 25–30 (1983)

Ülgen, O., Gunal, A.: Simulation in the Automotive Industry. In: Banks, J. (ed.) Handbook of Simulation: Principles, Methodology, Advances, Applications, and Practice, pp. 547–570. John Wiley & Sons, Incorporated, New York (1998)

Williams, E.J.: Downtime Data – its Collection, Analysis, and Importance. In: Tew, J.D., Manivannan, M.S., Sadowski, D.A., Seila, A.F. (eds.) Proceedings of the 1994 Winter Simulation Conference, pp. 1040–1043 (1994)

Williams, E.J., Orlando, D.: Simulation Applied to Final Engine Drop Assembly. In: Medeiros, D.J., Watson, E.F., Carson, J.S., Manivannan, M.S. (eds.) Proceedings of the 1998 Winter Simulation Conference, vol. 2, pp. 943–949 (1998)

Zeigler, B.P., Praehofer, H., Kim, T.G.: Theory of Modeling and Simulation: Integrating Discrete Event and Continuous Complex Dynamic Systems, 2nd edn. Academic Press, San Diego (2000)

4 Simulating Energy Consumption in Automotive Industries

Daniel Wolff, Dennis Kulus, and Stefan Dreher

Energy and resource efficiency emerge as strategic objectives in the operation of discrete manufacturing systems. In the future, energy consumption will have to be evaluated early during the planning phases, requiring the application of simulation technology. This has, until now, not been implemented into the tools of the digital factory supporting this phase of the product lifecycle.

The presented chapter discusses an approach to integrate energy efficiency aspects into an established software tool for discrete-event simulation. The basic principles of energy simulation are detailed along the successive phases of a typical pilot study, explaining challenges and restrictions for model building and calculation as well as subsequent experimentation. Standardized modeling and visualization plays a dominant role in these considerations. An outlook to possible future developments and challenges concludes the chapter.

4.1 Introduction

4.1.1 INPRO at a Glance

Innovationsgesellschaft für fortgeschrittene Produktionssysteme in der Fahrzeugindustrie mbH (INPRO) is a joint venture of Daimler, Sabic, Siemens, ThyssenKrupp and Volkswagen. The Federal State of Berlin, where the company is based since its founding in 1983, is also a shareholder. The joint venture aims to drive innovation in automotive production and transfer the results of its research to industrial applications. INPRO has approximately 100 employees engaged in developing new concepts in the fields of production technology, production planning and quality assurance for the automotive industry in close collaboration with a large number of shareholder's experts. INPRO's applications laboratory and

Daniel Wolff · Dennis Kulus · Stefan Dreher
INPRO Innovationsgesellschaft für
fortgeschrittene Produktionssysteme
in der Fahrzeugindustrie mbH
Hallerstraße 1
D-10587 Berlin
Germany
e-mail: Daniel.Wolff@inpro.de

testing facility is located in Berlin. More information on INPRO and its range of services is available at www.inpro.de.

Quick Facts:

- Headquarter: Berlin, Germany
- Founded in 1983
- Approximately 100 employees
- Collaboration of strong shareholders from automotive industry

Tools for material flow simulations are used globally today. Already in the 1980`s, INPRO developed a solution for the simulation of material flows in production, the simulation system "SIMPRO". INPRO's goal at the time was to establish the methods for material flow simulation in the planning departments of its shareholder companies. Today, more than 500 simulation projects using the tool SIMPRO have been carried out.

4.1.2 About the Authors

Dipl.-Ing. Daniel Wolff, born in 1977, studied mechanical engineering with the focus factory planning and operation at TU Chemnitz. Since 2004, he is project engineer at INPRO GmbH in the division „Production Systems and Intelligence Processes".

Contact

INPRO Innovationsgesellschaft für
fortgeschrittene Produktionssysteme
in der Fahrzeugindustrie mbH
Hallerstraße 1
D-10587 Berlin
Germany
Email: Daniel.Wolff@inpro.de

Dipl.-Kaufm. Dennis Kulus, born in 1981, studied economics with the focus logistics at TU Berlin. Since 2008, he is project engineer at INPRO GmbH in the division „Production Systems and Intelligence Processes".

Dr.-Ing. Stefan Dreher, born in 1964, studied manufacturing technology at TU Berlin. From 1996 to 2003, he was member of scientific staff at Fraunhofer IPK, after that head of the "Digital Factory" unit at the consulting company InMediasP GmbH. Since February 2008, he is managing the INPRO division „Production Systems and Intelligence Processes".

4.1.3 Motivation

Reducing cost, improving quality, shortening time-to-market, while at the same time acting and thinking sustainably pose a major future challenge for manufacturing industries. Until today, the monitoring of energy consumption and the improvement of energy efficiency did not play a dominant role in the operation of manufacturing systems. This is about to change as energy costs become subject to sharper focus of factory operators and machinery users, due to a more intensive analysis of lifecycle costs [4.7]. This is true both while preparing invest decisions and while securing operative production.

In the context of sustainability efforts of manufacturing companies, the entire complex "energy and resource efficiency" therefore emerges as a strategic objective. Classical target parameters in planning typically include invest figures, time demands and the number of workers required for manufacturing, further the area of floor space required for production. Jointly, these parameters constitute the planning objectives. They serve as a starting point to develop alternatives and to prognose production costs, while comparing these alternatives. In the future, next to the established criteria mentioned above, „energy efficiency" will constitute an additional aspect that is to be considered during planning (Fig. 4.1).

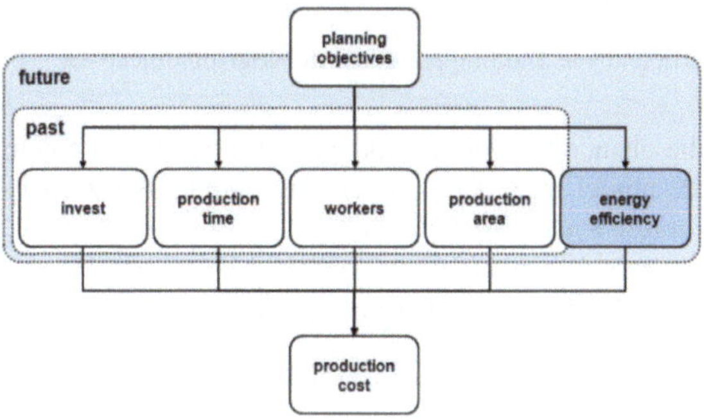

Fig. 4.1 Energy efficiency as a future strategic planning objective

While operative procedures are successfully establishing, the systematic consideration of system alternatives from an energy point of view especially during early planning phases - i.e. the concept stage - is still evolving. In this context, application of simulation technology holds great potential for energy-efficient manufacturing systems and processes.

Although in research first approaches are becoming available [4.10], the software tools of the Digital Factory so far do not include standard functionalities to calculate energy consumption. This is particularly true for material flow analysis based on discrete-event simulation. Therefore, augmenting the simulation tools with relevant energy flows seems like a promising approach. A review of commercially available simulation tools in this field reveals that these do not yet support those considerations.

The presented chapter discusses an approach to integrate energy efficiency into an established software tool (Plant Simulation, vendored by Siemens PLM Software), extending it in such a way that energy consumption values for an existing manufacturing system model can be calculated. Matched to logistic objectives such as minimization of inventory and throughput time or the maximization of utilization and adherence to schedule classically being evaluated in simulation, energy efficiency must therefore be valued against these (Fig. 4.2).

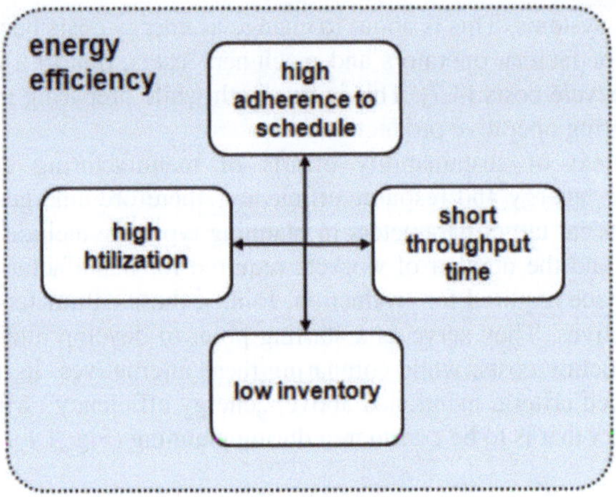

Fig. 4.2 Energy efficiency as framing parameter for logistic objectives

This offers the chance to evaluate changes of dynamic parameters and interacting effects in the model free of risk, deducing potentials for energy consumption reduction even before system realization.

4.1.4 Scope of the Proposed Approach

Figure 4.3 highlights the scope of the presented chapter, showing the characteristics of the taken approach.

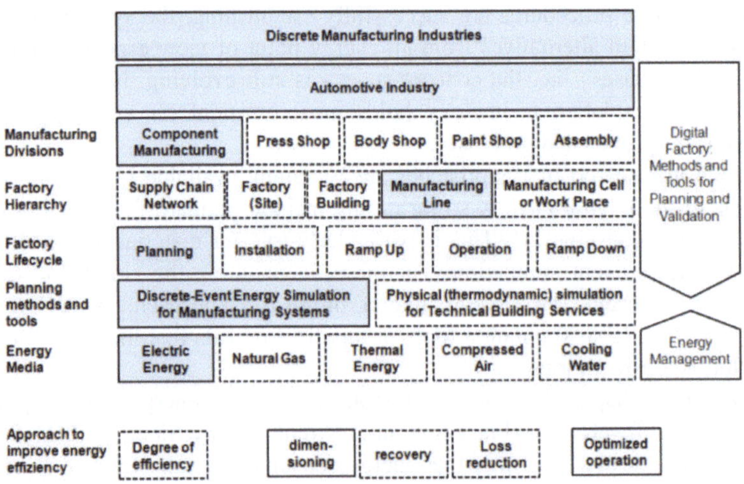

Fig. 4.3 Scope of discrete-event energy simulation, highlighting the described approach

Concentrating on the field that INPRO`s activities are primarily located in, the automobile manufacturing domain, selected crafts were subjected to sharper focus. In the production creation process, in the planning phase the essential strategic decisions are made. Foremost, consumption of electric energy was evaluated with the help of discrete-event energy simulation (energy simulation).

One of the drive manufacturing lines of the cylinder head “1.6 TDI common-rail” in a Volkswagen factory served as a pilot use case for energy simulation (discussed in 5.9). The component manufacturing processes performed in this use case require high amounts of electrical energy and other resources. Therefore, this production process represented a suitable pilot study.

Special focus was laid on the mechanical finishing processes after the foundry. These are located in the motor factory Salzgitter and can be divided into various steps for machining, assembly and cleaning. The machining workflow begins with drilling and milling operations and continues with washing to remove tension and cooling lubricant residues. After cleaning, the unfinished cylinder head is tested for leaks, followed by different assembly stations. The mechanical finishing is completed with the final cleaning and the manual inspection of each cylinder. Figure 4.4 shows an overview of the production system modeled in Plant Simulation.

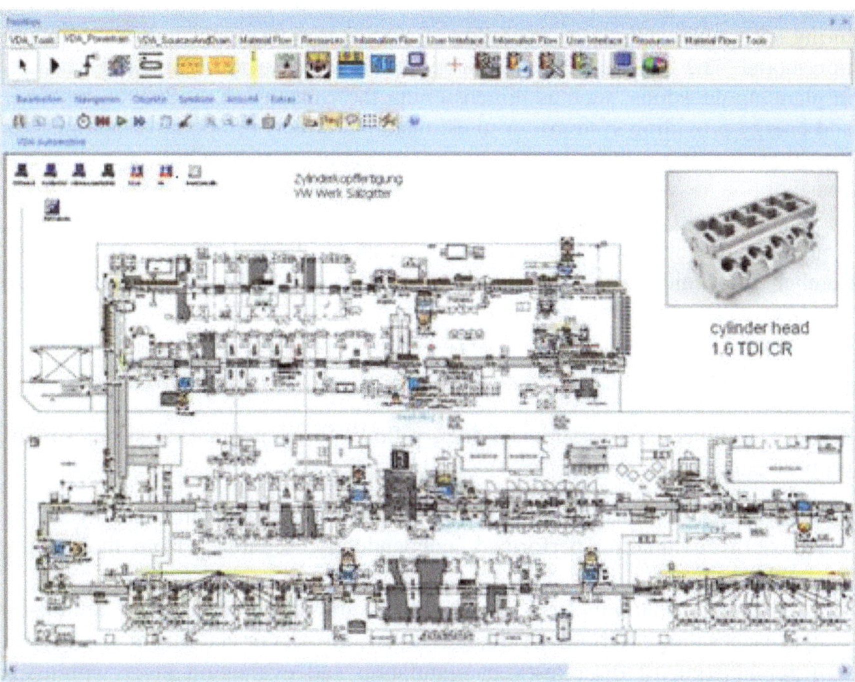

Fig. 4.4 Pilot study “cylinder head manufacturing”.

4.2 Energy Simulation

4.2.1 Definition

According to the VDI guideline 3633 [4.7], the term „simulation" refers to reproducing the dynamic processes in a system, with help of a model. This model must be capable of experimenting, so that knowledge can be gained and transferred to reality. When simulating energy and resource flows of manufacturing systems, the "system" may be perceived as the "traditional" material flow and manufacturing system, being extended to include a view on relevant energy sinks (consumers) and on the technical devices supplying energy and providing auxiliary materials, such as e.g. pressurized air, lubricants or technical gases.

The "dynamic processes" to be reproduced consist of the material flow processes that trigger the resulting electric energy consumption plus the flow of other energies and media. The latter may be modeled explicitly as moving objects, or may only be calculated on the basis of the material flow. Their dynamics result from the fact that consumption is directly influenced by the flow of materials and products. Additionally, both the technological manufacturing process itself and the operational state of the manufacturing resources influence energy consumption, which therefore varies over the course of time.

„Capable of experimenting"means that structural modifications of the manufacturing system as well as the operating strategies may be evaluated in the simulation model. The knowledge about the system`s behavior thus gained can be used in planning decisions, such as dimensioning the capacity of production resources and energy providing systems, or to estimate the effects of operative optimization measures. Foremost, this allows to exploit potentials to reduce both overall energy consumption at system level as well as energy per part produced.

Finally, „reality"can be understood as the designed planning solution if energy simulation is performed ahead of the realization phase, or the as an evaluated number of technical and organizational measures to be taken, if simulation is applied during the operation phase.

4.2.2 Simulating Energy in Discrete-Event Simulation Tools

Three paradigms exist to simulate energy flows in manufacturing systems.[4.10] propose either

- a „coupling of discrete-event simulation and external evaluation layer", i.e. an energy evaluation performed independently, separated from the existing simulation tool,
- a „dynamic coupling" of discrete-event simulation and other simulation approaches plus additional evaluation layers, internal or external or
- a combination of discrete-event simulation and evaluation layer within one application.

Acc. to [4.10], especially the last option (combined approach) is suitable to evaluate dynamic energy consumption on a system level. Thus, no further tools are required outside of the discrete-event simulation tool. This has advantages regarding model integration. Available functionalities, however, are limited by the simulation tool. Interaction with technical building services, for example, is limited, considering the restrictions of discrete-event simulation.

The approach presented in this chapter is based on this last option, the combined approach. Following, the basic functionalities as they were implemented in the simulation tool "Plant Simulation" are discussed in more detail.

4.2.3 Principle of Energy Simulation

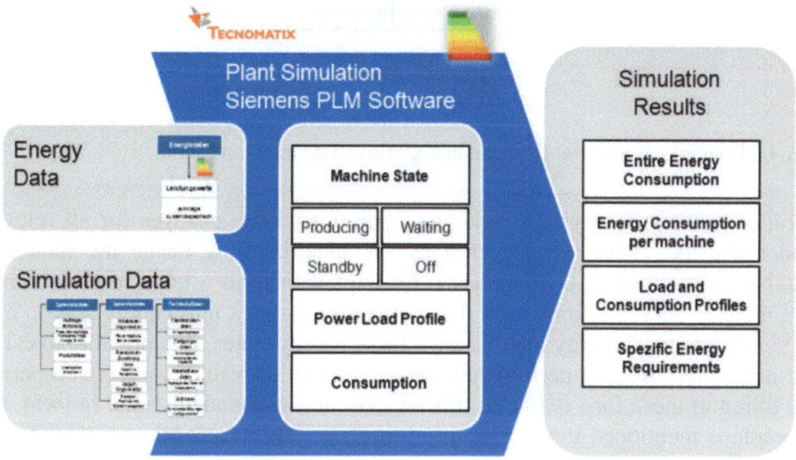

Fig. 4.5 Principle of energy simulation

A material flow simulation run will generate operational states for all model objects. These typically represent state such as Producing, Waiting, Failure, Setup etc. After a simulation run, time and utilization statistics provide information regarding the time share each object spends in the respective operating states.

To perform energy simulation based on these premises, is has to be assumed that the energy demands of the modeled resources vary according to their operating state (Figure 4.5). This behavior can either be constant or time-dependent. [4.4]

In the area of machine tools, [4.5] propose that power consumption during production can be distinguished into different levels. Praxis shows that energy consumption primarily depends on the type of operational state [4.4]. These states can be viewed as discrete segments, in combination representing a manufacturing task. To perform an energetic evaluation of dynamic load and consumption behavior of the modeled system, information about operational states has to be supplemented by information describing the energetic flows, thereby transforming the operational states into "energy states".

The principle of analyzing energy states in a material flow based simulation can be illustrated as shown in Fig. 4.6.

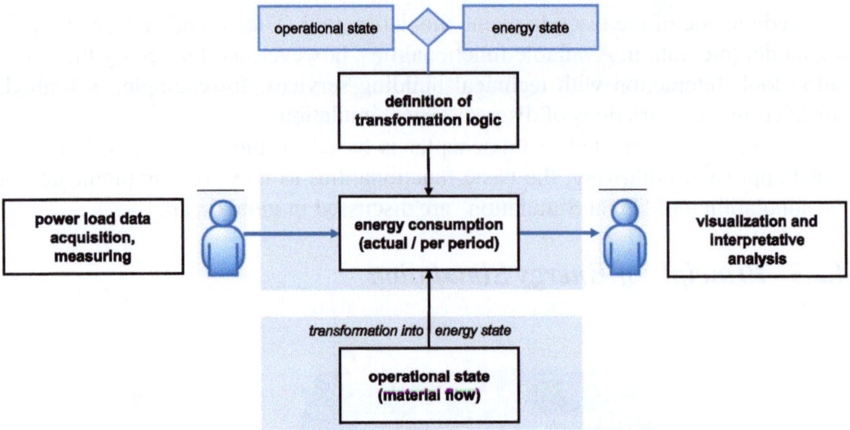

Fig. 4.6 Principle of material flow and energy flow state transformation.

First, a simulation system generates operational state changes for all relevant model objects. These are triggered by the material flow inside the model. A matching algorithm, in the simplest form implemented as a table or programmed as a method, serves to transform these operational states into energy states. With previously defined energy load data for these energy states, it is possible to calculate actual system load performance and consumption values for a given period, and to report these for - online or offline - visualization and analysis. In [4.5], this principle is mentioned in the domain of machine tools. According to this, energy consumption of a milling machine results from combining the energy load data for different operational machine states with the usage profile of a machine, representing the ordered sequence of states and their respective duration.

Regarding a general definition for the energy states required for state transformation, in practice various classifications are used. Literature review shows that currently no common definition for energy states of manufacturing systems exists. Also, energy states can differ according to application area and manufacturing craft (e.g. for body shop / robots, for component manufacturing / machine tools, for paint shop etc.):

- Typical is the distinction between four basic states with energetic relevance: "Off", "Standby", "Ready-To-Produce" and "Producing". [4.2]
- Alternatively, "Power Load during Start Up", "Base Load" and "Power Load during Manufacturing" are proposed. [4.1]
- Specifically for machine tools, aside from the state "Producing" the two states „Waiting in Manual Mode" and „Waiting in Automatic Mode" are distinguished, the last of which corresponds to the earlier mentioned state of "Ready-To-Produce". [4.2]

A practical classification system for energy states is proposed in [4.4] (shown in Figure 4.7). According to this methodology, production processes can be separated into segments with specific energy consumption, called "EnergyBlocks".

These segments are defined for the possible operational states of manufacturing equipment.

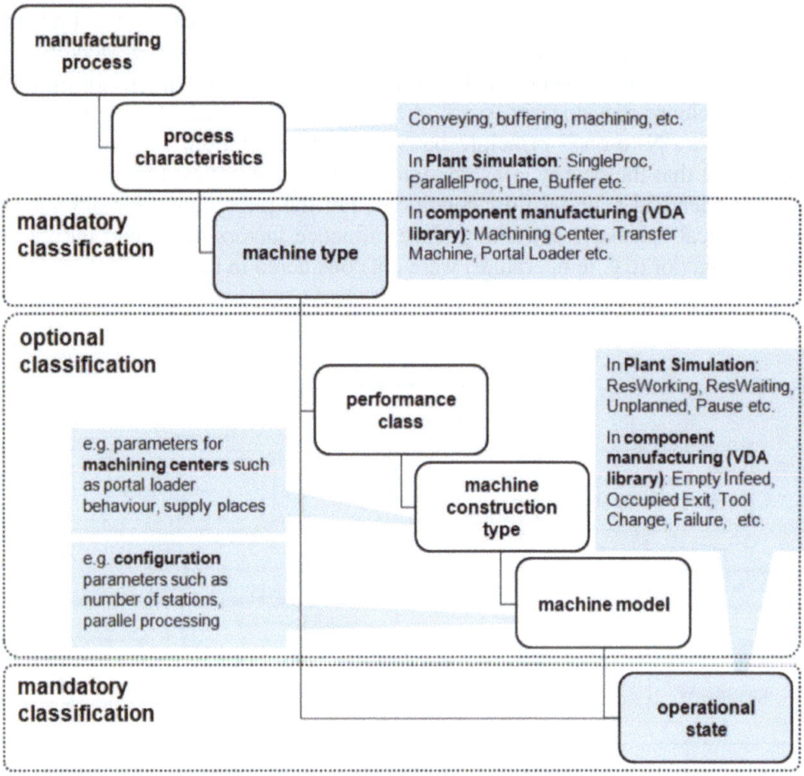

Fig 4.7 Classification system for operational states acc. to [4.4]

The actual transformation of an operational state into an energy state can be performed in different ways, as shown in Figure 4.8. In the pilot study, a deterministic approach was taken, defining exactly one energy state for every possible operational state, i.e. according to the "N-to-one" principle. This simplifies the matching process, since exactly one target state can be identified for each operational state. In contrast to this, a "one-to-N" principle implies that energy state changes cannot be calculated exclusively from material flow, because while the system assumes different energy states, no operational state change must necessarily occur at the same time. In reality, this may result from different product types or materials, requiring different amounts of energy on the same manufacturing step. Other reasons may be different manufacturing process parameters, such as milling speeds or feed rates, or special machine characteristics, or even external influences such as temperature.

Alternative ways to match states are a unified definition, valid for all relevant machine types, or a machine-specific definition. A unified definition for all model elements represents the most pragmatic approach, provided all types of machinery assume the same types of energy states. In the pilot study a combined approach was required: Buffer systems, for example, do not assume the same typical manufacturing states as machine tools, but rather consume energy during "loading" and "unloading". Therefore, specific definitions for each machine type, if not for the individual machines, became necessary. A static matching algorithm is the simplest solution for a "N-to-one" principle based, unified definition. Dynamic matching would imply that the energy consumption of a machine varies during a simulation run, given identical material flow conditions. For the pilot study, this did not seem like a practical approach, especially since influence factors that could induce such machine behavior (e.g. temperature) were not considered in the model.

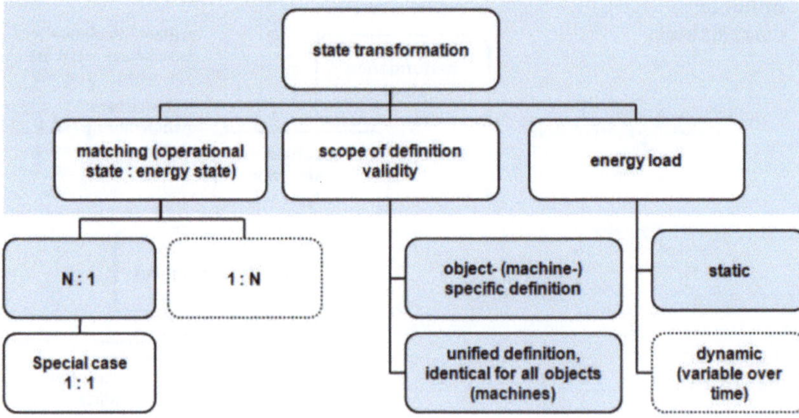

Fig. 4.8 Classification of state transformation strategies.

Figure 4.9 shows the calculating logic to implement the functional principles discussed above. Three basic steps are performed in a calculation cycle inside the simulation model to calculate energy consumption.

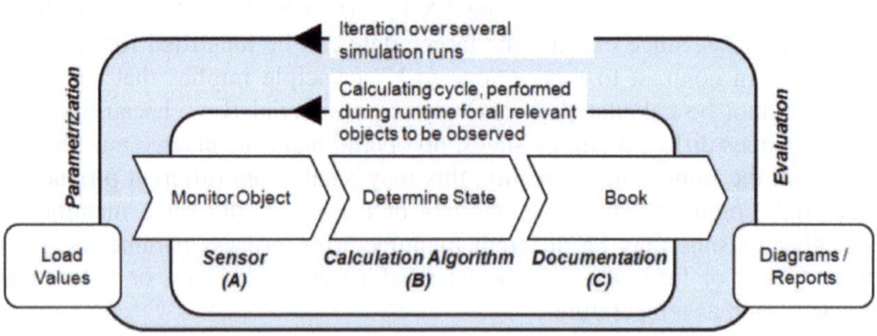

Fig. 4.9 Calculation cycle.

As elementary step, a state sensor (A) is introduced into the model. It monitors state changes in all relevant model elements. This sensor can either be implemented as a method or as an observer in Plant Simulation. It detects changes in the object attributes or in the variables that are used to describe material flow states. For example, at a conveyor modeled with a "line" object in Plant Simulation, different attributes such as ResWorking, Pause etc. can be observed. For machine objects that are modeled as network objects due to their complexity (as implemented in the VDA library, cp. section 4.1.4), status variables exist that internally translate material flow into operational state information for this object. Based on the above discussed principle of transformation and with the knowledge of load values provided as input parameters, in each cycle the current energy state can be determined (B). Finally, the results are booked to logging tables in a documentation step (C). This provides the basis for visualization (in diagrams) and later statistic evaluation (in tables and reports).

Implementation of this logic must take into account that at any point in time, current power load can be calculated, documented and visualized, however for the calculation of resulting energy consumption the elapsing of the current state has to be waited for. Thus, two steps are required:

- Step 1: Determine current power load, valid during the current cycle.
- Step 2: Determine current power load, valid during the new cycle and determine consumption for the elapsed cycle.

To realize this logic, additional functionalities are required that have to be implemented as model elements. In the pilot study, this was done by programming specific methods. In doing so, programming state sensor methods dedicated to single machines turned out to be practicable. Methods to determine operational and energy states as well as the booking steps, however, could be implemented as universal methods to be used with different machines. The implementation is described in more detail in Sect.4.1.4.

4.2.4 Process-Oriented Approach to Energy Simulation

In order to establish energy simulation in the envisioned sense, a procedural model is required. VDI guideline 3633 [4.7] offers a reference process to carry out simulation studies. Structured into the preparatory phase, the execution phase and the evaluation phase, this procedure is well suited to serve as a basis for simulating energy consumption of discrete manufacturing systems, if adapted to the specific aspects of energy simulation. This is shown in Fig.4.10. Subsequently, each of the procedure steps will be discussed regarding their significance in energy simulation.

In the pilot studies in component manufacturing, the simulation studies were carried out according to this adapted model.´

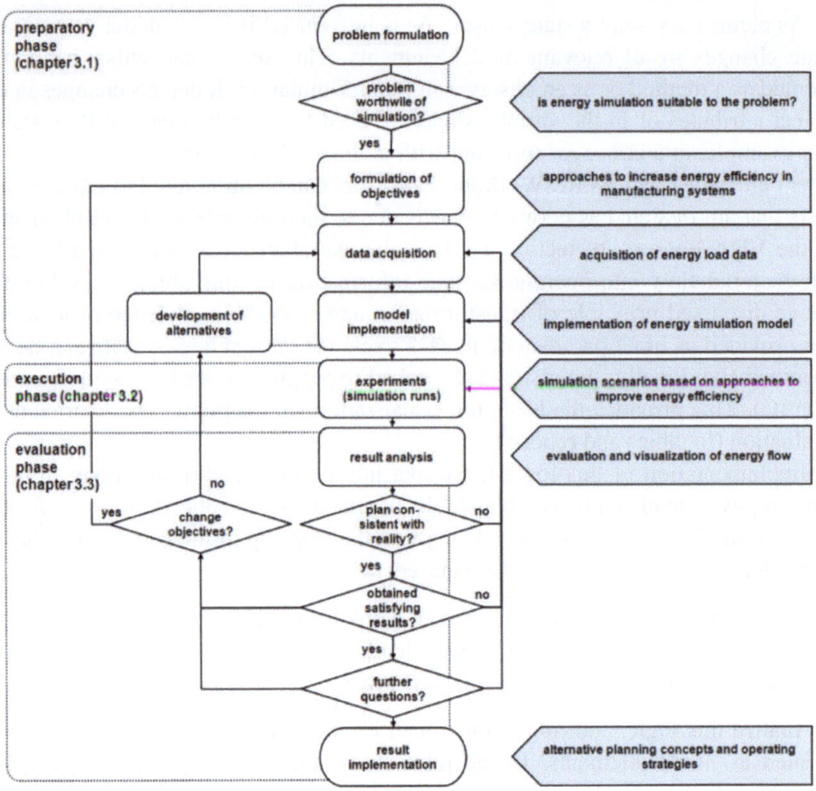

Fig. 4.10 Simulation procedure for material flow simulation, based on VDI 3633, extended with energy aspects

4.2.4.1 Preparatory Phase

The preparatory phase starts with the first step of problem formulation. A range of potential uses can be envisioned by the systematic application of energy simulation, as e.g.

- To prognose the energy consumption of manufacturing systems;
- To generate performance indicators describing the energetic behavior of manufacturing systems, e.g. according to VDI 4661 [4.11];
- To assess interdependencies between energy consumption of a system and the basic structural and parametric design decisions, in order to deduce options to influence planning and operation of these systems;
- To visualize energy flows (e.g. load and consumption profiles) inside the modeled systems, showing the dynamic properties of the flows and their correlation to production profiles;
- To differentiate value-add and non-value-add energy consumption;

- To evaluate and validate technical and organizational approaches to increase energy efficiency and the actual measures to be taken, ahead of the realization phase;
- To quantify these optimization potentials;
- To calculate (hitherto unknown) specific consumption, i.e. consumption in relation to the volume (number) of parts produced;
- To assess goal conflicts and problem shifts, to subsequently address and avoid these.

With regard to a specific manufacturing system, these potential uses illustrate the way that problem formulation for an energy simulation study can be detailed.

4.2.4.1.1 Suitability of Simulation

The fundamental decision to apply material flow simulation to a defined planning problem should contain a reflection if simulation is suitable to the problem and if the problem is worthwhile of simulation [4.7]. Also, costs for conducting the simulation study as well as the potential benefits should be considered. When proposing energy simulation, therefore, these questions similarly should be raised.

Criteria in this decision include, first, the complexity of the problem analyzed: Can the system be analyzed using analytical methods? In analogy to a „regular“ material flow simulation, energy simulation can be judged as at least equally complex since the energy flows logically result from the operational machine states induced by material flow. A machine object assuming the operational state “Producing” (reflected by a „resWorking“ attribute in Plant Simulation, for example) must automatically incur an energy demand according to this state. Therefore, the additional energy state view on the system`s behavior adds an extra complexity layer, justifying a simulation approach.

Another criterion is the accessibility for experimentation. Since in the planning phase the manufacturing system to be realized typically does not yet exist in the envisioned configuration, uncertainty regarding the realistic energy load values can be assessed using statistical experiments. During the operation phase, boundary conditions regarding the operating strategies can be tested free of risk.

A further aspect will be the evaluation of the energetic system behavior over longer time periods. Additionally, energy consumption itself can exhibit time-dependencies, as e.g. in a “Waiting” state that is changed to a “Standby” state with reduced energy consumption after a defined period of time.

Against the background of these listed reasons, among others, application of simulation to the problem of energy consumption analysis in material flow systems seems suitable.

4.2.4.1.2 Formulation of Objectives

Potentials to improve energy efficiency in manufacturing systems exist in different areas. 4.12 distinguishes between six principal approaches: Higher degrees of efficiency, reducing energy losses, energy recuperation, energy substitution, optimal dimensioning and optimized operation. Foremost, the latter two approaches seem the most promising to be evaluated and quantified by energy simulation. [4.12]

"Optimal dimensioning" relates to the danger of oversizing reserves, installed to handle failure situations, which in turn leads to low degrees of efficiency at manufacturing stations. Additionally, energy infrastructure is installed based on the energy demand prognoses, so that oversized capacities in this area incur further idling losses, aside from unnecessary invest. To focus on the second approach, "optimized operations" are to optimize the load profile of a manufacturing system, avoiding non-productive operation times, and to adapt the energy absorption of the machines to the actually required power demand (secondary media etc.) [5.12]

If not taking into account aspects like energy provisioning or the transformation and transmission of energy to the final point of consumption, optimized energy use therefore represents the most reasonable approach for increased energy efficiency in manufacturing (see Fig. 4.11). While production volume must always satisfy the requirements (representing a basic planning premise) the reduction of

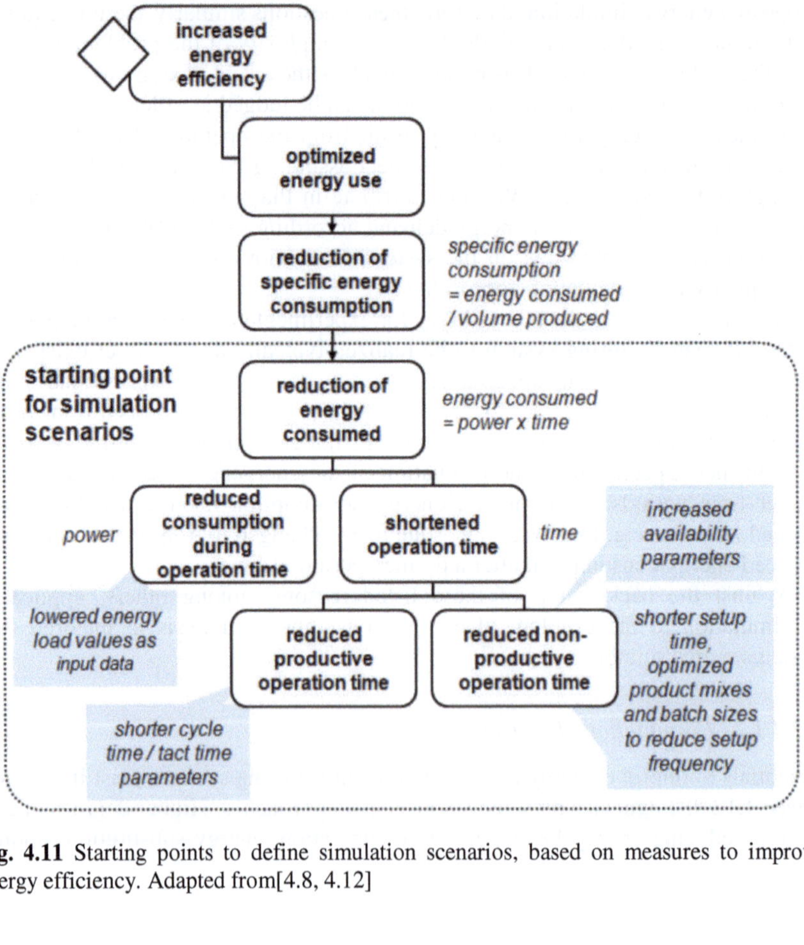

Fig. 4.11 Starting points to define simulation scenarios, based on measures to improve energy efficiency. Adapted from[4.8, 4.12]

energy consumed can be achieved either by reducing consumption (power load) during operation time, or by shortening the duration of the operation time. Applied to simulation, this corresponds to calculating a model with lowered energy data input values, or to evaluating alternative operating strategies to control the material flow in the system. The second approach can either focus on the reduction of productive operation time, i.e. shorten the cycle time on selected machine objects, or aim at reducing non-productive operation times, which relates to identifying operating strategies resulting in low time percentages for waiting times, setup and inspection times and failures.

4.2.4.1.3 Data Acquisition

The simulation of energy consumption is based on a material flow model. Therefore, the basic system load, organizational and technical data required to configure and initialize the model are also required for energy simulation. The consumption and performance data for all modeled machines and operating states must complement this, as shown in Fig.4.12.

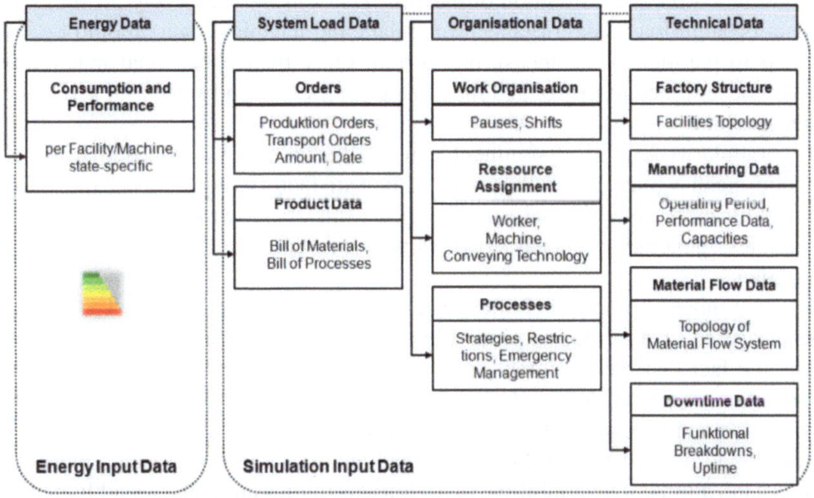

Fig. 4.12 Data inputs for energy simulation, adapted from VDI 3633 [5.8].

The starting point for data acquisition is the measurement of electric power in the field. In the pilot studies, mobile technology was used. With a data logger (e.g. from Company "Janitza") the electric measurements (such as power, voltage, current, cos φ, etc.) can be logged on the central power supply of each machine. The logged data is then transferred to a PC and analyzed, e.g. using the Software "GridVis".

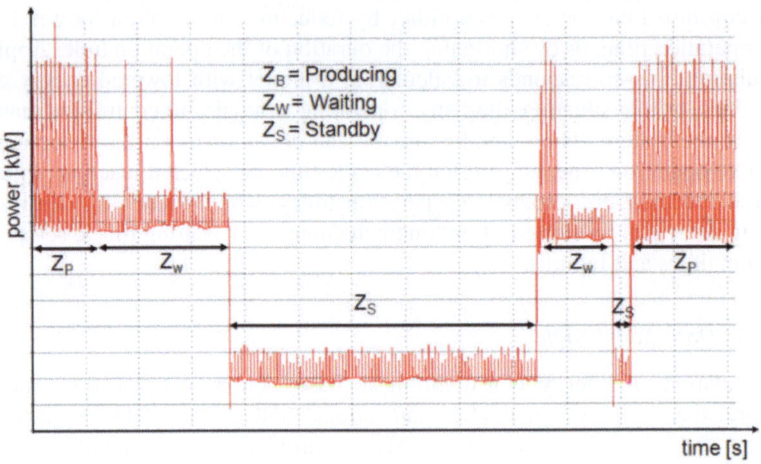

Fig. 4.13 Logged measuring profile (see [4.10]).

The actual power load of a machine largely depends on the current operating state. For the correct identification of machine states in the measuring profile, various data (system load data, organizational data and technical data) should be documented in parallel to the measuring period. Only then operational machining states can be assigned to the logged measuring profiles, as shown in Fig. 4.13 for a transfer machine.

The granularity of this assignment can be discussed. As proposed in [4.4], arithmetic mean values generally prove to be satisfactory, considering the effort necessary for more detailed analysis. Therefore, this approach was followed in the pilot studies, generating energy load values for representative periods in the measurement.

Also, a number of practical challenges exist during data acquisition. The tracing of the measured energy loads to individual machines might not always be possible, due the fact that measurement opportunities may only exist on central power supplies. Access to the electrical cabinets in praxis is restricted, requiring maintenance personnel to assist in the measuring process. Under certain circumstances, this can delay or even hinder long-term readings to acquire representative data, due to organizational unavailability. Also, long-term measurements quickly generate very large amounts of data. Overall, the effort involved in measuring and analysis must not be underestimated and therefore represents a critical step in the setup of an energy simulation.

To acquire energy data during the production creation process, a number of principal options exists (Fig. 4.14). Today, a continuous lifecycle of energy load data is not defined in praxis. In the planning phase, load values can be approximated from the knowledge about installed power supplies, considering simultaneity factors or correction factors. This results in rather imprecise data. Another principal option is to use reference values from previous experience, based on expert knowledge, or from past simulation studies. A definition of reference machines and processes should support this. More exact are laboratory values gathered from machine manufacturers, or values acquired during acceptance tests in the commissioning phase of machine installation. Processes to collect this kind of information are still to be established. During the operation phase, finally, measuring under field conditions becomes possible. Manual analysis to gain representative load data for all operational states from the measurements will still be required, as described above. Alternatively, data might be gained from Energy Monitoring Systems, where installed.

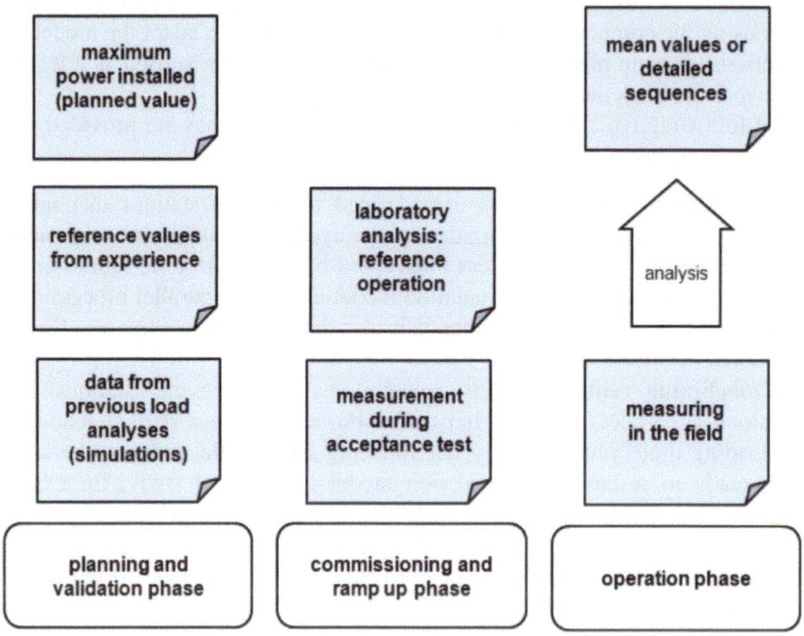

Fig. 4.14 Principal options to acquire energy load data during the production creation process.

4.2.4.1.4 Model Implementation

Implementation of the envisioned approach was based on certain premises. An established simulation tool should be used in order to assure conformity with existing processes of automobile manufacturing. The modeling and simulation system "Plant Simulation", produced by Siemens PLM Software, fulfills this requirement. It is established as a standard tool for material flow simulation with major German automobile manufacturers and suppliers, among others, BMW, Daimler, Ford, Opel and Volkswagen. For this reason, the tool was applied to the pilot study on which this chapter is based, the manufacturing craft „component manufacturing".

Plant Simulation is built on an object-oriented approach. As one possible way to build models in Plant Simulation for the Automotive manufacturing domain, a standardized component library published by a working group of the German Association of the Automotive Industry (Verband der Automobilindustrie e. V. - VDA) exists. This library, called "VDA Automotive Bausteinkasten" (VDA automotive component library), facilitates model building and parameterization for tool users by offering pre-defined components for typical machines of component manufacturing, as well as other crafts. The configuration of simulation models that are based on these components can thus be done more quickly. Input data is documented in a standard way (by using pre-formatted tables and import/export functions) so that users can focus on critical programming issues and experiments. When using the component library, individual extensions to adapt the model components still remain possible. This VDA library therefore was used as a technical framework for implementation.

The following typical component manufacturing machines are provided by the VDA library (Figure 4.15):

- A **transfer machine** consists of a defined number of stations that must be traversed in series. After expiration of the cycle time, all component parts are moved ahead one station, unless interrupted by e.g. inspection, failure or setup events. Each station in the simulation model allows for parallel processing and empty cycles. Transfer machines can also be used for continuous flow, for instance washer.
- A **machining center** typically contains several processing stations. These stations can process two parts in parallel. For each process, defined loading and unloading times must be set. After finishing one parallel process, the stations are ready to restart. The simulation model in the pilot study, for example, contains two machining centers, each with five parallel processing stations.
- **Portals** are conveyors for the loading and unloading of processing stations. Each portal typically consists of two different loaders. One loader supplies stations with blanks or raw parts. The other loader removes the machined parts from the station and transfers them to the unloading point. During the loading/unloading process a certain strategy is implemented, so that the loaders do not block each other in a deadlock. Different settings, such as speed, acceleration or positions of the loaders can be applied.
- **Buffers** are used for decoupling work sequences by storing machined parts. Typically a part is moved to the storage when parts on the main conveyor line can pass on and adequate buffer capacity still exists. The simulation model in the pilot study, for example, includes three buffers positioned between transfer stations. Each buffer can store or deliver two parts at the same time. The user can choose between two different strategies (FIFO or LIFO) according to which the parts from the buffer return into the material flow. Different settings, such as buffer capacity, time to store in/out or delay time, are possible.
- The **inspection station** allows to channel defective parts out of the process. The inspection station can be placed between conveyors and/or machining stations such as transfer machines or machine centers. With each new batch, an inspection of the first part of the new batch is performed, during which the material flow is blocked. During the inspection any number of parts can be tested and sorted out, in case the quality does not meet the requirements.

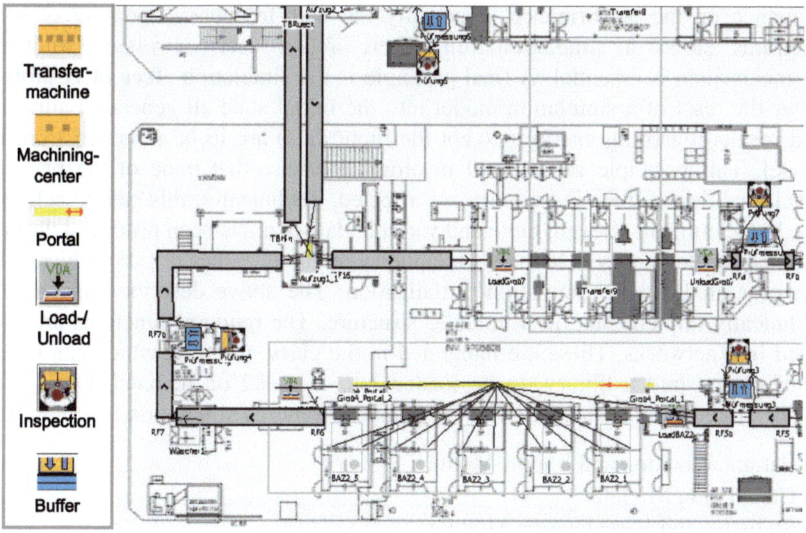

Fig. 4.15 Component manufacturing machines.

As a first premise, therefore, the modeling principles of this library had to be complied with in order to ensure the future integration of the developed energy efficiency modules into the standard.

One of these principles is **modularity**: The required functions should be clustered into modules that can be loaded as an extension to the basic VDA library. Plant Simulation offers a standard functionality ("Update Library") for this import. Thus it is ensured that the imported functionalities can be removed again from a model at any time, for example if a "standard" simulation (without the energy calculation) should be performed or if a model should be transferred to other business partners. Modularity is therefore implemented both in the global functionality as well as in single functions, as Fig. 4.16 shows.

A second consideration affects the **control** aspect: The user should be able to activate or deactivate selected energy calculation functions according to his need. Therefore, switch-on/switch-off boolean variables were implemented for different functionalities, such as global calculation, statistics documentation, specific consumption and others. Preferably these control variables should not be changed during a simulation run, in order to generate valid results. Still, total user control over the performance of a simulation model thus is assured.

Simultaneity: Looking at runtime, the energy calculation should be performed during the simulation run, to facilitate debugging and communication of results. This should generate a dynamic view onto energy consumption during simulation. It is important to recognize that the implementation of this principle has negative influence on the performance. Especially the high-frequency updating of some diagrams slows a simulation run. Therefore, a user-controlled switch-off/on-mechanism is essential. A final principle is the simulation „**free of residues**". After the reset of a simulation model into the initial state all generated attributes and results regarding energy (except the input data) are to be removed from the model. This principle is followed in order to ensure that none of the principal functionalities of the VDA library are affected. Technically, this can be achieved e.g. by deleting all relevant table and variable data that has been produced during a simulation run and by deleting all temporary methods, tables or other objects that are generated during simulation initialization. The above described approach is technically implemented in a modular structure. The required functions are clustered into networks. These are integrated into a class structure, which can in turn be clustered into a library, to be loaded into a model on demand. During this import, three modules are sufficient to realize the energy simulation functionality:

- Parameterization and Import Module
- Calculation Module
- Statistics and Visualization Module

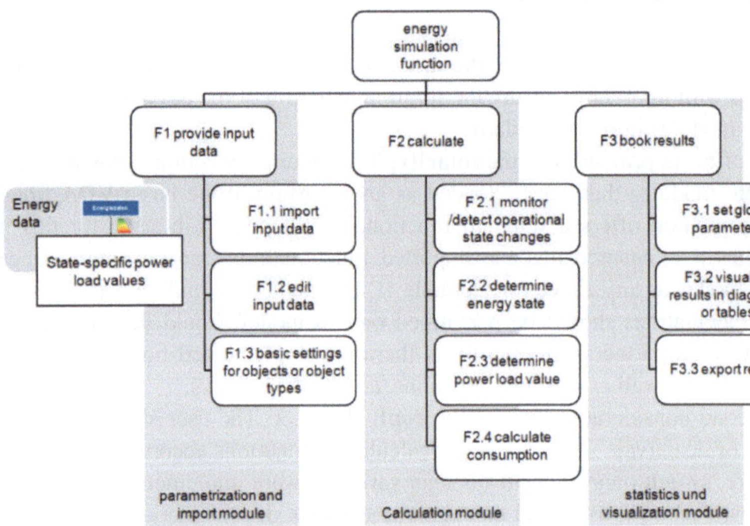

Fig. 4.16 Schematic overview of required functions to integrate energy consumption into material flow simulation [4.10].

Figure 4.16 shows the elementary functions required to realize the approach. Following is a short technical description of these:

Providing the necessary input data (F1) deals with the import of prepared energy values, i.e. the state-specific energy values, into parameter tables inside the model (F1.1). These tables should be accessible by the user in order to edit or update them if necessary (F1.2). Also, basic parameter settings should be available, as e.g. simulating only certain types of model objects (such as the object type "SingleProc" or

"Line") or only typical components of the VDA library, that modeled as networks representing machine types. In this way, simulation can be performed with focus only on specific model objects, or with focus on the entire system, according to the specific aspects that are to be examined. In the pilot study in drive manufacturing, for example, the model consisted of a significant amount of conveyor belts modeled using the "Line" element in Plant Simulation. Since, however, the conveyor systems were responsible only for a limited share of energy consumption in the system, it was not desirable to focus strongly on the conveyors. By eliminating them from the energy monitoring mechanism, model complexity could be reduced.

The calculation module (F2) contains a state monitoring function (F2.1). Here, the selected model objects have to be monitored to detect operational state changes. These can be changes in material flow, observable via object attributes like "ResWaiting" (e.g. on a "SingleProc" object) or operational status variables like "Occupied Exit" (as used by the VDA library). With the knowledge of the current operational state, the corresponding energy state can be determined (F2.2) and the matching power load value can be read from input data (F2.3). Finally, after the current state is elapsed, consumption results from power load and state duration (F2.4). To keep it simple and accessible, the matching algorithm to assign energy states to certain object attributes can be modeled statically in a two-dimensional table. This leaves flexibility to change assignments if necessary, should additional states be required.

The documentation of the calculated values is implemented the statistics and visualization module (F3). Here, global parameter settings allow to determine if certain booking operations are to be performed or not and which type of table or diagram should be used. This allows to use special documentation issues that are calculation-intensive, such as specific energy consumption per part (requiring the parallel logging of throughput) or a regular arithmetic mean calculation; e.g. for power load, consumption per part or for each energy state. In a simulation run, the user can access different diagrams and tables to visualize calculated consumption. After a simulation run, the results can be exported to a spreadsheet application (e.g. MS Excel).

4.2.4.2 Execution Phase

In the pilot studies in drive manufacturing, both analyses regarding the principal behavior of model elements and the influence of practical measures in production system operation were evaluated. This included the modification of energy load values under the premise of different technical measures taken to optimize energy consumption at single machines. Variation of input data can easily explore these scenarios effectively.

More significant changes of the existing manufacturing process were performed by modifying the process order in the manufacturing line:

- Where technologically practicable, consecutive machining steps can sometimes be integrated into one single process, or can even be assigned to the same equipment or machinery. This allows for analyses of the resulting energy demand now occur-ring during longer periods at the occupied resource, while at the same time reducing setup and waiting times in the surrounding machinery.

- Machining processes typically produce chippings that necessitate repeated washing steps. Eliminating these washing steps in the process sequence holds great potential for saving energies required for heating, pumping capacity and the auxiliary media flows such as coolants and washing solution. However this option remains technologically delicate for danger of spreading chippings into successive machining processes. The conducted experiments were therefore performed with reservation, perceived as case studies to determine the principal potential of eliminating certain washing processes.

To evaluate possibilities for energetically optimized operation, simulation scenarios can focus on control strategies. Classical product mix and batch size variation experiments can show how these parameters influence both absolute and specific energy consumption [4.9]

- Increase in batch size in the pilot scenario, for example, quickly showed that this typically results in higher throughput and therefore increases overall energy consumption. Since the increase in volume output was even higher, however, specific energy consumption per part actually could be improved. This results from a reduction of setup times and associated idle-running consumption.
- The evaluation of availability scenarios follows typical material flow simulation procedures. Failures and non-productive time intervals for setup and maintenance negatively influence throughput and induce idle-running consumption. The main question in this area is how specific energy consumption per part will correlate with falling absolute consumption, giving reduced availability of the manufacturing system.
- The effect of defective goods and scrap produced is another essential parameter to be evaluated in experiments. Depending on the positioning of inspection stations in the manufacturing processes, combined with the absolute amount of defective goods produced, the energy input into defective products influences energy consumption per part negatively. The lower the scrap rate and the earlier the detection takes place, the higher the increase in energy efficiency is to be expected. Experiments can show these correlations and bring them into context with the earlier mentioned parameters.

4.2.4.3 Evaluation Phase

4.2.4.3.1 Verification und Validation

Validation is the process of confirming adequate consistency between model and reality. Understanding of the system and of model behavior are thus increased. A particularly crucial activity in simulation, this step is to assure that the model reflects the real system`s behavior accurately and correctly. This adequacy can only be evaluated based on previously defined result precision: No fixed rules for this however exist, so that validation has to be conducted problem-specific, involving individual analysis. [4.7]

An important validation step is to check the input data used. In the case of energy simulation this relates to power load values. To validate the results of an energy simulation run, different types of information prove to be particularly helpful:

- Overall energy consumption during a period,
- Machine specific energy consumption during a period (e.g. operating cycle),
- Energy consumption plotted over time, i.e. the load profiles of energies consumed.

When planning a not yet realized manufacturing system, reference values for the above mentioned information will be hard to obtain. During operation phase, the simulated results can therefore be matched with the real load data measured in the factory at power feed-in points, allowing comparison of simulated and real load profiles, as well as the cross-checking of calculated and operatively arising consumption values.

Another validation step is the visual observation of simulation runs. [4.7] this proved to be particularly helpful in validating the correct parameterization of the portal strategies and the behavior of the machining centers. Appropriate interactive visualizations will subsequently be discussed.

The duration of a simulation run and the initial adjustment period are further aspects in validation. An option to handle the adjustment period is to log and subtract the overall consumption values as well as the throughput generated during this period from the overall calculation. Fig. 4.17 shows this behavior of the system during adjustment period, the cumulated waiting consumption rising until a realistic filling degree of the system has been reached, then lessening to more representative levels, when more energy is consumed in the producing state.

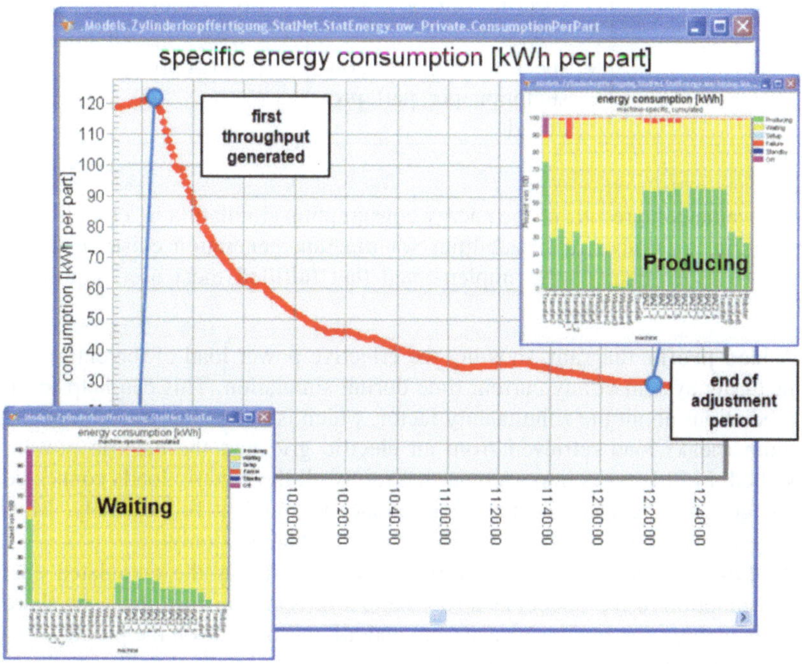

Fig. 4.17 Specific energy consumption before and after the adjustment period

4.2.4.3.2 Visualization and Documentation of Results

The energy consumption data that is generated in an energy simulation model can be presented to the user in different ways. Table 4.1 shows principal options to display energy data, either as raw data, i.e. in the quality generated by the state monitoring and transformation mechanisms or further processed into aggregated views. Both can be displayed in static views, being calculated at discrete time intervals or only at the end of a simulation run, or continuously over time, generating a dynamic "live" view.

Table 4.1 Options to visualize energy consumption data in a simulation model

	Static view	Dynamic view
Raw data	• energy consumption for last state completed • cumulated energy consumption	• load profile for single machine
Aggregated data	• sum of state-specific energy consumption (e.g. waiting-consumption in system) • percentages of state-specific energy consumption • specific energy consumption per part produced	• percentage of power load • load profile for machine type or cumulated for entire system • energy consumption per part (as profile)

To evaluate the results of the energy consumption simulation in a tool like Plant Simulation, a number of possibilities for diagram generation exist. In the pilot studies, various charts were implemented that fulfilled most user requirements (Fig. 4.18). These include:

- A **load profile** diagram, showing the effective power load of the entire manufacturing system at any current time during simulation. This can help to make predictions about the simultaneity factor, which is defined as the ratio of maximum (peak) load retrieved from an electric grid and the electric power installed. [4.8] It takes into account that rarely all the electric loads connected to the grid require electric energy simultaneously and at full capacity. Mostly, they are based on the experience of the electric planners only.
- A **state-specific energy consumption** diagram shows the cumulated energy consumption of the system as a bar chart. It informs the user about the share of each energy state in the entire system and allows to focus on the productive and non-productive energies consumed.

- A diagram for **machine-specific energy consumption** additionally displays the share of each energy state in total consumption on each machine, also implemented as a bar chart. The user is free to choose a percentual view or an absolute representation, helping him to focus on either the most intensive energy share (e.g. the waiting-state-consumption on machine 1) or the most intensive energy-consuming machine in general.
- A plotter diagram for **cumulated energy consumption** shows the resulting overall consumption, calculated over all machines over the period simulated so far. Here, the user can quickly determine if the system is producing at high or low capacity, recognizably when energy consumption at current simulation time is of high or low intensity.
- Another plotter diagram for **specific energy consumption** (per part) shows if non-productive machine states currently influence energy consumption negatively. If the curve gradient is steep, e.g. setup, waiting and failure times currently add a lot of non-productive consumption to the system`s energy consumption.

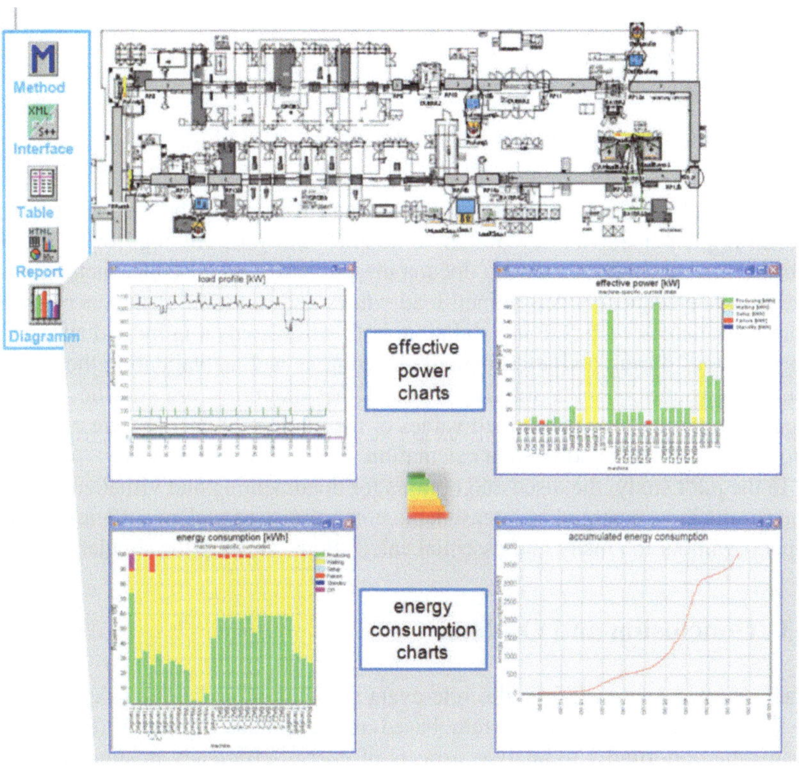

Fig. 4.18 Selected energy charts.

The simulation results can also be expressed in term of certain key performance indicators (KPI) programmed into the model. Basically, two types of KPI can be meaningful (Fig. 4.19). Examples for absolute KPI are minimum and maximum power load during simulation or the average energy consumption. Relative KPI are the ratio of two values such as consumption and throughput. For comparing energy-related machining, equipment and processes, one of the most important indicators is the specific energy consumption. [4.11] This KPI is typically applied in automotive industry as consumption per automobile or per part, measured in kWh or kJ. [4.12].

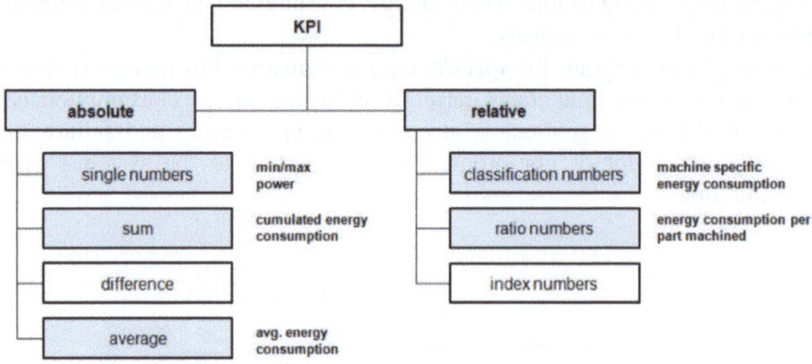

Fig. 4.19 Different types of KPI, acc to GOLDMANN/SCHELLENS 1995 [4.14]

In parallel to these visualization options and KPI, a number of statistic tables can be implemented to allow for more detailed analysis. Among others, the load profile (per machine) should be documented in a structured table, including data like the state sequence, determined load values and duration of the energy states. This plays an essential role in validation, both while analyzing individual machine state changes during and after a simulation run, as well as validating the resulting energy consumption of a simulation run. Other examples are statistic tables for logging minimum/maximum load values or the number of energy state changes, to allow for final mean and median calculation.

In the pilot study, the discussed options for documenting and visualizing simulation results have emerged as a reasonable compromise to realize sufficient analysis support without too many non-essential calculations, slowing the simulation run.

4.3 Conclusion and Outlook

The presented approach for a discrete-event simulation model that calculates energy flows in a manufacturing system, based on the simulated material flow, demonstrates the opportunity to analyze aspects of energy efficiency in simulation tools of the Digital Factory. Validation and performance analysis of such a manufacturing system can be conducted in such a model, as required by production system planners. It follows that the experimental approach of material flow simulation can be applied to energy simulation as well, facilitating the analysis of the effects of process changes, or the analysis of efficiency measures to reduce energy consumed. This can be evaluated in experiments ahead of the system realization and thus contributes to establishing "energy efficiency" as a planning objective during the concept stage of planning.

A number of technical developments remain to be addressed. Among these, the integration of complex standby strategies into the energy calculation logic seems promising, assuming that in the future machines will increasingly be controlled by intelligent energy control systems. This requires a more detailed modeling approach for energy states and state change sequences that cannot be triggered by material flow signals alone. Further, the integration of cost and value analysis aspects into the simulation model might strengthen deployment in hands-on environments, making the potential returns on invest immediately obvious.

An essential upcoming issue, after proving the technical feasibility of the energy simulation method and procedure, will be the integration into existing planning organizations and into the planning processes. Practical implementation of new and innovative approaches however are complex and risky endeavors. This is obvious in the case of energy simulation when looking at the number of organizational actors involved in a potential project. While technical and methodical support by a company`s IT organization is an essential, just as the technical vendor support that should aim to enhance current simulation tools, actual users of the energy simulation (i.e. the simulation experts) must be convinced of the quality of results and must be qualified to apply this method. As they perform simulation services for the actual planners internally or externally, they must be committed to integrate energy simulation aspects into their simulation studies. Planners, on the other hand, who receive the analyses evaluating the energetic performance of their concepts, must seriously consider the energy efficiency aspects in those results, must be able to interpret these correctly and must finally assume the responsibility to convince the realizing subcontractors (i.e. the machine manufacturers) and the operating divisions (i.e. manufacturing) to implement those energy efficiency measures deemed feasible. During the operation phase, finally, energy efficiency as a complementary planning objective will compete with throughput and quality targets, an issue that in operative manufacturing today still impedes the swift implementation of e.g. standby strategies. Support can be called from „energy officers" establishing themselves in operative production.

Therefore, an energy simulation method must be systematically integrate into a company`s production system development processes. The specific aspects of realizing this are still to be addressed: Collecting use cases for practical benefits and showing financial gains for practical planning scenarios; consideration of the risks during introduction and application of the energy simulation method; finally proper milestones to locate energy simulation procedures in the planning processes.

Concluding, a need for action remains to practically establish energy simulation as innovative approach to future manufacturing systems planning:

- Integrating energy simulation into the planning and validation processes during production creation process.
- Establishing valid cost-benefit relations for the involved company divisions.

- Encouraging people to use energy simulation and establish specific use cases, learning from implementation and developing standard scenarios based on practical alternative solutions (alternative components, flexibility models for operation etc..)
- Standardization of modeling energy aspects.
- Coupling of models for larger-scale analyses.

The use case presented above demonstrates the great potential that exists in the use of simulation technology when planning and operating energy-efficient manufacturing processes. The technical modules developed in implementation will be integrated into the VDA library for standardized application. In the future, discrete-event energy simulation will thus become an established part of the Digital Factory in Automotive manufacturing.

References

[4.1] Rudolph, M., Abele, E., Eisele, C., Rummel, W.: Analyse von Leistungsmessungen. ZWF Zeitschrift für wirtschaftlichen Fabrikbetrieb, Seite, 876–882 (October 2010)

[4.2] Beyer, J.: Energiebedarfsarme intelligente Produktionssysteme. 1. Internationales Kolloquium des Spitzentechnologiecluster eniPROD, Chemnitz (2010)

[4.3] Eisele, C.: TU Darmstadt. In: Conference Talk at Effiziente Produktionsmaschinen Durch Simulation in der Entwicklung, AutoUni., February 16 (2011)

[4.4] Weinert, N.: Vorgehensweise für Planung und Betrieb energieeffizienter Produktionssysteme. Dissertation, TU Berlin (2010)

[4.5] Dietmair, A., Verl, A., Wosnik, M.: Zustandsbasierte Energieverbrauchsprofile. wt Werkstattstechnik online, Jahrgang 98, H. 7/8 (2008)

[4.6] Neugebauer, R., Putz, M.: Energieeffizienz. Potentialsuche in der Prozesskette. In: Conference talk at ACOD Kongress, Leipzig, February 18 (2010)

[4.7] VDI-Richtlinie 3633 Blatt 1: Simulation von Logistik-, Materialfluss- und Produktionssystemen –Grundlagen. Verein Deutscher Ingenieure, Düsseldorf (2010)

[4.8] Müller, E., Engelmann, J., Löffler, T., Strauch, J.: Energieeffiziente Fabriken planen und betreiben. Springer, Heidelberg (2009)

[4.9] Kulus, D., Wolff, D., Ungerland, S.: Energieverbrauchssimulation als Werkzeug der Digitalen Fabrik. Bewertung von Energieeffizienzpotenzialen am Beispiel der Zylinderkopffertigung - Berichte aus der INPRO-Innovationsakademie. ZWF Zeitschrift für wirtschaftlichen Fabrikbetrieb, JG 106, S585–S589 (2011)

[4.10] Herrmann, C., Thiede, S., Kara, S., Hesselbach, J.: Energy oriented simulation of manufacturing systems – concept and application. In: CIRP Annals Manufacturing Technology, pp. S45–S48. Elsevier (2011)

[4.11] VDI guideline 4661 "Energetic characteristics. Definitions – terms – methodology". Verein Deutscher Ingenieure, Düsseldorf (2003)

[4.12] Engelmann, J.: Methoden und Werkzeuge zur Planung und Gestaltung energieeffizienter Fabriken. Dissertation, TU Chemnitz (2008)

[4.13] Goldmann, B., Schellens, J.: Betriebliche Umweltkennzahlen und ökologisches Benchmarking, Köln (1995)

5 Coupling Digital Planning and Discrete Event Simulation Taking the Example of an Automated Car Body in White Production

Steffen Bangsow

Abstract. MAGNA STEYR aims to establish digital planning in all important areas. In the field of BIW (Body In White) the digital process planning is already a reality. Starting from a digital product model, welding process are planned completely in digital form. Process simulation and offline robot programming safeguard the planning. With the connection of the digital process planning and discrete event simulation MAGNA STEYR took an important step towards realizing the digital factory.

5.1 The Task

The task of the project was modeling an automated body in white production with more than 170 robots. Important demands of the model were:

- Easy to use and customizable by the planning engineers
- Reusability of the library elements
- Sufficiently fast experiment runs
- No redundant data storage (using data from digital process planning)
- Import of availability data from the real production system
- Use of real production job data

In the future the simulation model should give planners the opportunity to verify changes in the process only by pressing a button in the production line simulation (for example regarding a possible change in the total output within a given time). Building and maintaining the model must be possible without changing the underlying programming. For the digital process planning MAGNA STEYR uses Process Designer, for process simulation and offline robot programming Process

---

Steffen Bangsow
Freiligrathstrasse 23
D 08058 Zwickau
Germany
e-mail: steffen@bangsow.net

Simulate and for material flow (discrete event) simulation Plant Simulation, all are applications of Siemens PLM Software.

5.2 Data Base in Process Designer

MAGNA STEYR is a leader in the field of digital production planning. For the area to be modeled digital planning is used starting from the product, through production processes to the major equipment. This way the body in white planners can react quickly to changes like construction modifications. To date, however, a link to material flow simulation was missing. Although a simple simulation model already existed, it was decided to create a new model from scratch, custom-tailored to the specific requirements. In principle the following data for the simulation exist in Process Designer and are also used for process simulation:

- Process steps (in different detailing, starting from weld point and the movements of the robot between the weld points)
- Sequence of process steps (stored in so-called flows)
- Estimated and simulation-checked (offline robot programming) process times
- Resources allocated to the process steps

The data for modeling of dependencies (shake hands) between the robots are missing in digital process planning. The resources are only partially included.

Digital process planning is very limited when it comes to evaluating the effects of dependencies between the elements of the line. Robots for example have to wait for the completion of previous steps of other robots, or are dependent on available places in the conveyor system. Also there are many processes which are executed by several robots together. To avoid collisions, the robots have to sidestep or wait within their processes, which affects the process time. The relatively static process simulation does not offer sufficient hold for these aspects.

5.3 Selecting of Level of Detail for the Simulation

Solutions already exist for the automated creation of simulation models using data from process simulation. For the present task, this approach is not feasible. The automatic export generates one item per process. For representing the dependencies within the process (especially if more than one resource is involved in a process), it is necessary to model the processes, "one level down".

A process is stored in Process Designer in several levels of aggregation (Figure 5.1).

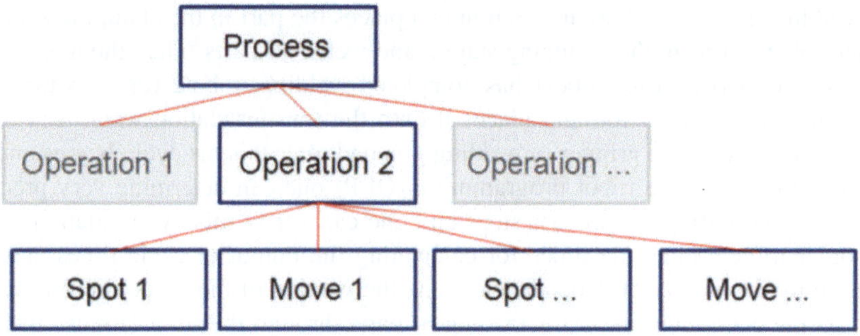

Fig. 5.1 Process Designer process aggregation levels

Each process consists of about 2 to 10 compound operations, which in turn consist of a large number of individual operations. Modeling the activities of the welding robot at process level (for each robot one station with one process time) was insufficiently precise for creating the material flow simulation. Modeling at the lowest operating level in turn is "too exact" for the material flow model, since the amount of data would explode through the inclusion of this level. The choice of the "Compound Operations" as a data base for the simulation however, entailed that it would be impossible to write back material flow simulation data to the digital planning process.

The following scenario is typical (Figure 5.2):

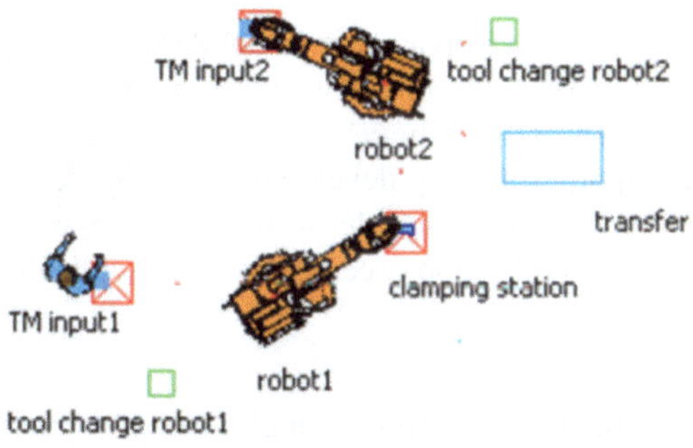

Fig. 5.2 BIW scenario

Two robots are working together within a cell. The worker puts parts on the station "TM input1" and "TM input2" and confirms this. Next robot1 welds the parts together. Then he changes the tool from welding gun to gripper. He takes the part from the loading station, turns to the clamping station and places the part there. Then, the robot robot1 makes another change from gripper to welding gun and waits. In parallel, the worker places parts in TM input2 and sends a release signal for robot2. Robot2 welds the parts together, changes from welding gun to gripper and removes the part from TM input2. Robot2 now waits until the robot1 has placed his part in the clamping station and places the part in the clamping station. Then robot1 turns to the clamping station and welds all parts. Then the next cycle begins for robot1. After robot1 has completed welding, robot2 removes the part from the clamping station and places it onto the transfer station when it is free. Robot2 changes from gripper to welding gun and after this his cycle begins anew. By employing offline robot programming (OLP) one can determine very precise times for the individual process steps and one can verify this by simulation runs. In order to determine the times for calculating the output or cycle times, delays caused by the variety of dependencies (e.g. the waiting of the robot1 before welding on the clamping station for loading of parts through the robot2) must also be taken into account. The impact of these dependencies is in reality often estimated by the line planners. Digital planning offers the possibility to use so-called line studies to simulate the cooperation of several robots. Creating these simulation models is very complex though.

Three different dependencies were to be considered in the present project:

- Dependancy on other robots (insertion, welding, gluing ...)
- Dependancy on workers (e.g. insertion of parts)
- Dependancy on materials handling equipment (e.g. free space for storing a part, which in turn depends on the following work stations)

Several dependencies per process usually exist.

5.4 Developing a Robot Library Element

The first challenge was to develop a robot model that can handle process tables as input and displays a similar (chronological) behavior as a real welding robot. Therefore a data model was initially developed into which the data from Process Simulate could be imported. During development it became clear that it is necessary to categorize the operations in order to realize a universal programming approach. The robots in the body shop execute the following main activities:

- Load parts
- Place parts
- Welding, gluing, hemming, ... (processing)
- Shake hand operations (a robot holding a part while another robot processes the part)
- Tool placing
- Tool loading
- Turning, positioning
- Maintenance activities (cap milling and cap changing)
- (waiting)

This information can mostly be extracted from Process Designer or can be entered directly as an additional attribute value in Process Designer.

For the robot (and the worker) a process-oriented behavior model was developed. The behavior of the robot is based on 100% of the process from Process Designer. The robot waits before each operation step in his waiting position until the condition is met for starting his next operation. Then, he turns into the processing position, remains there until the operation time is over (which affects all operations except for the transportation of parts) and, after finishing the operation, he possibly sends a release signal for the next process step. Next, he turns back to his waiting position. Then, the next operation step is determined from the operation list. This approach ensures accuracy of the modeling of the processes up to a fraction of a second compared to the process simulation. Each part has been considered in the simulation to ensure a future connection to the logistics processes. For this reason the robot loads parts and places them at their destination in the operations "Load parts" and "Place parts".

To reach the goal of ease of use, the configuration of the robot and its peripherals is solely accomplished by drag and drop. The user of the simulation model does not have to change the underlying programming to model the different processes.

5.5 Linking, Shake Hands

Links are implemented via release bits. For this purpose some library elements (clamping and insertion stations and skid-stopping places) were equipped with a set of control bits. Within the process it has to be entered which control bit must be set to start an operation, and which control bit is set when the operation is completed (Figure 5.3).

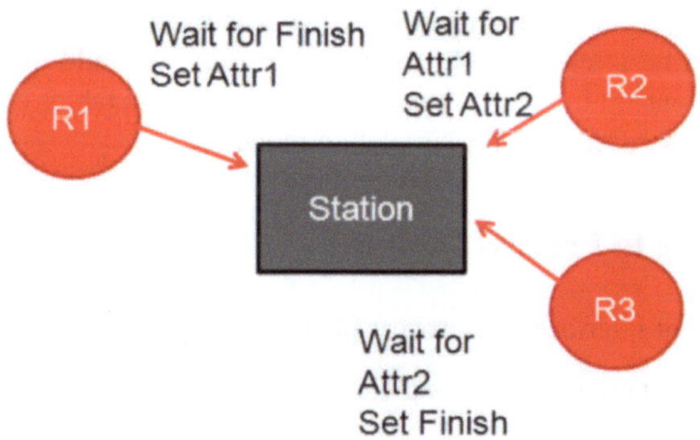

Fig. 5.3 Release Bits

Figure 5.3 shows a typical situation. A robot 1 waits for the end of the previous cycle (finish). He performs his work and sets a release bit (Attr1). The robot R2 is waiting for this release, begins his part of the process and in turn sets a release bit (Attr2). The robot R3 is waiting for this release, starts his operation, and at the end of his operation sets a bit to indicate the end of the process. The simulation model required up to 7 different release bits.

Initially only the manual input of the linking information (location and symbolic name) was intended. But it became clear that this approach was too time-consuming and error prone. In order to avoid input errors and to improve maintainability of the simulation model a network of relationships for modeling the dependencies was developed. It is generated automatically and can be edited with the instruments of Plant Simulation (connectors, Figure 5.4).

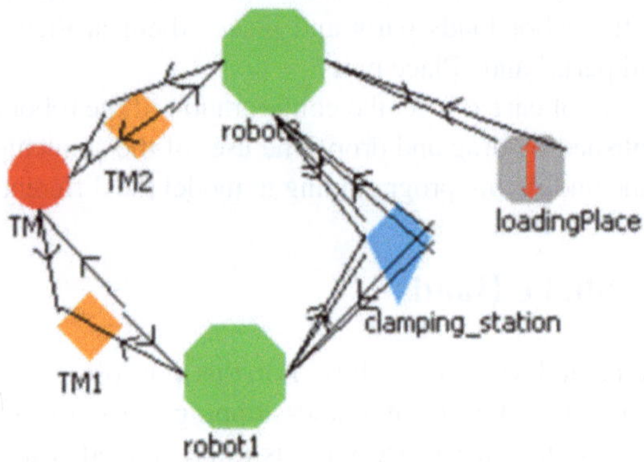

Fig. 5.4 Network of relationships

The animation of the robot (turn to the relevant operation positions) allows a very simple graphical debugging when relationships are set incorrectly. Temporary drains can be integrated into the frame in order to build the model step by step. The relationship information is stored in the process tables of the simulation and is in this way saved with the simulation model.

5.6 Interface to Process Designer

Different types of interfaces for process planning are possible:

- Automatic generation of a complete material flow model based on the process planning data
- Automatic transfer of processes from process planning into an existing material flow (plant) model
- Linking/updating of individual data (e.g. processing times) from process planning

5.6.1 Automatic Model Generation

A prerequisit for automatic model generation is that all required resources are modeled in Process Designer and these are inserted in the plant layout. It is then relatively easy to read the resource and its coordinates from Process Designer and create a discrete event model based on that. The inclusion of all line elements, e.g. loading places for the workers, connecting conveyor systems and the entire periphery of the robot would cause an extreme increase of the modeling expense within process planning. The added value of including the entire resources is smaller than the expected problems with the extreme extension of the planning process model. Creating a robot cell using the library elements developed in this project only takes a few minutes. Body in white production does not require fast and frequent layout changes. Therefore it was decided to abstain from automatic model generation and to manually create the simulation model in Plant Simulation (DES).

5.6.2 Transfer of Processes from Process Planning to Material Flow Simulation

Manually entering process data into the material flow simulation is a relatively large amount of work. This is not necessary if the processes already exist within digital process planning. Therefore, the processes in this example were not created and maintained in the material flow simulation model but imported from digital process planning. To increase the performance, the processes are cached for each robot in the material flow simulation. A direct connection of the material flow simulation to the production environment of the process planning could in reality imply a significant limitation of working ability of the complete process planning system, if the material flow simulation generates a large number of queries per second for the process planning database to determine the next process steps. Therefore it makes sense to work with an export file (e.g. XML). The export file however contains a lot of information, which is not necessary for the material flow simulation. The extraction of the data is computationally intensive, thus is not useful to work directly with the XML file in the simulation runs. In this example, the interface was developed as a kind of middleware. Another reason was that within Magna Group process planning is partly performed in the Excel format, which is to be modeled with the same library. Importing process data encompasses four steps:

1. Selection of the robot (and product version)
2. Correction of process steps if necessary (deleting and changing the order)
3. Assigning the resources, if it does not exist in the process planning
4. Graphical creation of the Shake Hands

To 1) It has been shown that it is not useful to have a completely automatic process. The corrections needed are too great if all processes for all robots will be imported at once, especially since it is necessary to adjust the processes of individual robots in the ongoing maintenance of the model. Importing the data per robot only takes about a minute. Assignment of resources per robot while importing data is also extremely user-friendly.

To 2) Processes need to be corrected under certain circumstances, before they can be inserted into the material flow simulation. If the process planner for example inserts comments into the process flow, then these are created as operations and are read in by the interface. Also, there may be problems if the individual process steps are not fully connected by flows. Then, the order cannot be clearly identified during importing. This step can be executed with minimal effort with a simple configuration menu in which parts of the process can be deleted and the order may be modified during the import process. For better orientation flows are displayed in the respective menu (with appropriate ID, Figure 5.5). Unconnected objects (without flow IDs) usually can be deleted.

Fig. 5.5 Interface process configuration

To 3) In the simulation model all components of the robot cells can be assigned by simply dragging and dropping them on the robot. These assigned resources (clamping stations, tool change positions, transfer positions to the workers and for materials handling equipment, etc.) are made available during import via a menu. The resources can be assigned to the operations via simple mouse operations (Figure 5.6). This step is also used to verify the completeness of the modeled robot cell.

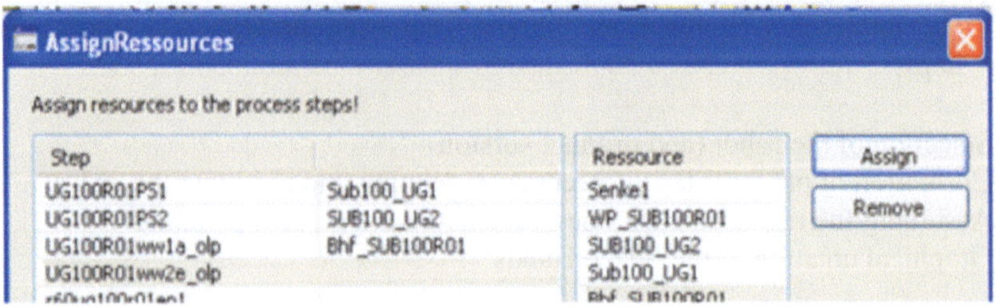

Fig. 5.6 Interface resource assigning.

The operations in of the material flow simulation are equipped with references to the planning process operations. This way a simple update of the processing times is possible by just clicking a button.

To 4) After importing data is completed, the processes will automatically generate a Shake-hand frame (network of relationships). In this frame, symbols are located analogous to the position of elements in the material flow simulation layout. By setting the connecting lines (connectors) the dependencies between individual operations can be defined with the instruments of material flow simulation.

5.7 One Step Closer to the Digital Factory

The Association of German Engineers defines the Digital Factory as follows [5.1]:

> "... A generic term for a comprehensive network of digital models and methods, including simulation and 3D visualization. Its purpose is the integrated planning, implementation, control and continuous improvement in all major plant processes and resources associated with the product "

Connecting digital planning with the material flow simulation enables digital planning starting with product planning via the production process to the production line using one integrated data base. Figure 5.7 shows the data model implemented in the simulation of the body shop.

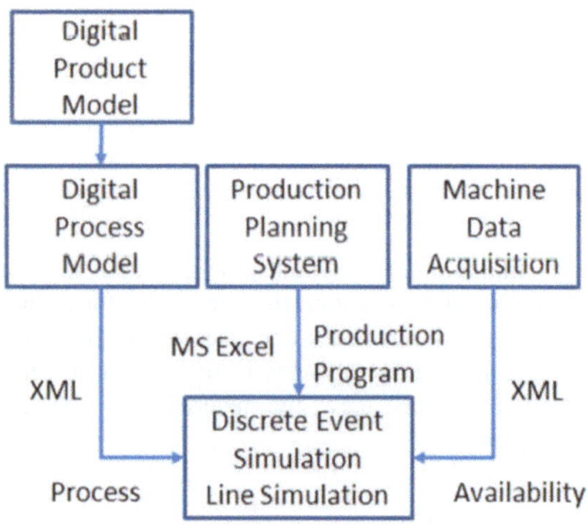

Fig. 5.7 Simulation data model

After a design change the following procedure now is possible (see Figure 5.8):

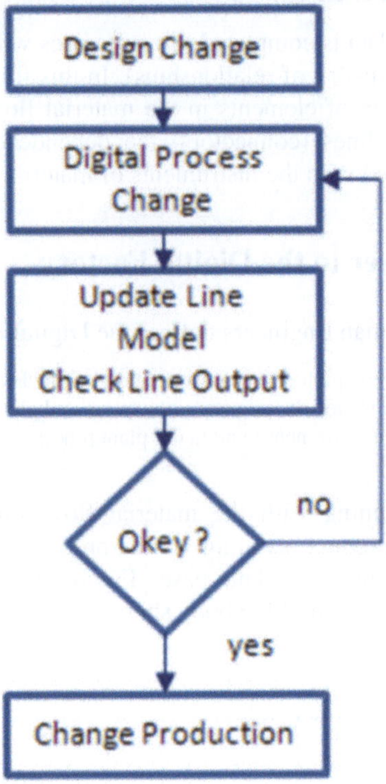

Fig. 5.8 Possible design change process

The process planner changes the welding process according to the design and possibly creates a new robot simulation. Then he loads the changed process times in the material flow simulation and examines the impact of the changes on the total output. If the result is not satisfactory, then he might, for example, move welding points to another process and re-test the line output. If the line output of the simulation meets the expectations, changes in real output are made. Only when processes are created completely anew, the material flow simulation needs to be changed (reload operations).

5.8 Result of the Simulation

A highly accurate representation of the body shop could be realized through the high detailing of the simulation and by using real process data (times, availability). The results of the simulation on average only deviate about 2% from the results that were reached in reality. The simulation of a production day takes 1.5 minutes (Dual Core, 2.6 GHz), which allows for the simulation of longer production programs. The maintenance of the simulation model is possible without having to intervene in the underlying programming so that the planners themselves, who are not simulation experts, can take over this task in the future.

5.9 Outlook and Next Steps

Detailed process modeling within the material flow simulation opens up a series of other fields of study for the body shop:

- Energy use: A consumption profile will be assigned to each robot (current consumption for welding, turning, waiting, ...), the consumption will be recorded and summed up over the simulation (Figure 5.9 shows the simulated (current) consumption of electricity within the material flow simulation of 172 welding robots).

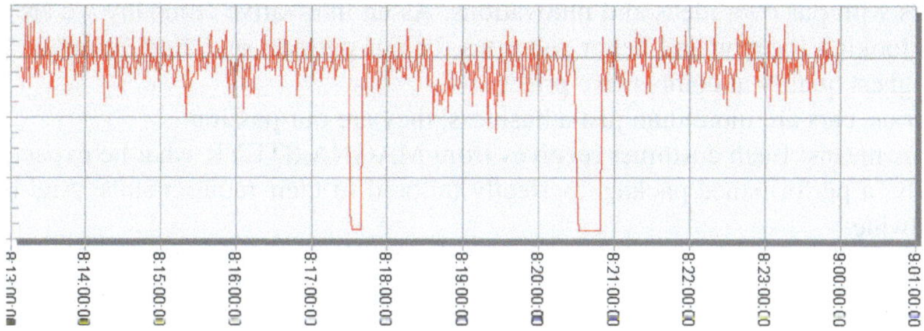

Fig. 5.9 Simulated power consumption

- Output optimization, for this detailed resource statistics will be generated, that breaking down the utilization data of the robot into welding, loading, unloading, tool changing, process caused and idle waiting time. Identifying the idle waiting time can serve as a basis for optimizing capacity utilization
- Workers, the study of staffing with various number of workers and the impact on the line output
- Buffer allocation and failure concepts

5.10 Company Presentation and Contact

5.10.1 Magna Steyr Fahrzeugtechnik Graz (Austria)

Over 100 years of experience in the automotive industry and the comprehensive range of services of the company make MAGNA STEYR the leading global, brand-independent engineering and manufacturing partner for automotive manufacturers.

Our comprehensive range of services covers four divisions:

- Engineering: Development services from systems and modules to complete automotives
- Automotive Contract Manufacturing: Flexible solutions from niche to volume production
- Fuel Systems: Innovative fuel systems made of steel, plastic and aluminum
- Roof Systems: Entire range of roof systems such as soft tops, retractable hard tops and modular roofs

As a contract manufacturer, we have produced to date 2.5 million vehicles, divided into 21 models. In addition to our competence in the area of fuel systems and roof systems, we increasingly intend to offer customized solutions in the fields of aerospace and non-automotive.

Behind all these performances stand 10,200 people worldwide. Through our global network of 37 locations on three continents, we are close to our customers. Partnership for us means to strengthen and expand the market position of our customers with our own ideas and innovations. As an innovative company we are always looking for new and better solutions for our partners and are committed for the highest quality at competitive prices.

For us, cars are more than just a business, they are our passion.

This means: Each customer receives from MAGNA STEYR what he expected: namely, a performance package perfectly tailored to their requirements. And this worldwide.

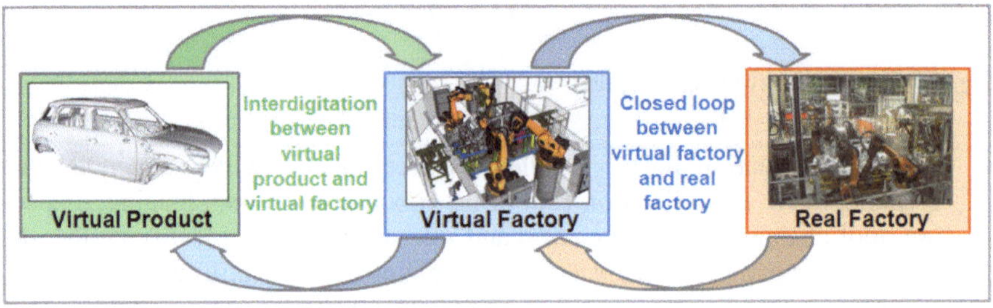

Fig. 5.10 Digital factory at MAGNA STEYR

Target of the digital factory at MAGNA STEYR is the cost and time optimization of the planning, implementation and ramp-up processes.

It is essential to make the right products at the right price in the desired quality at the defined time available. With the "digital factory" planning approaches are described that create even before the construction of a factory a realistic image of the future reality. This opens up the possibility to define an optimum overall system.

The digital factory contains at MAGNA STEYR Fahrzeugtechnik:

- The close integration with product development for the joint development of product and process
- Planning for the body shop, paint shop, assembly, logistics and plant infrastructure
- securing of planning through simulation of the processes and equipment

Components of the digital factory at MAGNA STEYR are:

- Alphanumeric planning and 2D Plant Layout
- 3D-Process simulation
- Material flow simulation
- 3D Plant layout
- Offline programming
- Facility management
- Integration into the Quality Planning
- Integration in the serial planning

Results and objectives are:

- Early influence of product development with regard to manufacturability and process optimization
- Early planning based on the virtual factory independent of the location decision
- Reducing of the investment and production costs
- Accelerate the planning and commissioning processes (shortening time-to-market)
- Improved quality reduces start-up costs and allows steeper ramp up curves
- Better integration of product development and production planning in the form of common data base for product structure, components and technology
- Standardization of planning processes to eliminate redundancies and interface problems
- Decision support through visualization and simulation
- Consistent change management and continuous project time tracking
- Integration platform for product, process and resource data within a unified MSF system architecture
- Faster and easier access by all process participants in current product/process and resource data

Contact

Walter Gantner
Magna Steyr Fahrzeugtechnik
Liebenauer Haupstraße 317
8041 Graz
Austria
Email: walter.gantner@magna.com

5.10.2 The Author

Steffen Bangsow is working as a freelancer and book author. He can look back on more than a decade of successful project work in the field of discrete event simulation. He is author of several books relating to simulation with the system Plant Simulation and technical articles on the subject of material flow simulation.

Contact

Steffen Bangsow
Freiligrathstrasse 23
D 08058 Zwickau
Germany
Email: steffen@bangsow.net

Reference

[5.1] VDI: Digitale Fabrik Grundlagen VDI-Richtlinie 4499, Blatt 1, VDI-RICHTLINIEN, S. 3 (Februar 2008)

6 Modeling and Simulation of Manufacturing Process to Analyze End of Month Syndrome

Sanjay V. Kulkarni and Prashanth Kumar G.*

Manufacturing industries across the globe face numerous challenges to become 100% efficient but each and every industry has its own constraints / problems with their functional system to achieve 100% excellence. End of the month syndrome is one of the major problems almost all manufacturing industries face with the ever growing demand and the competition around. Manufacturers find it really difficult to achieve their potential if they produce more than 25% of their monthly shipment plan in the last week of the month or more than 33% of their quarterly shipment plan in the last month of the quarter. Companies that live with the "end-of-the-month-crunch will be burdened with premium freight, internal expediting, overtime costs, and production inefficiencies that will crush their bottom line goals. But effective upfront planning and timely execution can make the "end-of-the-month-crunch" a bad memory and eliminate those profit killers. The causes for end of the month syndrome are raw material constraints and production inefficiencies, last minute product changes, stoppage and machine down time in manufacturing line etc. Manufacturing industries can analyze these challenges through the application of modeling and simulation technique with the existing system and try out various "what-if scenarios" (sensitivity analysis) without any physical changes to the existing process & thus find a solution to all those problems leading to End of the Month Syndrome.

6.1 Introduction

Manufacturing industries across the globe face numerous challenges to be 100% efficient but every industry has its own constraints / problems with its functional

Sanjay V. Kulkarni
Industrial and Production Engineering Department,
B.V.B CET,
Hubli - 580021, Karnataka, India
e-mail: skipbvb@gmail.com

Prashanth Kumar G.
Student – Industrial and Production Engineering Department,
B.V.B College of Engineering and Technology,
Hubli - 580021, Karnataka, India

* Co-author.

system to achieve this 100% excellence. To overcome these constraints / problems with their existing systems is again a challenge because of unidentified bottlenecks or lack of insight into their processes. If one attempts to resolve an existing bottleneck a similar / different kind of bottleneck would surface out and will be shifted to another area of the manufacturing process. Manufacturing industries can overcome these bottlenecks by studying the gaps / short come that occur in the existing system with lean manufacturing tools / techniques and with the technology available. Modeling and Simulation techniques best suit such applications since various "what-if" scenarios could be tried virtually along with knowing the impact of lean manufacturing tools and techniques on the process even before they can be implemented.

Some of the challenges, that the manufacturing industries face, are listed below.

1. Throughput – Under Average & Peak Load.
2. Unidentified Bottle Necks.
3. End of the Month Syndrome.
4. Unreliable Suppliers.
5. Excess Inventory.
6. Exceeding System Cycle Time.
7. Queuing at Work Station.
8. Capital Investment Justification.
9. Production Planning and Scheduling.
10. Line Balancing.

Manufacturing industries can analyze the impact of these challenges through Modeling and Simulation of the systems in question and find answers using various ***"what- if scenarios"*** (sensitivity analysis) without any physical changes to the existing processes.

6.1.1 End of the Month Syndrome

Manufacturers find it very difficult to achieve their full potential if they produce more than 25% of their monthly shipment plan in the last week of the month or more than 33% of their quarterly shipment plan in the last month of the quarter. This is a phenomenon called "End of the Month Syndrome". Companies that live with "end-of-the-month-crunch" are burdened with premium freight, internal expediting, overtime costs, and production inefficiencies that will crush their bottom line goals. But effective upfront planning and timely execution can make the "end-of-the-month-crunch" a memory of the past and eliminate those profit killers.

Most companies are using ERP systems coupled with lean manufacturing techniques to plan and control their business processes. These have eliminated the end-of-the-month-crunch to some extent for some companies but there are many more still burdened by it. If we spend time observing certain MRP scheduled factories during the last weeks of a financial quarter, we can't appreciate the extent of the remaining problems. We will typically observe profit draining in overtime, internal/external expediting, last minute product changes and production inefficiencies. The inevitable scrap and rework too add to the profit drain. Then there are the long-term consequences of quality problems in the field, warranty costs and not to forget the resulting customer dissatisfaction.

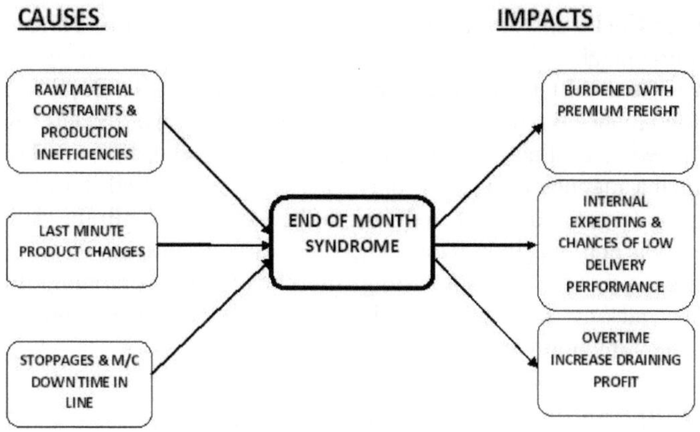

Fig. 6.1 End of month syndrome.

6.1.2 Objective

- Modeling and Simulation of Manufacturing Line to Analyze End of the Month Syndrome.
- Reduce bottlenecks.
- Prevent under utilization of resources.
- Optimize system performance.
- Inclusion of new orders / customers.
- Capacity improvement.

6.1.3 Problem Statement

This case is the result of project work carried out in one of the automobile parts manufacturing company. The Company manufactures two wheeler parts and special purpose machines and supply to worldwide customers.

After a series of discussion with the shop floor managers and the production heads it was found that the plant faced *end of the month syndrome*. The plant produces mainly gear shifter fork component for the various types of two wheelers. They are overloaded with orders; however the plant finds it difficult in fulfilling the orders "on time" every time. The actual problem in this plant is, they have three gear shifter fork manufacturing lines and they are expected to produce seven components for seven different customers which are listed below:

Fig. 6.2 Gear Shifter Fork

Gear Shifter Forks customers: Honda, Ducati, Bajaj, Piaggio, Yamaha, Motorai Miner, Motto Guzzi.

The three Gear shifter fork-manufacturing lines are: Honda, Bajaj & Yamaha.

The Honda and Bajaj lines are busy with their own models as they are completely dedicated lines. The plant needs to produce all the remaining models in the Yamaha manufacturing line only, due to this they find it very difficult in producing the targeted quantity and in turn face problems with the delivery dates of those models and thus month end syndrome starts developing.

The case study aims at suggesting alternatives in overcoming this end of the month syndrome after a thorough analysis of the existing processes using modeling and simulation techniques.

6.1.4 Modeling and Simulation Concepts

Simulation is a technique to evaluate the performance of a system, existing or proposed, under different configurations of interest and over long periods of real time. When used for a new system, helps to reduce chances of failure to meet specifications, eliminate unexpected bottlenecks, prevent under or over-utilization of resources, and optimize system performance.

Modeling and Simulation (M&S) is one of the tools, which help managers in decision making by using various sensitivity analysis (What- If). Modeling and Simulation must become the method of product and process design, it plays major role with the processing time but does not deal with the dynamics of machines.

M&S is a numerical technique for conducting experiments on a computer, which involves logical and mathematical calculations that interact to describe the behavior and structure of a complex real world system over extended period of time. Simulation is a process of designing a Model of a Real or an Imaginary System for:

- Conducting Experiments with it.
- Understanding its behavior.
- Evaluating Various Strategies.

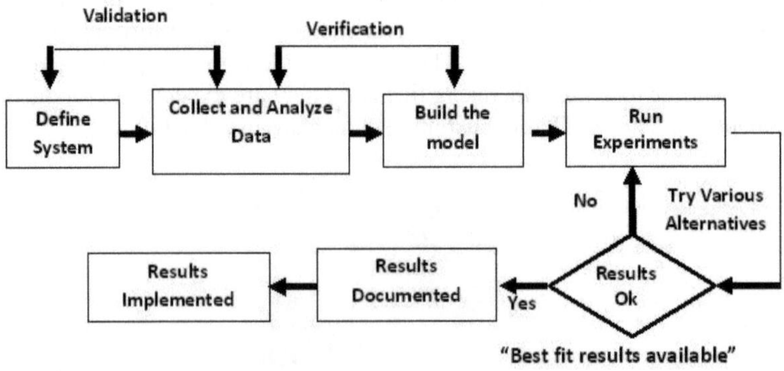

Fig. 6.3 Simulation process.

6.1.5 Software Selected for the Project Work

Many Modeling and Simulation software packages are available in the market, but ARENA software is used for the proposed work. This software provides more accurate average values for the analysis rather than the theoretical values and uses statistical average distribution as input / output for analysis.

6.2 Study of the Process to Be Modeled

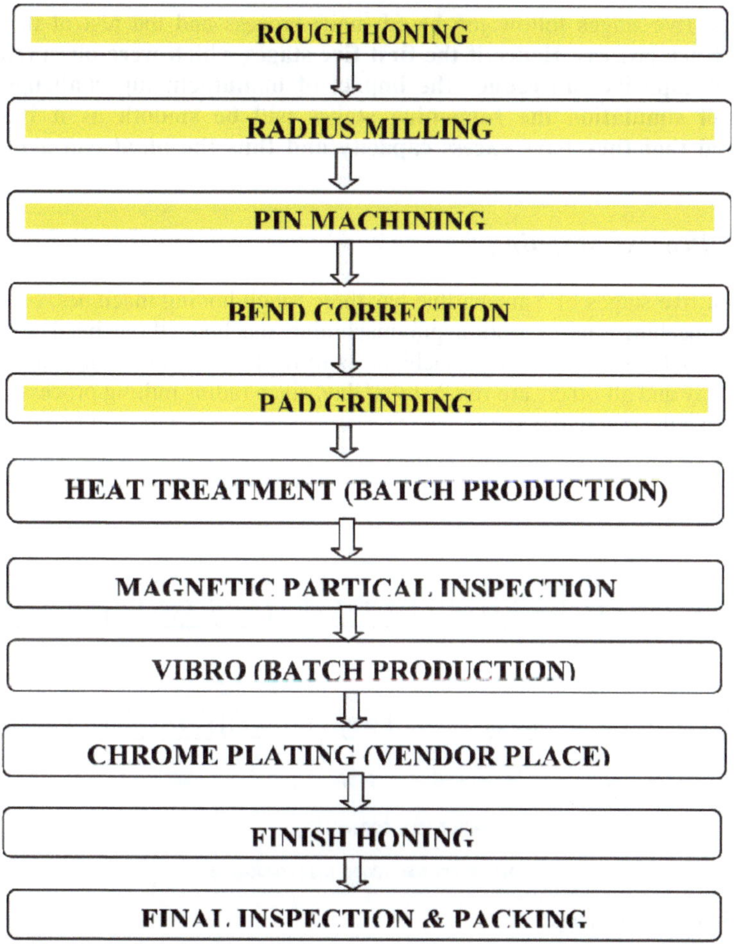

Fig. 6.4 Yamaha Line Processes

The detailed study of all the processes was conducted along with the discussions with the concerned production heads and line managers. Finally it was decided to focus on Yamaha gear shifter fork manufacturing line (YMG Line) which was the subject of interest contributing in the End of the Month Syndrome. Different stages of Yamaha gear shifter fork processes are listed above as in Fig. 6.4.

Further study pointed out that the most contributing stages which lead to end of the month syndrome were the first five as mentioned below:

1. Rough honing
2. Radius milling
3. Pin machining
4. Bending and
5. Pad grinding.

The first five stages follow job based (unit) process and the rest of the stages follow batch process. **Hence if the first five stages which were observed to be short of capacity can reduce the impact of manufacturing challenges as a result of simulation the remaining stages will be smooth as it has been observed that they have excess capacity and thus the plant can achieve its target**.

6.2.1 Process Mapping

The first five stages of Yamaha line has three rough honing machines, one radius milling machine, one 8 station pin machining machine, three bend correction machine and two pad grinding machines as in fig.1.5. The radius milling is run 3 shifts / day and all others are run 2 shifts/ day, since radius milling processing time is less than the rest in the line.

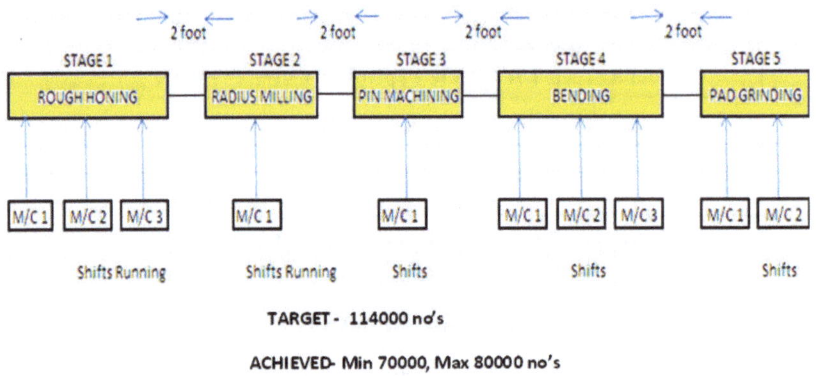

Fig. 6.5 Process Mapping.

The above line has a target to produce 1,14,000 units per month however it has been observed that the achieved target is around 80,000 units per month only.

Entire plant runs 3 shifts per day with a shift time of 480 minutes however the effective utilization is 390 minutes only which means 90 minutes would be the standard loss in the line which is as shown below:

1) 2 Tea time 10 minutes = 20 minutes
2) Lunch time = 30 minutes
3) Inspection = 20 minutes
4) Start & End up = 20 minutes

6.2.2 *Data Collection*

The data was collected and tabulated following the stages after carefully filtering and verifying the same with the concerned. Tables below show the collected data and the distribution of collected data (refer appendix) as required to build the model in ARENA software.

6.2.3 *Machine Wise Data Collection*

Table 6.1 Machine wise data collection.

	MACHINES				
Contents	Rough Honing	Radius Milling	Pin Machining	Pads Grinding	Bend Correction
No of Machines	**3**	**1**	**1**	**2**	**3**
Standard Cycle Time	**5**	**10**	**10**	**10**	**5**
Man power	**3**	**1**	**1**	**2**	**3**
Start Up loss(min/shift)	**10**	**10**	**10**	**10**	**10**
End up Loss(min/shift)	**10**	**10**	**10**	**10**	**10**
Target output (units /shift)	**900*1**	**2600**	**3000**	**2600*1**	**800*1**
Achieved output/shift	**700*1**	**1900**	**2500**	**1900*1**	**800*1**
Setting time (hrs) for component to component	**1/2**	**1- 2**	**1-2**	**1-2**	**1/2**
Rework(No's)/shift	**4**	**3**	**4**	**2**	**50**
Rejection(No's)/shift	**0**	**0**	**0**	**5**	**3**

6.2.4 CYCLE TIME (Seconds)

Table 6.2 Cycle times – with legend

No.	A	B	C	D	E	F	G
1.	8.71	11.50	10.87	25.38	15.20	10.65	10.71
2.	9.60	25.44	17.70	25.54	16.10	24.33	12.12
3.	19.20	1.7.70	9.50	24.94	15.02.	27.05	12.93
4.	15.50	30.84	11.31	25.22	16.16	24.21	12.21
5.	8.41	5.75	10.31	25.14	17.00	13.81	13.40
6.	11.07	25.02	9.86	25.00	14.00	24.31	9.94
7.	16.75	20.04	8.47	25.18	17.00	18.36	10.68
8.	10.61	6.98	11.85	24.88	15.00	27.03	12.39
9.	9.61	10.87	12.61	25.44	20.00	22.65	10.52
10.	9.50	8.08	11.63	25.14	22.10	44.50	12.77
11.	11.31	17.82	25.13	25.00	13.32	7.27	11.30
12.	10.31	22.07	11.20	25.59	13.52	16.11	12.25
13.	9.86	9.71	13.40	24.89	13.24	9.37	13.39
14.	8.47	8.98	9.50	25.02	10.20	16.37	12.00
15.	11.85	11.23	11.31	25.42	10.40	35.70	11.79
16.	12.61	26.48	10.31	26.10	16.10	17.08	13.09
17.	11.63	26.92	9.86	25.83	14.10	12.92	11.74
18.	25.13	8.08	8.47	25.16	15.20	25.08	11.42
19.	11.20	8.75	11.85	24.52	8.11	14.28	14.00
20.	13.40	9.50	12.61	25.72	15.10	21.27	11.45
21.	20.61	38.98	11.63	25.2	14.00	15.16	10.20
22.	21.22	25.60	25.13	20.40	16.00	39.84	12.35
23.	13.33	21.56	11.20	15.70	12.00	30.24	12.93
24.	10.50	22.00	13.40	22.30	10.10	34.97	13.80
25.	8.10	17.20	9.50	24.00	11.00	42.57	10.50
26.	9.50	12.80	11.31	26.00	15.00	30.39	12.00
27.	12.00	17.11	10.31	30.00	17.12	25.19	11.30
28.	15.10	21.00	9.86	33.55	12.40	42.15	10.80
29.	16.10	9.00	8.47	26.20	14.12	56.34	9.60
30.	20.10	22.18	11.85	23.15	15.20	34.79	10.50

A	ROUGH HONING M/C 1	**E**	PIN MACHING M/C
B	ROUGH HONING M/C 2	**F**	BENDING M/C
C	ROUGH HONING M/C 3	**G**	PAD GRINDING M/C
D	RADIUS MILLING M/C		

6.2.5 Dispatch Plan for the Yamaha Line (GSF-Gear Shifter Fork)

Table 6.4.

SL NO	GSF PART NO	6TH WEEK	7TH WEEK	8TH WEEK	9TH WEEK	10TH WEEK	TOTAL Quantity
1	DUG01	0	600	500	0	0	1100
2	DUG02	0	1200	1000	0	0	2200
3	DUG03	0	2200	2200	0	0	4400
4	DUG04	0	1100	1100	0	0	2200
							8900
1	PIG02	0	0	0	1800	0	1800
2	PIG03	0	0	0	0	0	0
3	PIG04	0	0	0	2600	0	2600
4	PIG05	0	0	0	5400	0	5400
5	PIG06	0	0	0	638	0	638
6	PIG09	0	0	0	375	0	375
							10813
1	YMG07	2000	3000	3000	2000	0	10000
2	YMG08	2000	3000	3000	2000	0	10000
3	YMG09	5000	6000	6000	5000	7000	29000
4	YMG10	5000	6000	6000	5000	7000	29000
5	YMG11	5000	6000	6000	5000	7000	29000
							10700

Out of the above weekly dispatch schedule it can be seen that Yamaha Gear-shifter Fork (YMG) has higher production rate compared to DUG and PIG models, hence the same has been considered for the further analysis.

6.2.6 Delay Timings in the Processing Line

Machines	No. of Shifts	Total (Min) Working	Total Time Lost (Min)
RH	22	8360	360
RM	34	12920	1145
BM	22	8360	545
PM	22	8360	1980
PG	22	8360	4735
TOTAL		**46360**	**8765**

Table 6.4 Delay timings.

RH - Rough Honing.
RM - Radius Milling.
BM - Bend Correction Machine
PM - Pin Machining.
PG - Pad Grinding.

6.3 Building a Virtual Model and Achieving "AS IS" Condition

This is the most crucial step where it is required to build a virtual model of the process under study and fine tune it till the "AS-IS" condition is achieved. The output of the model confirms the achievement of as-is if the output of the virtual model matches with the monthly output of the YMG line. However the entire exercise is carried out in consultation and continuous interaction with the process owners. This "AS-IS" model can further be used to carry out various "WHAT-IF" (Sensitivity) analyses without any physical changes to the line. As-Is model and a sample reports are shown in fig 6.6 and fig 6.7.

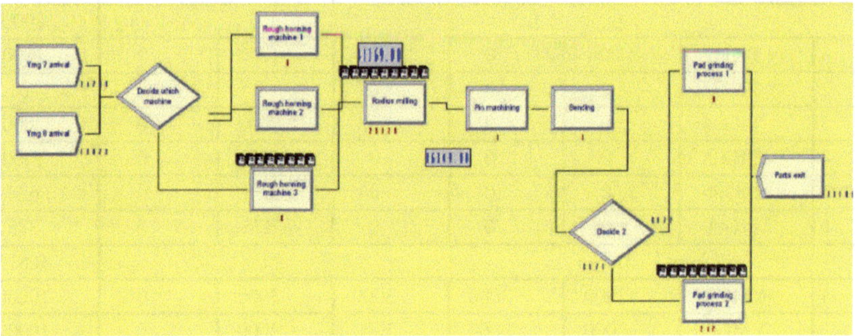

Fig. 6.6 As-is model.

Failure - Advanced Process

	Name	Type	Up Time	Up Time Units	Down Time	Down Time Units	Upti
1	RH 1 Failure	Time	EXPO(390)	Minutes	EXPO(19.85)	Minutes	
2	RH 2 failure	Time	EXPO(390)	Minutes	EXPO(19.85)	Minutes	
3	RH 3 failure	Time	EXPO(390)	Minutes	EXPO(19.85)	Minutes	
4	RM failure	Time	EXPO(390)	Minutes	-0.001+EXPO(34.7)	Minutes	
5	PM Failure	Time	EXPO(390)	Minutes	-0.001+EXPO(66)	Minutes	
6	PG 1 Failure	Time	EXPO(390)	Minutes	NORM(71.8 , 66.4)	Minutes	
7	PG 2 Failure	Time	EXPO(390)	Minutes	-0.001+EXPO(71.7)	Minutes	
8	BM 1 Failure	Time	EXPO(390)	Minutes	WEIB(0.022, 0.21)	Minutes	
9	BM 2 Failure	Time	EXPO(390)	Minutes	WEIB(0.022 , 0.21)	Minutes	
10	BM 3 Failure	Time	EXPO(390)	Minutes	WEIB(0.022 , 0.21)	Minutes	

Double-click here to add a new row.

Fig. 6.7 Schedule Assigned in ARENA (sample).

The Data sheet shows the "up-time" and "down-time" of the line.

6.3.1 *Report - As Is Condition*

The output of 15 day production run of YMG-7 & YMG-8 models were recorded with the AS-IS model (Fig. 6.8). The number out was 42123 which matched with the existing output of the process line under study confirming the model to be right for the study. Further various "What-If" or "Sensitivity Analyses" were conducted and the results were recorded.

Table 6.6 Results.

Component	Number in	Number out	Work in Process
Ymg 7	64519.00	25153.00	19610.35
Ymg 8	43310.00	16970.00	13083.54
TOTAL	107829.00	42123.00	32693.89

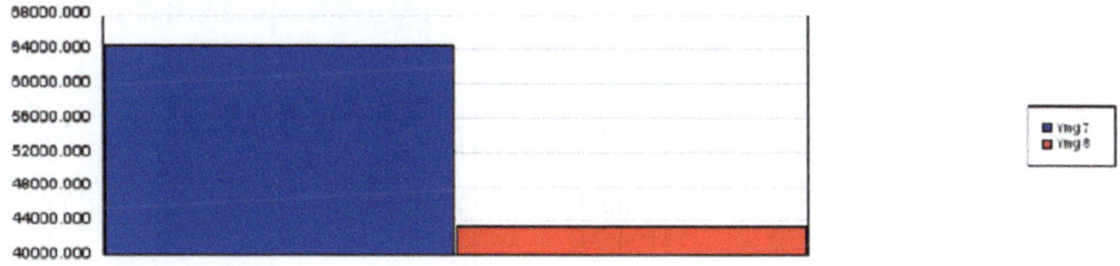

Fig. 6.8 Result diagram

Table 6.7 Utilization

SCHEDULED UTILIZATION

Machine	Utilization in %
BM 1	0.3060
BM 2	0.3053
BM 3	0.3110
PG 1	0.2141
PG 2	0.2148
PM	0.2808
RH 1	0.3991
RH 2	0.4023
RH 3	0.4101
RM	0.9130

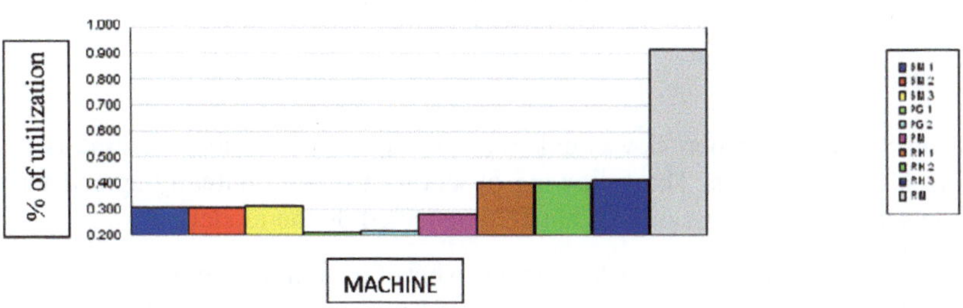

Fig. 6.9 Utilization.

6.3.2 Reports and Analysis

Analysis of report for the 15 days of production runs of Yamaha Ymg-7 & Ymg-8 models is as given in table 6.8.

Table 6.8 Output results & analysis.

ELEMENTS	AS-IS	AS-IS 50% DOWN TIME REDUCTION	AS-IS 75% DOWN TIME REDUCTION	RM CAP INCREASED to 2 SHIFTS	RM CAP INCREASED to 2 SHIFTS & 50% DT REDUCTION	RM CAP INCREASE D 3 SHIFTS
1.Number Out	42123	44054	44691	73332	74192	86474
2.Average Wait Time	Ymg 7 108.75, Ymg 8 108.36	Ymg 7 108.20, Ymg 8 107.90	Ymg 7 104.74, Ymg 8 104.40	Ymg 7 56.88, Ymg 8 56.77	Ymg 7 57.19, Ymg 8 57.62	Ymg 7 38.59, Ymg 8 38.58
3.WIP	Ymg 7 19610.35, Ymg 8 13083.54	Ymg 7 19346.77, Ymg 8 12858.85	Ymg 7 18897.01, Ymg 8 12572.09	Ymg 7 10278.48, Ymg 8 6783.32	Ymg 7 10326.55 Ymg 8 6900.32	Ymg 7 6878.14 Ymg 8 4555.14
4.Waiting Time	108.18hr	107.95hr	104.52hr	56.34hr	57.20hr	37.95hr
5.Number Waiting	32632	32179	31447	16948	17175	11271
6.Instantaneous Utiliza-tion	0.8169	0.8533	0.8651	0.8568	0.8603	0.8380

A detailed study of the analysis results after discussions with the concerned managers led to the final conclusion that **by increasing radius milling machine capacity to 2 shifts and 50% of down time reduction on delay time in the line results into achieving the production schedule target on date thus reducing the effects of end of month syndrome substantially.**

6.3.3 Results

Based on Simulation report it was evident that the radius-milling machine is the bottleneck in the process. The following observations were mutually agreed with the end users during the various "What-If" conducted on the model.

1. Number Out - Increases for various what-if as in the table 6.8.
2. Average Wait Time - Waiting time of Entities shows gradual decrease in the system.
3. WIP – Work in Process decreases for various what-if as in the table 6.8.
4. Waiting Time - Waiting time of an entity in front of the Resource de-creases.

5. Number Waiting - Decreases with various what-if as in the table 6.8.
6. Resource Utilization – Decreases.

6.3.4 Conclusion

After seeing the analysis and results we can conclude that the plant can achieve targets with the existing line by increasing the capacity of radius milling machine for 2 shifts and in turn plant can also save / reduce one shift of production to achieve monthly target.

If the plant increases the capacity of radius milling machine for 3 shifts, then the plant can achieve targets within 15 days of production run. Another 15 days plant can run for different models and concentrate on adding new customers to the existing line.

Analyzing & comparing the down time of all machines it was evident that the pad grinding and pin machining have it more than the radius milling machine as shown in the delay time column of the table 6.4 but those machines are running for 3 shifts as compared to the radius milling machine. If the plant can reduce the down time of those machines then the output of the line will increase and they can also reach their target quantities of production before the target dates and can reduce the end of the month syndrome.

Authors Biography, Contact

About the College (www.bvb.edu)

The versatile manifestations of engineering have had a profound and lasting impact on our civilization. From the grandeur of the pyramids and mans journey into the space, to the recent information revolution, engineering continues to fascinate and enthrall. The B. V. Bhoomaraddi College of Engineering and Technology (BVBCET) believes in kindling the spirit of this unique and creative discipline in every student who enters its portals. Preparing them for a world in which their contribution truly stands apart.

Established in 1947, BVBCET has achieved an enviable status due to a strong emphasis on academic and technical excellence. From a modest beginning when the college offered only an Undergraduate program in civil engineering, the college has indeed come a long way. Currently college offers 12 UG and 8 PG programs affiliated to Visvesvaraya Technological University, Belgaum and is recognized by AICTE, New Delhi and accredited by NBA. Current annual student intake for Undergraduate & Post Graduate programs is in excess of 1200. The faculty consists of extremely qualified and dedicated academicians whose commitment to education and scholarly activities has resulted into college gaining Autonomous Status from the University and UGC. The college has adopted Outcome Based Education (OBE) framework to align the curriculum to the needs of the industry and the society. Innovative pedagogical practices in the teaching learning processes form the academic eco system of the institution. The active involvement of faculty in research has led to the recognition of 8 research centers by the University.

Spread over a luxurious 50 acres, the picturesque campus comprises of various buildings with striking architecture. A constant endeavor to keep abreast with technology has resulted in excellent state-of-the-art infrastructure that supplements every engineering discipline. To enable the students to evolve into dynamic professionals with broad range of soft kills, the college offers value addition courses to every student. Good industrial interface and the experienced alumni help the students to become industry ready. The college is a preferred destination for the corporate looking for bright graduates. There is always a sense of vibrancy in the campus and it is perennially bustling with energy through a wide range of extra-curricular activities designed and run by student forums to support the academic experience.

Author: Sanjay Kulkarni
Graduated as a mechanical engineer in the year 1995, Sanjay worked for various engineering industries as a consultant in and around India. He started off as a consultant introducing "Clean Room" concepts to various engineering industries when the technology was very nascent in the Indian region. He had a great opportunity coming across as a software consultant after two years of his first assignment after which he never had to look back. As a software consultant Sanjay had best opportunity to learn various technologies relevant to engineering industry right from Geographical Information Systems, Geographical Positioning Systems, CAD and CAM solutions, Mathematical modeling, Statistical modeling and Process modeling tools to various hardware associated with the above technologies. He spent 14 years serving the engineering industry before he quit and began his second innings with academics.

Presently Sanjay is a professor with one of the oldest and leading engineering colleges of North Karnataka – B V Bhoomaraddi College of Engineering and Technology Hubli, Karnataka, India. He is associated with Industrial and Production department handling subjects like – System Simulation, Supply Chain Management, Organizational Behavior, Marketing Management, and Principles of Management. A rich industry exposure of Sanjay has given an edge while delivering the lectures to the students and it has been a memorable experience to experience both the worlds of engineering profession and engineering academics.

As a consultant he has handled challenging engineering projects in the past for various engineering industries and delivering the results successfully. As a professor he is learning new things every day from his students – actually learning **never ceases**.

Co-Author – Prashant
After completion of engineering in mechanical Prashant worked in an engineering industry as quality engineer for about 30 months till he took up M-Tech in production management from BVB College of Engineering and Technology, Hubli – Karnataka.

As a part of his academic project in M-Tech, Prashant had an opportunity to work closely with an automobile parts manufacturing export unit. The unit was facing end of the month syndrome and they were relying on their past experience and knowledge of the employee to overcome the syndrome. However Modeling and Simulation technique was employed to solve such problem which was appreciated by the company and the results were much better than their conventional approach.

Presently Prashant is employed with a high precision manufacturing unit and is responsible for the profit and loss of the company. Prashant has a keen interest in solid modeling and has learnt many related software's from CAD modeling to Analysis.

7 Creating a Model for Virtual Commissioning of a Line Head Control Using Discrete Event Simulation

Steffen Bangsow and Uwe Günther

The increasing mastery of the instrument Discrete Event Simulation and increasing detailing of the simulation models open up new fields for the simulation. The following article deals with the use of discrete event simulation in the field of commissioning of production lines. This type of modeling requires the inclusion of sensors and actuators of the manufacturing facility. Our experience shows that it is well worth the effort. Essential coordination with the development of automation can be integrated in the planning process. The simulation helps to find a common language with all people involved in the development.

7.1 Introduction and Motivation

HÖRMANN RAWEMA takes on the role as a general contractor for many projects. As a general contractor, one task is to coordinate and supervise the construction of the entire system. The construction of highly automated plants abroad is an especially big challenge for the project manager. To ensure the proper functioning of the plant components, we typically use a multi-stage acceptance concept (Figure 7.1):

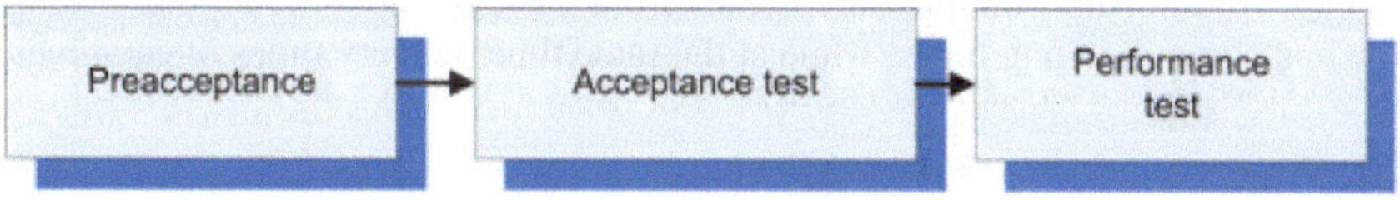

Fig. 7.1 Acceptance sequence

Steffen Bangsow
Freiligrathstrasse 23
08058 Zwickau
Germany
e-mail: steffen@bangsow.net

Uwe Günther
HÖRMANN RAWEMA GmbH
Aue 23-27
09112 Chemnitz
Germany
e-mail: uwe.guenther@hoermann-rawema.de

During the pre-acceptance phase, the ability of the machinery and equipment is tested to meet the agreed-upon requirements. Pre-acceptance may include cold tests (without machining of parts) or sample processing. Deficiencies during the pre-acceptance phase will be recorded. Shipping of plant components and equipment takes place only after eliminating all significant deficiencies and possibly after repeated inspection. This way repair or improvement at the customer site is avoided. Functional tests of line sections examine the function of the machine (with and without workpieces) and the function of the automation technology used to transport materials. For this purpose, after installation of all related technology, the line segments manually "will be fed" with parts. These workpieces are transported either in automatic mode or by manual operation through the line segments. All important operating states are examined (acceptance test). The performance test of the entire system is used to detect the contractually specified performance parameters to the client. The performance test in general consists of a certain production time under full load. Within this context, the performance of the head control components is also tested.

Between readiness for operation of the individual machines and the functional tests of the line segments usually a lot of time passes in practice. This has, among others, the following reasons:

- The integration of automation typically begins only after all system components and machines are set up and functioning. Normally, the construction of automation begins only when the individual machines are installed.
- The programming/customization of the control will start only after finishing the construction of the automation hardware.
- Poorly prepared programs lead to long trial and error phases.

During the software adaptation phase, the system shows a state that is hard to understand for the client. All machines are operational, but the production facility, often worth tens of millions of Euros, doesn't produce one single part for months on end. Additionally, the pressure comes from customers to shorten the installation and commissioning times, while at the same time delivery times of equipment

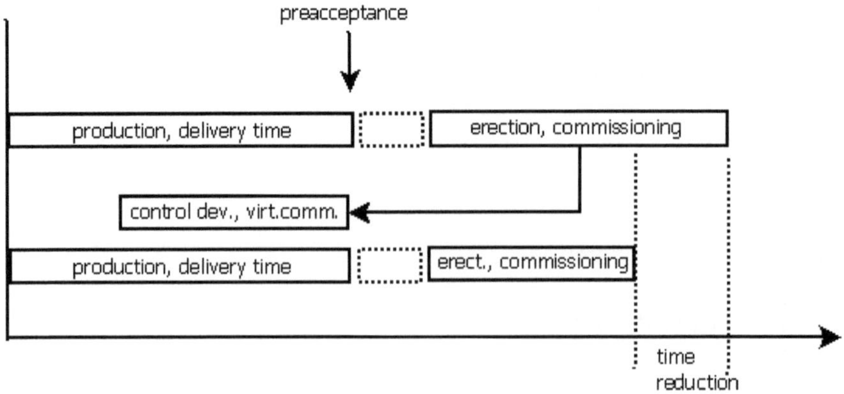

Fig. 7.2 Project time reduction through virtual commissioning

manufacturers are extended. One way we see to achieve this, is using virtual commissioning of the line (head) control. With the help of virtual commissioning, it is possible to bring forward a part of the line software development and software testing in the project process and to shorten the execution time of the project (Figure 7.2).

7.1.1 Definitions

7.1.1.1 Commissioning

In operational practice, commissioning has the task to put the mounted products on time in readiness for operation, to verify their readiness for operation and, if readiness for operation is not given, to establish it [7.1].
Regarding controls commissioning activities include:

- Correction of software errors
- Correction of addressing failures, possibly the exchange of signal generators
- Teaching of sensor positions
- Parameter adjustments (for example, speeds)

The correction of software errors in highly complex manufacturing facilities takes up most of the time (see also [7.2]).

7.1.1.2 Virtual Commissioning

The basic idea of virtual commissioning is to provide a large part of the commissioning activities of the controls before installing the system (eg, parallel to the construction of the facilities) with the help of a model. The concept of virtual commissioning describes the final control test based on a simulation model that ensures the coupling of real and virtual controls with the simulation model with a sufficient sampling rate for all control signals [7.3]. According to our understanding virtual commissioning can be realized on three different levels.

- Virtual commissioning at machine-level or individual equipment level
- Virtual commissioning at line level
- Virtual commissioning at production system level

7.1.1.3 Virtual Commissioning at Machine Level

Due the increasing complexity of machines and through the integration of additional tasks in the machines the development of controls needs to start at an earlier stage. For virtual commissioning at machine-level a variety of proven approaches and instruments exist. A 3D model of the machine or equipment will be extended by the relevant individual sensors and actuators. Accordingly, there are, for example, solutions are provided as add-ons to 3D CAD systems. Very high demands are made to the simulation regarding sampling rates and response times, to reach a behavior that is as close to reality as possible. It can be controlled relatively easily with "virtual machines" collisions and the function of the control.

7.1.1.4 Virtual Commissioning at Line Level

Through the coupling of machines and equipment with suitable materials handling equipment production lines result. To produce an overall function, it is necessary that the individual components communicate in an appropriate manner. In many cases, protocols and regulations exist, in some cases however; special software needs to be developed. Complete lines usually cannot be modeled as a 3D model before the technical hardware development is finished, because all individual components are necessary to build the complete model. Subject of a virtual commissioning at line-level is the communication of the individual machines and equipment with the line control. With a simulation at a higher level (machinery and materials handling), it is possible to model all necessary operating states of the production system and the associated sensor and actuator signal exchange. The response times are less demanding than at the machine level, which gives rise to a big amount of opportunities for couplings. Due to the longer response times, the models can be tested in fast motion (software in the loop) or in real time to validate a coupled PLC.

7.1.1.5 Virtual Commissioning at Production System Level

The control of a production system (ERP, MES, head control) requires a lot of information from the machine and line level. Many control systems also provide important information for the line control, which, for example, are stored in databases. When new lines are integrated into existing production control systems, a lack of adequate preparation may lead to a failure of the entire production system, which can cause huge costs. A 3D model is completely unnecessary at this level. A discrete event simulation for modeling the operating states and system responses can provide important impulses for error handling, especially since discrete event simulation models can be created hierarchically and in this way contain complete production systems. Virtual commissioning on production system-level would simulate the input and output signals of the production control (and all higher-level systems) and test the appropriate response of the system elements (machine, material handling and equipment). According to our experience, virtual commissioning on line level can be combined with virtual commissioning on production system level.

As a system supplier we are dealing with virtual commissioning on production lines and system level. At line level, we are testing the sensor-actuator communication of all major components. Our objective is to combine virtual commissioning with the pre-acceptance.

7.1.2 Software in the Loop and Hardware in the Loop Approaches

According to the type of coupling of the controller to the simulation, one distinction is between hardware-in-the-loop simulation (HIL) and software-in-the-loop simulation (SIL) [7.3]. At HIL simulation the model is directly connected to the control hardware. For this purpose the simulation computer must have interfaces to the automation system and must be connected directly to it (Figure 7.3).

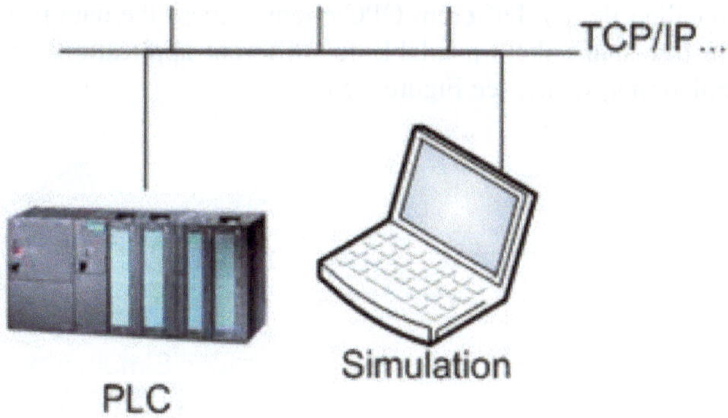

Fig. 7.3 Hardware in the loop simulation

In software in the loop (SIL) approaches the simulation computer is not connected directly to the controller hardware, but is connected with simulation software that simulates the control hardware again (such as software-PLC, Figure 7.4).

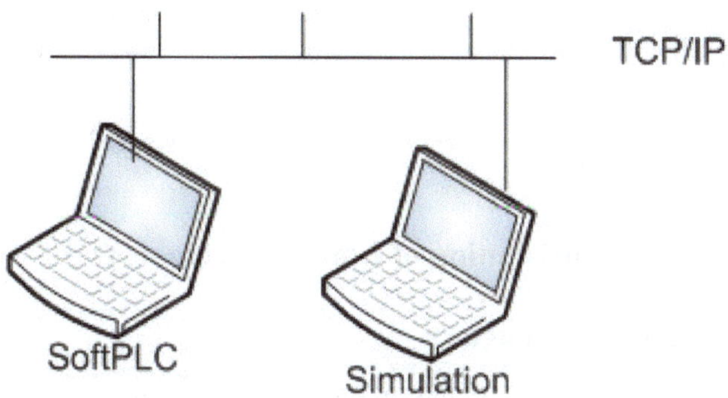

Fig. 7.4 Software in the loop simulation

SIL approaches are not readily suited to virtual commissioning of equipment in which a high sampling rate of the signals is necessary (machine level).

7.1.3 OPC

OPC is a standard for manufacturer-independent communication in automation technology. It is used where sensors, actuators and control systems from different manufacturers must work together [7.4]. For each device only one general OPC driver for communication is required. The programmers of OPC servers are testing their software according to a specified procedure (OPC Compliance Test) for compatibility. A major part of the automation technology used today is OPC-compliant. In a standard constellation an OPC server receives data via the proprietary field bus of the PLC/DDC controller and makes them available in the OPC server (as so called items). Different OPC clients access the data provided by the server and in turn make them available for different applications (e.g. graphical console, simulation systems; see Figure 7.5).

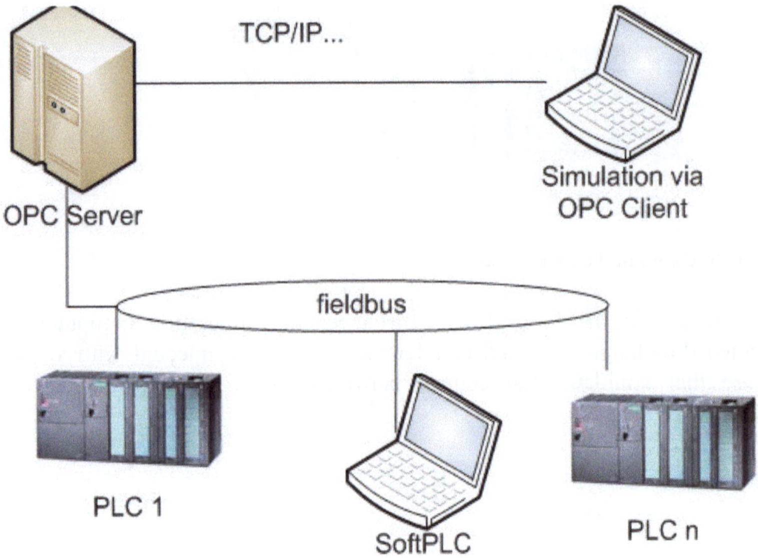

Fig. 7.5 OPC communication

7.2 Virtual Commissioning of Line Controls

7.2.1 Task and Challenge

In the commissioning phase of linked automated production systems the same type of problem came up time after time in the past:

- Incompatibility of the installed components (e.g. analog output - digital input)
- Uncoordinated address spaces (e.g. not enough inputs or outputs, too complex programming for the built-in control)
- Data is not delivered or delivered to the wrong address
- The implementation deviates considerably from the planned (and possibly simulated) processes, which leads to problems in the output of the equipment

One reason is the wanting or too late integration of the automation developer in the project schedule. Normally, the work of software developers does not begin until the production automation hardware is finished, often only after finishing the construction of the complete production line. A major challenge is to move their involvement forward in the project process to an earlier stage. For testing purposes it is necessary to provide a model of the production line, with the same basic behavior as the future equipment. Another problem is the large number of suppliers.

All suppliers must provide and process accurately defined data. Even small errors (such as in the programming of the interfaces in the machine control) practically lead to large delays if a programmer has to come on site for the software test. The poor or nonexistent coordination between the customer and the control development also results in an often inadequate design of the programming. Since the time pressure at the end of the project is the greatest, the project often goes into operation with the first working control variant because there is no time for an elaborate optimization. The performance parameters of the system can be affected to a significant extent.

7.2.2 Virtual Commissioning and Discrete Event Simulation

7.2.2.1 Experience Background and Capabilities

For more than 12 years discrete event simulation has been used by HÖRMANN RAWEMA for planning support. Over the years a highly skilled base of simulation specialists has been established, who realize simulation projects, sometimes integrated into plant implementation projects. A basic idea of virtual commissioning at HÖRMANN RAWEMA is the integration of virtual commissioning in the planning process. We developed a methodology by which virtual commissioning can be integrated into the material flow simulation from a certain progress of the planning process.

Discrete event simulation is used within planning to prove the contractual parameters (e.g. output in a given time, overall plant availability, strategies for changes in operating conditions). For this purpose, we simulate the plant at a high level of detail. We found that especially in the area of the line control the completion of the simulation models with sensors and actuators is possible with acceptable effort. For these reasons we decided to develop virtual commissioning as part of the plant simulation. Especially for lines and head controls DES, in connection with OPC, provides a sufficiently high sampling rate. The simulation allows to define and to simulate all necessary test cases. The OPC interface allows connecting the discrete event simulation with a large number of automation technologies.

7.2.2.2 Suitability of DES for Virtual Commissioning

A large number of techniques and software packages for virtual commissioning are on the market. Within many programming packages the control is developed in the same manner as later in the controller (e.g. PLC-programming), so that the entire simulated software can be transferred to the original control hardware at the end of development. After careful consideration, this variant for line and head control has proven to be insufficient. In many cases, head controls are so complex that they no longer run on PLCs. For connecting to higher level systems (which in turn are also simulated) a higher programming language with support for all required interfaces is required, such as ODBC, XML or socket communication. That's usually beyond the scope of most specialized systems for virtual commissioning, but can easily be made available by many DES systems. Many PLC development systems can be coupled to the DES via OPC, so that the PLC-program can also be developed in a DES system.

7.3 Use Case

7.3.1 Virtual Commissioning Simulation Methodology

The task was to design an automated handling system for bars. The handling system consists of equipment for storage and transport from different suppliers. For developing virtual commissioning models we use Plant Simulation by Siemens. Figure7.6 illustrates the basic structure of our simulation approach:

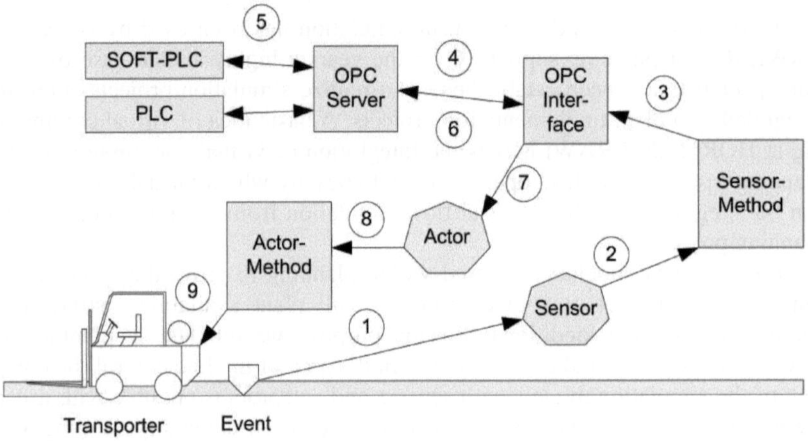

Fig. 7.6 Simulation methodology

Events in the simulation trigger changes of sensor states (1). The sensor is connected to a sensor method, which is called, when the sensor value changes (2). The sensor method transfers the changed sensor value to the OPC server using a method of the OPC interface (3 and 4). The PLC control, which can either be a real PLC or a Soft PLC, reads the changed sensor value as input value and then sets an output value (5). The OPC server then updates its memory. The OPC interface detects the modified output value to the OPC Server (6) and sets the value of an assigned actuator-variable in the simulation (7). The change of the actor value triggers a method call (8). The method initiates the actions required to create an appropriate response to the change of the actuator value within the simulation (9). Both actuator and sensor controls can be used together with a suitable parameterization for a wide range of sensors and actuators, so that the number of elements required keeps manageable. During the project process we found, that the high level of detail of the virtual commissioning model allows detailed studies of the control strategies. Within the simulation we used a control bypass. The control bypass acts like a PLC within the simulation (Figure 7.7).

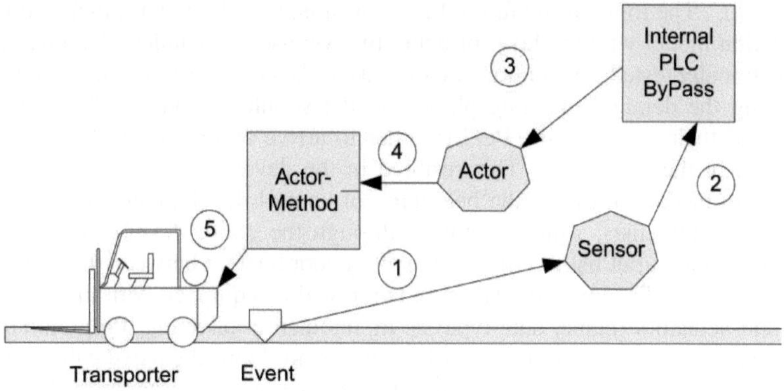

Fig. 7.7 Control bypass

The internal control bypass scans the sensor values (2), processes these and sets the actuators accordingly (3). The control system is modular, so that certain parts of the control can be enabled and disabled. This way a hybrid control is possible. While most of the control is realized through the bypass and the simulation delivers events and sensor changes, an external connection can be established for each supplier and PLC and the control signals via the OPC interface can be supplied. This way each and every PLC can be tested step by step (Figure 7.8).

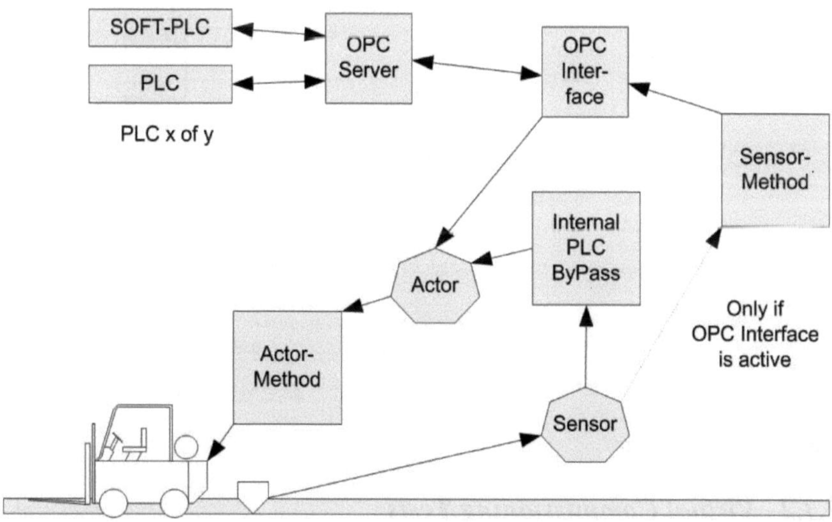

Fig. 7.8 Hybrid control

This constellation had a startling side effect. We've often been confronted with the question how to pass the logic of a simulation model to the automation developers. The control-bypass is working with the same input and output values, as the future PLC. The logic of the future line control is, for a large part, included in the simulation model with the level of detail that we use for detailed planning, and it is functionally tested. In addition, we optimized the control of the simulation model during the detailed planning phase and the simulation phase. These changes must find their way into the PLC in order to arrive at similar results in the real world as in the simulation. This resulted in the development of a specific programming methodology. At the beginning of control development we coordinate the input/output lists, which continues through the entire development process. The input and output lists are the first level of coordination between the simulation and automation development. The simulation will be equipped with the same sensors and actuators (name, data type) as in automation planning. Programming of the simulation is similar to that of a PLC in a main loop (recursive call). All program-specific commands are omitted in the simulation; it is programmed with only the instruction set, which is also available in the PLC. The communication with the simulation is exclusively controlled by the sensors and actuators. Only the actuator control includes direct access to the objects of the simulation model. The result is code which is very similar to the PLC programming. The program code can be handed over to the PLC programmer as pseudo code or it can be transferred with very little effort to the PLC (Figure 7.9).

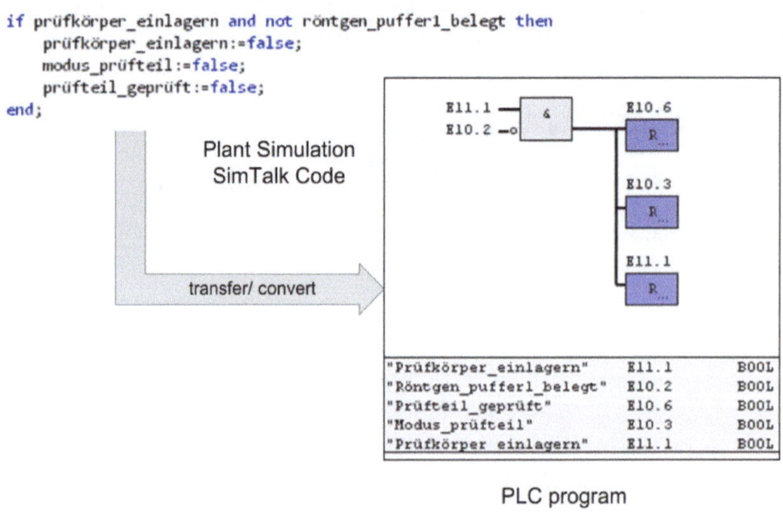

Fig. 7.9 Code hand over

7.3.2 Virtual Commissioning Tests

As part of virtual commissioning a number of tests can be run.

7.3.2.1 Addressing Tests

For virtual commissioning the control must be connected with the OPC server. In practice we realize the connection with the help of alias lists. Within the lists, addresses of the PLC program are assigned to alias names. The server reads the values from the PLC and makes them available for the OPC clients using the alias names. The alias list is prepared on the basis of automation planning (it defines the addresses for the communication between the elements). In a first step we check whether all of the required addresses are "serviced" or if there are errors in the assignment (which particularly affects the addressing within data blocks). This is accomplished through logging the data traffic on the OPC server. Only after full conformity has been reached, functional tests can be run (Figure 7.10).

Fig. 7.10 Addressing tests

7.3.2.2 Function Tests

Within the simulation the different operating states of a system can be modeled (machine, line, plant). The function tests produce combinations of sensor states and other data and the PLC program must respond adequately, so that the behavior of the system matches that of the expected or planned behavior. Operating states to be tested could be for example:

- Ramp up(the line is empty, the first part arrives)
- Shutdown, empty line
- Remove of test pieces (either automatically or by request)
- Feed in the tested part
- Lock lots and remove them
- Machine failure, maintenance
- Handling of nio-parts
- Lot change and set up

All system states to be examined in the simulation can be easily prepared and be triggered by pushing a single button. This simplifies a systematic review. The modular design of the virtual commissioning model allows individual tests with all suppliers involved. So we are a big step closer to our goal of integrating virtual commissioning into the pre-acceptance phase.

7.3.3 Problems during Virtual Commissioning

The interaction of PLC, OPC and discrete event simulation is problematic in some areas. In principle, all parties involved are working independent of each other. The main problem is to reset the simulation without restarting the PLC. All initial system states must be written to the PLC via the OPC server within an initialization routine to avoid false responses of the PLC if different states of the variables exist. Within the simulation accelerated execution must be put aside to meet the response times of a real PLC or of a Soft PLC. This makes virtual commissioning models unsuitable for some purposes. Long-term studies (for example for detecting of system parameters) or the use of random functions for displaying availabilities could be very time-consuming with virtual commissioning models, depending on the size of the model. The high level of detail also causes an increase in the complexity of modeling. In many cases it is necessary to work more closely with realizing authorities to obtain a realistic picture in the model. This force everybody involved to think along the requirements of the control of the manufacturing system even during the planning phase.

7.3.4 Effects of Virtual Commissioning

A major goal of the virtual commissioning at HÖRMANN RAWEMA was the involvement of the automation development at an earlier stage during project implementation. This goal has already been reached in the stage of development and optimization of the simulation model. This required the following coordination tasks:

- Precise coordination of I/O – lists
- Precise specification and adaptation of the sensor equipment
- Coordination of the addressing among all parties ("up to the individual bits")
- Accurate handover of the control strategy as a template for programming the control

The control of the simulation could be handed over to the automation developer without loss of information and communication problems. Coupling the control with the simulation model in the pre-acceptance phase verified that the PLC matched the specifications. Unrewarding control strategies have already been identified in the simulation phase. This also shortened the time required for commissioning. I/O conflicts could be eliminated as well as, allowing for the networkings of the individual components to go more quickly. Without a question the commissioning time was reduced by the higher input quality of the control. However, this effect cannot be quantified, since a reference value is missing.

7.4 Outlook

The next logical step is to expand virtual commissioning to the communication with the higher-level production control systems. This may, in the simplest case, be a machine data acquisition system, in the most difficult case, a corporate manufacturing execution system (MES). These systems don't exist in an early phase of the project, the exchange of signals is usually defined in comprehensive functional specifications. For virtual commissioning the signal exchange between these systems and conveyor systems or machines has to be modeled. Using suitable interfaces to database systems this issue can be realized with reasonable effort.

7.5 Summary

Virtual commissioning provides a solution for many problems that occur in the implementation of complex automation projects. It significantly improves the communication with the automation developers and leads to a mutual understanding of problems and solutions. Virtual commissioning forces the planning executive unit to deal with the logic of production controls early and in detail. The greater maturity of planning and the better coordination of installation of the system in advance significantly reduce the commissioning times. This leads to an increased planning and modeling effort.

Company Profile and Contact

HÖRMANN group and its 20 subsidiaries with its two business segments: "Industrial Services" and "Communication" is a diversified company, which offers, in particular the customers in the automotive industry, comprehensive over-all and customized components solutions. The range of services in the business area "Communications" includes the fields "Traffic & Control", "Automotive" and "Security". The business field of "Industrial Services" covers the spectrum of "Automotive" with component production and production of sheet metal molds, "Energy and Environment" covers renewable energy, the area "Plants" for the supply of logistics and assembly systems and the field "Engineering" stand for planning, project management and delivery of turnkey production systems. A significant activity of HÖRMANN RAWEMA, a subsidiary in the field "engineering", is the application of tools for the digital factory. Early in the planning stage the goal is to provide customers with safeguarding the future production and therefore to avoid high costs through the efficient use of these tools. It is also possible to optimize processes and resources already during the planning phase, not just during operation. HÖRMANN not only uses this effect in its own group, but offers this knowledge to other companies as a service as well.

Steffen Bangsow works as a freelancer and as book author. He can look back on more than a decade of successful project work in the field of discrete event simulation. He is the author of several books about simulation with the system Plant Simulation and of technical articles on the subject of material flow simulation.

Contact

Steffen Bangsow
Freiligrathstrasse 23
08058 Zwickau
Germany
Email: steffen@bangsow.net

Uwe Günther is at the company HÖRMANN-RAWEMA hired as project manager. In particular, he working of the following main topics: project management, factory planning and material flow simulation.

Contact

HÖRMANN RAWEMA GmbH
Dr. Uwe Günther
Aue 23-27
09112 Chemnitz
Germany
Email: uwe.guenther@hoermann-rawema.de

References

[7.1] Eversheim, W.: Die Inbetriebnahme komplexer Produkte in der Einzel- und Kleinserienfertigung. In: Inbetriebnahme komplexer Maschinen und Anlagen, (VDI-Berichte 831), p. 9. VDI-Verl., Düsseldorf (1990)

[7.2] Wünsch, G.: Methoden für die virtuelle Inbetriebnahme automatisierter Produktionssysteme, pp. 1–2. Herbert Utz Verlag, München (2007)

[7.3] Wünsch, G.: Methoden für die virtuelle Inbetriebnahme automatisierter Produktionssysteme, p. 33. Herbert Utz Verlag, München (2007)

[7.4] Internet: Wikipedia, `http://de.wikipedia.org/wiki/OLE_for_Process_Control`

8 Optimizing a Highly Flexible Shoe Production Plant Using Simulation

F.A. Voorhorst, A. Avai, and C.R. Boër

This paper explores the use of simulation for the optimization of highly flexible production plants. Basis for this work is a model of a real shoe production plant that produces up to 13 different styles concurrently, resulting in maximum 11 different production sequences. The flexibility of the plant is ensured by organizing the process in a sequence of so-called work islands, using trolleys to move shoes between them. Depending on production needs one third of the operators are reallocated. The model considers the full complexity of allocation rules, assembly flows and production mix. Analyses were performed by running use cases, from very simple (providing an insight in basic dynamics) up to complex (supporting the identification of interaction effects and validation against reality). Analysis gave insight in bottlenecks and dependencies between parameters. Experiences gained distilled in guidelines on how simulation can support the improvement of highly flexibly organized production plants.

8.1 Introduction

Discrete event simulation has been widely used to model production line (Roser et al. 2003) and to analyze its overall performances as well as its behavior (Boër et all. 1993). For the most part, past models have concentrated on the mechanical aspects of assembly line design and largely ignored the human or operator component. (Baines et all. 2003). The simulation model, presented in this paper, was developed in Arena (Kelton et all. 2003) and it augments the standard production system model to include labor movements and its dynamic allocation many times per shift. This paper describes the experiences and findings in using discrete event

F.A. Voorhorst
HUGO BOSS Ticino SA, Coldrerio, Switzerland
e-mail: Fred_Voorhorst@hugoboss.com

A. Avai
Technology Transfer System, Milano, Italy
e-mail: Antonio.Avai@ttsnetwork.com

C.R. Boër
CIM Institute for Sustainable Innovation, Lugano, Switzerland
e-mail: Claudio.Boër@icimsi.ch

simulation as tool to better understand a plants dynamic behavior prior to optimization and further improvements The remainder of this paper is organized as follows: in section 8.2 a short description of the problem is presented and section 8.3 gives an overview about the actual system to produce men shoes. Section 8.4 provides a description of all the modeling and implementation issues to be faced in order to get a simulation model with a correct detail level. In section 8.5 the results are presented and conclusions follow.

8.2 Problem Description

The challenge we face is to better understand the dynamic behavior of the shoe production plant in order to be able to predict the daily volume and as basis for improvements to obtain a more fluent production. Actually there are many factors influencing these aspects, such as labor availability and allocation of operators, availability of lasts and, clearly, the composition of the daily production plan, the so-called production mix. The production process has almost 40 different operations, grouped in work islands, to which approximately 70 operators are allocated. The production plant can work on more than 100 shoe variants, each one different in production routing and/or cycle times for operations. The main goal of this project is to identify the scenarios under which the system breaks down (production target is not achieved) in order to evaluate the impact of key factors such as production mix and labor allocation on the overall performances. The theoretical target productivity is about 1.700 pairs of shoes per day. However in the real system, daily through-put is not constant and shows large variations, sometimes 25% below target value.

8.3 System Description

The actual production plant assembles high quality man shoes in various colors, mainly of 3 different families:

1. Shoes with glued leather sole
2. Shoes with stitched leather sole
3. Shoes with rubber sole

From the 3 families the production processes of 50 shoe styles were modeled, amounting in 11 different process sequences (differences due to color are not included). The organization in work islands makes the production a very flexible system, both in terms of product types and capacity allowing the possibility to maximize through-put while minimizing investment such as the total number of lasts per style needed. At any point in time there are up to 13 different styles in production, each needing a specific last model with a significantly different (between families) or a slightly different (within one family) production sequence. In addition, shoes of the same style can have different colors, such as black, ebony, brown, grey, white, etc. which have an additional impact on the production sequence.

The production plant, organized in a circular fashion, is split in 2 main departments:

• The assembly department where shoes are assembled by means of last starting from upper, sole and insole, as it is displayed in Figure 8.1.

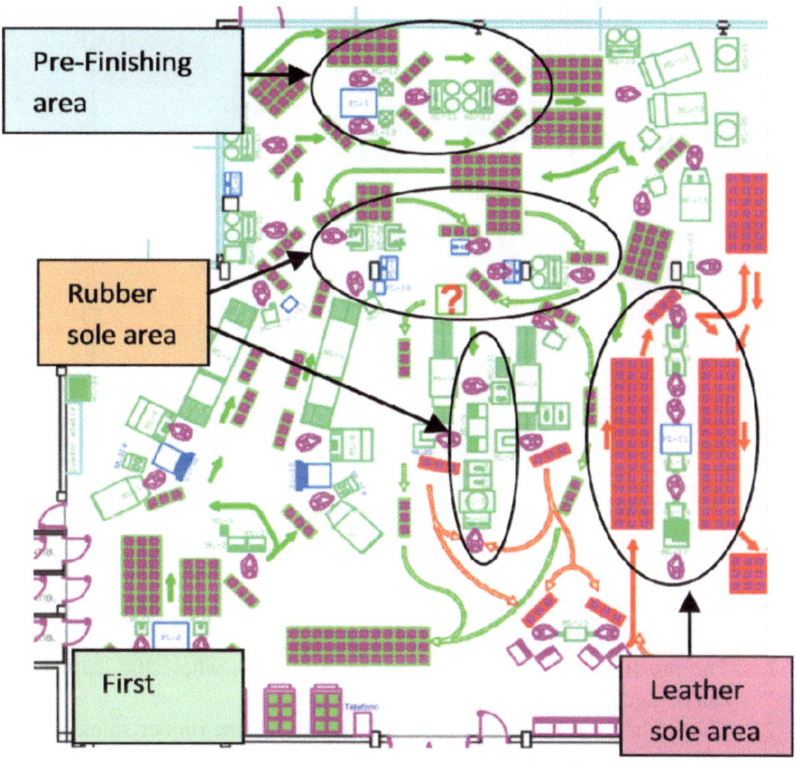

Fig. 8.1 Layout of assembly department.

• The finishing department, see Figure 8.2, where shoes are creamed, brushed, finished and packaged.

Each department is organized in different working islands, by grouping one or more machines and working positions. Furthermore, as shown in Figure 8.1, in the assembly department, 3 macro areas, to which a single team is assigned, can be identified:

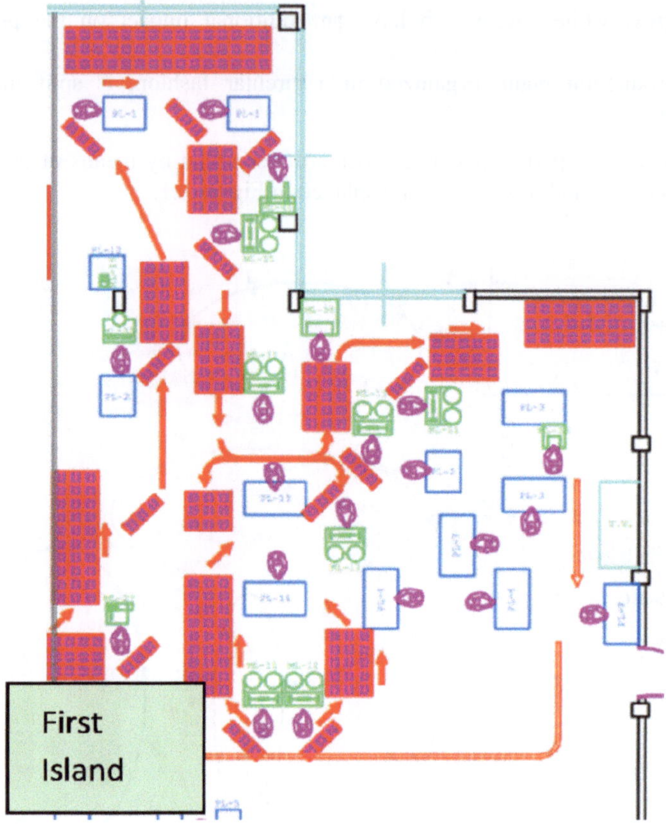

Fig. 8.2 Layout of finishing department.

1. The pre-finishing area, composed by 3 islands, where the leather upper can be aged, daubed of cream and brushed.
2. The rubber sole area, formed by 7 islands where rubber soles are glued and coupled with shoes.
3. The leather sole area, composed by 4 islands where shoes, with leather soles, are stitched.

The rubber and leather sole areas are crossed only by some shoe articles, so workers are allocated only when some trolleys are waiting to be worked. Shoes move from one island to the other by means of trolleys, moved by workers. In general an operator takes a waiting trolley, performs an operation to each shoe on the trolley and pushes the processed trolley to the waiting area for the next island. There are 2 trolley types:

- Assembly trolley: each one holds uppers, with the respective lasts, soles and insoles. They are used only in the assembly department.
- Finishing trolley: it transports shoes trough the finishing department.

The number of assembly and finishing trolleys is limited in order to keep constant the flow of shoes but, on the other hand, it can have a negative impact on the through-put. If many trolleys are stacked up in different positions, there are none available to be loaded with new shoes. Better production fluency is achieved when the length of trolley queues are minimal.

8.4 Modelling Issue

This paragraph describes the simulation architecture as well as all the relevant aspects analyzed during the modeling and simulation model deployment phases. The applied methodology follows a top down approach: first, the flow of shoes in the production plant has been simulated, and it has been refined adding details and rules by means of several meetings and interviews with foreman and production responsible. Then, the rules dealing with the production batches composition and dispatching have been modeled and tested. Last, the dynamic behavior of labor allocation between different islands and inside the 3 macro areas has been simulated. It was assumed that operators have equal skills and are interchangeable. Furthermore, an extensive campaign to measure cycle times by direct observations was carried on.

8.4.1 Simulation Architecture and Input Data Analysis

The simulation model is driven by 3 Excel files with the following input data:

1. The production mix, in terms of shoe articles, quantity and color to be produced
2. The assembly sequence per style along with stochastic cycle times for each operation
3. Several parameters related to the process together with the distances between islands

All these data are automatically imported in the simulation model at the beginning of each run. For the stochastic cycle times a triangular distribution was used. (Chung C. A. 2004)

8.4.2 Simulation of Shoes Flow

A particular attention was kept to simulate the following issues that are described in the next 2 paragraphs:

- The input buffer policy in each island
- The trolley selection and dispatching rules at the roughing island

8.4.2.1 Input Buffer Policy

Every island has an input buffer where trolleys are stacked up if they cannot be processed immediately. These buffers are simulated as queues following the same policy except for the last removing island. The defined policy for a queue is as following: each coming trolley is ranked based on its order number and, then, it is released following the FIFO rule (first in – first out) when the machine is free. In this way, each island tries working together all trolleys with the same order number.

At the last removing island, lasts are taken out from the shoe and put back into baskets. To minimize the number of baskets being filled in parallel, the last removing island does not follow the FIFO rule. Instead, trolleys are worked on last codes. This ensures a minimal change in baskets as large numbers of the same last are processed in one batch.

8.4.2.2 Trolley Selection at Roughing Island

All worked shoes have to be roughed in the roughing island then they pass through a reactivation oven where the cement is reactivated and, eventually, sole is applied to shoe bottom and pressed. There are 2 reactivation ovens for shoes with leather soles and one for rubber soles. In order to reach the productivity target and to keep the number of workers involved in the mentioned processes as small as possible, the worker at roughing island follows some rules of thumb to decide which trolley to take out from his/her queue, work it and move to the right reactivation oven.

The main issue in the modeling phase was just to understand the basic lines followed in this decision process and then to clearly define the several rules of thumb.

By means of direct observations and interviews with foreman and workers staffing the roughing island as well as reactivation ovens, it was found out the second reactivation oven for leather sole is switched on when

- The amount of stacked up trolleys at first oven for reactivating leather sole is greater than a certain threshold
- The oven for activating rubber sole is switched off.

Once it's switched on, it should work for about an hour and then it is switched off again.

Generally, more than 10 trolleys with different shoe articles are staked up at roughing island. Many times during a shift, the worker in this island has to decide when the second oven for leather sole has to be switched on, which and how many trolleys sent to it, or, vice versa, when the oven for rubber sole has to be activated.

The selection process is triggered by 2 events:

1. If some trolleys, holding shoes with rubber sole, are waiting at the roughing machine, they will be worked if the queue at oven for rubber sole is very short. This kind of process goes on until the queue at first oven for leather sole is long enough to avoid its stopping.
2. If no trolleys, holding rubber sole, are waiting and the queue at first oven for leather sole is too long then the selection process is a little bit complex. The basic idea is to try to work at roughing machine a certain amount of trolleys holding the same last in order to reduce the number of set up at roughing machine and to keep on the second oven for leather sole for an hour, at least. This area could become a

candidate to be investigated by means of simulation to improve system performances. Furthermore, when too many trolleys are staked up at this island, another manual roughing machine is activated for about an hour staffed by an operator to reduce the length queue of waiting trolleys.

8.4.3 Production Batches Composition

A production batch represents a single lot put in production at the same time in order to use the available lasts efficiently. It can be composed by one or several orders of different shoes but using the same last code to be produced. The batch size represents the amount of lasts used for each production batch. At the very beginning of the simulation, the whole production plan is examined to aggregate sequential items with the same last code and to disaggregate items with ordered quantities greater than the number of available lasts. In the first case, the aggregation mechanism is mainly based on homogeneous batch concept: the basic idea is to create batches, using the same last code, with a similar size. In the latter case, orders with big quantity are split based on

- Available last
- Homogeneous batch as mentioned before.

A split order is put in production again when, at least, there is a certain percentage of available last in the stock again compared to the batch size.

8.4.4 Simulation of Dynamic Labor Reallocation

Workers are re-allocated many times during a shift mainly because:

- The amount of available labor is less than the actual working positions
- Some shoe articles have long cycle times for some operations/islands and the number of workers allocated to these islands have to be increased to avoid queues

The decision on how to allocate labor takes into account many factors such as:

- Batch size
- Assembly sequence and cycle times
- Work already in process
- Last availability
- Labor availability
- Skill of each worker

By changing the schedule it is possible to influence the labor need. In the real system, the production responsible can modify the schedule based on the actual situation in production. This is done in order to increase flexibility in labor management, and to avoid trolleys being stacked up in front of some islands. This supervisory behavior is discarded as it is beyond the scope of this project and the simulation strictly follows the schedule.

The first step to simulate the dynamic labor reallocation was to understand the general principles and rules applied by the production responsible and model them in a formal way. In particular, the following items were defined:

- The decision events: when decisions on labor reallocation have to be taken
- The worker allocation or de-allocation rules for each decision moment

In general, labor allocation rules can be applied during these four specific decision moments:

1. When a new item arrives to an island with no worker available
2. When a queue of an island is getting too long
3. When an island has no item to be worked
4. When a worker has completed a certain number of trolleys

In the first two moments, an available worker has to be moved to the needed island, in the third case, an operator becomes available to be moved and in the last case a worker is eligible for transferring.

8.4.5 Labor Allocation Modeling

About 65% of available workers have a fixed position. In both assembly and finishing some work islands are continuously staffed whereas others are not. The remaining flexible workers are assigned depending on the production needs. In the simulation this is modeled by grouping the flexible workers in a single pool, and allocating them according to rules reacting to the first or second event, as mentioned before. An operator, if available, is taken from the pool immediately when an island 'requests' an operator, for example when the number of waiting trolleys exceeds a specific amount. When there are no workers available in the pool, 2 different situations have been simulated:

1. If the requiring island belongs to a macro area, as mentioned in the paragraph 3, an operator, working in the same area of the empty island, can be shared: he/she can work in 2 different positions alternatively.
2. If the requiring island does not belong to a macro area, it has to make a "reservation". This mechanism will be described in the next paragraph.

8.4.5.1 Reservation Mechanism

The reservation mechanism simulates the request of labor dynamic reallocation when all available workers are busy and some trolleys are waiting for being worked in, at least, one island. In the actual system this mechanism represents the moment when some trolleys reach an empty island and the foreman has to wait until, at least, a worker can be moved.

A reservation is triggered when some trolleys are staked up in an empty island and when no workers can be moved to this position. This situation can become critical because many trolleys could pile up. In order to avoid this scenario, a worker has to start working in this empty island as soon as possible.

When a reservation is made, the first worker becoming available (either free or candidate for transferring) is reallocated. The simulation model calculates the travelling time based on the starting and arrival positions.

8.5 Simulation Results and Performances Evaluation

After having concluded the validation, the simulation model was ready to make several runs and analyze of production performance and through-put under different conditions, aiming to identify bottlenecks and main important process drivers. To help the validation and analysis an animation was provided as shown in Figure 8.3.

The simulation was tested against different production mixes. Production mixes defines the combination of shoe families produced and for each family the quantities (batch sizes) produced. Both the combinations of families as well as the batch sizes were systematically changed.

The following variables were measured:

- The overall performances, mainly daily through-put
- The labor utilization
- The production fluency indicated by the staking trolley in some key islands.

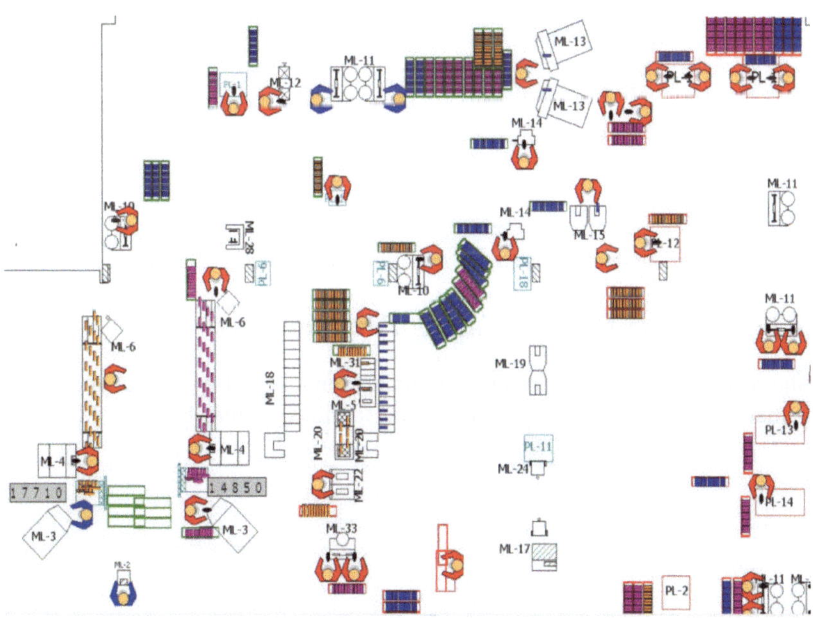

Fig. 8.3 Screenshot of simulation model animation

To obtain a good understanding of the production dynamics the analysis was based on use-cases of different complexity, first simulating simple production plan composed by only one shoe family and then adding the other two families changing the mix and the batch size, and finally using production mixes composed of three types of shoes. Furthermore, first the performances of the two departments were assessed separately and then the whole production system was analyzed. In addition, a specific analysis was carried out to investigate some input parameters dealing with labor management.

8.5.1 *Use-Case One for Assembly Area: Producing Only One Family of Shoes*

In this first use-case the production mix is composed only by a single shoe family to identify the family specific bottlenecks. Figure 8.4 shows an example of the trough-put for the different shoe families in the assembly area. Such a result demonstrates the large difference in produced quantities depends on the shoe family. Similar differences were found for resource allocation and production fluency. As expected, through-put is determined by the produced shoe family and not influenced by the batch size.

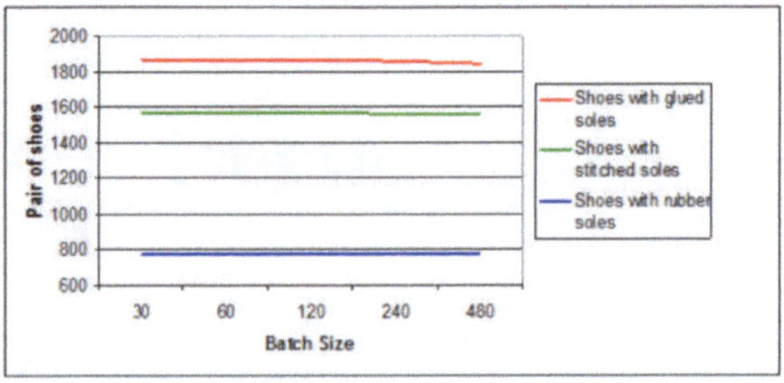

Fig. 8.4 Daily through-put vs. batch size for each shoe family in the assembly area.

8.5.2 *Use-Case Two: Producing Two Shoes Families*

In the second use-case the production mix is composed of two shoe types to identify main interaction effects between shoe families. Figure 8.5 and 8.6 show an

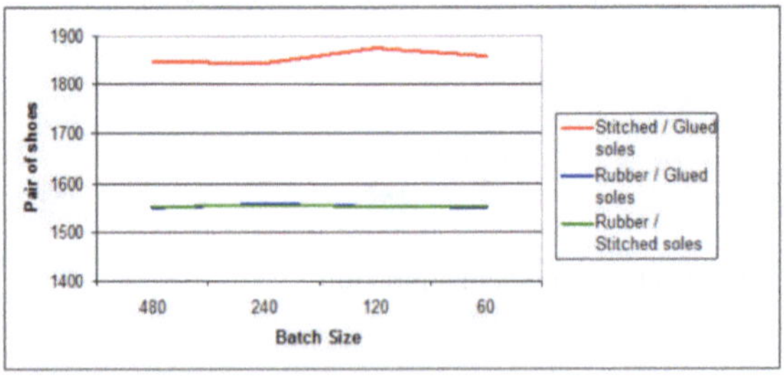

Fig. 8.5 Through-put vs. batch sizes when combining two shoe families.

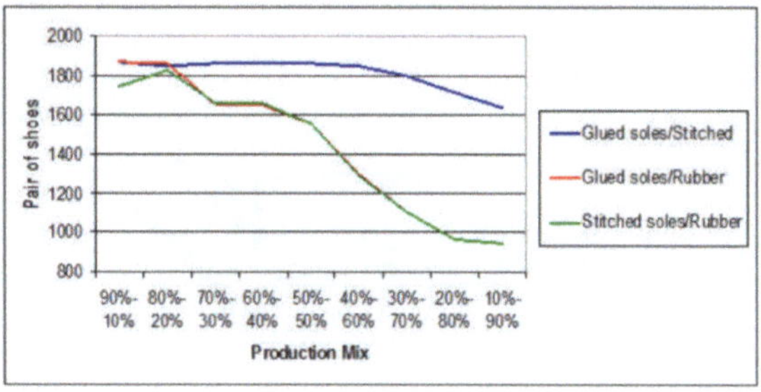

Fig. 8.6 Through-put vs. production mix when combining two shoe families.

example of through-put when combining two shoe families in the assembly area. Results show how produced quantities are impacted by the production mix i.e. the combination of families/styles produced, and not influenced by the batch sizes. As expected, through-put is determined by the production-mix.

Similar differences were found for resource saturation (see Figure 8.7) and production fluency.

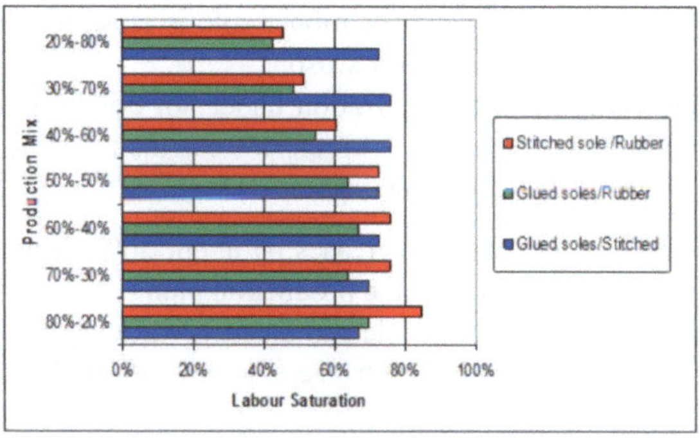

Fig. 8.7 Labour saturation VS production mix.

8.5.3 Use-Case Three for Assembly Area: Producing Three Shoes Families

In the third use-case the production mix is composed of three shoe types. To limit the number of simulation runs the production mixes followed the strategy used in production. Typically, half of the production capacity is assigned to one shoe family while the second half is shared by the remaining families. Looking at Figure 8.8, the productivity is influenced minimally by the batch size if between 120 and 240 and if the ratio of stitched/rubber shoe families is between 1 and 2. Rubber shoes have a significant impact on productivity that is lower about 5% then the target if the daily percentage of produced shoes with rubber soles is bigger than

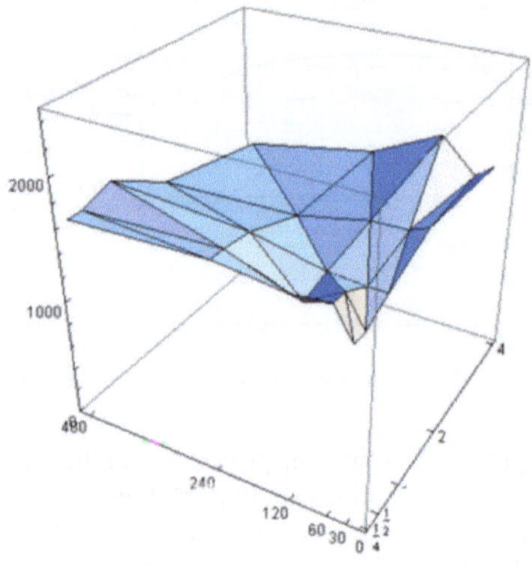

Fig. 8.8 Through-put vs. batch size when combining three shoe families, for the assembly area.

30%. Currently, the annual demand for rubber sole is close to 20-25%, although demand changes with every year and/or season.

As far as concerning the labor utilization under not critical production mixes, its overall saturation ranges from 64% up to 76% for the assembly area and most variations were found at the following areas:

- The cream island, its utilization increases by about 30% rising the quantity of shoes with stitched leather sole in the production plan
- The reactivation oven for rubber sole and the last removing island, their utilization is largely influenced by the batch size of shoes with rubber sole

Although these more complex production-mixes allowed for a validation of the simulation against the real production, we did not find a clear relationship between the production mix and through-put.

8.5.4 Finishing Area Overall Performances

The finishing area performances are not directly related with shoe families/styles, but with finishing sequence as well as cycle time of each shoe article. Based on this consideration, all the shoe articles were grouped in three macro categories i.e. easy, normal and difficult to finishing, and specific production mixes were defined with different compositions.

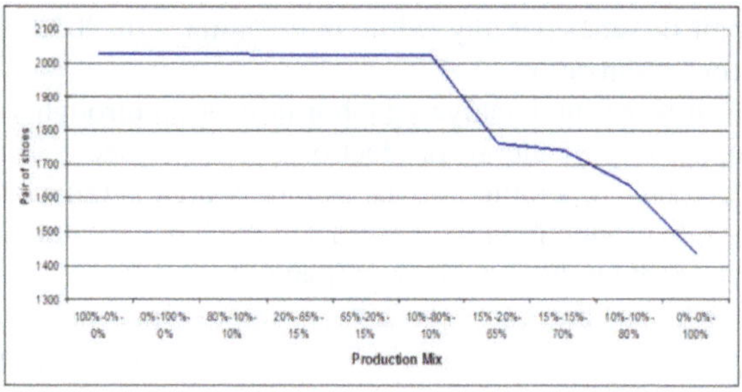

Fig. 8.9 Through-put vs. production mix

The daily through-put considering only finishing department, see Figure 8.9, ranges from about 1400 up to 2050 pairs of shoes and it is not influenced by batch size. Brushing and cream islands are the main bottlenecks and most of the finishing trolleys are staked up in these key position. As far as concerning labor utilization, its saturation ranges from 72% to 95%, as shown in Figure 8.10, simulating only the finishing area.

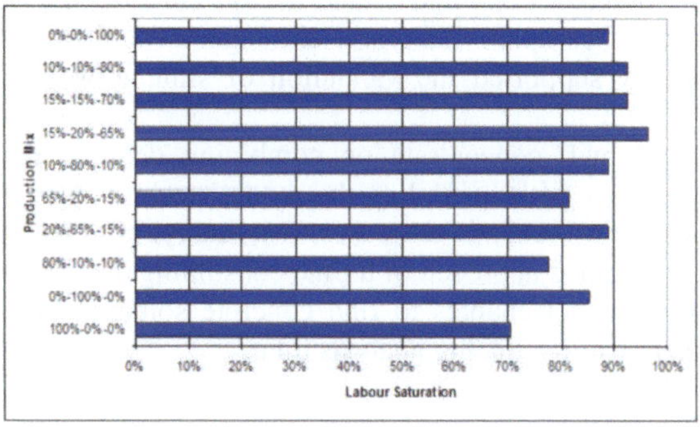

Fig. 8.10 Labor saturation in the finishing area vs. production mix.

8.5.5 Production Plant Overall Performances

Based on the previous results, the target productivity for the plant can be reached under scenarios with the following constraints:

- The rubber sole percentage in the daily production mix is lower than 30%
- The percentage of shoe articles with long cycle times in the finishing area is lower than 65%

In the first case, the assembly area is the bottleneck for the production plant, while in the second case the finishing department cut down the productivity.

Finally, a real production mix of two weeks was tested simulating the whole production plant as well as only the finishing area. In the first case the through-put is about 1863 pair of shoes per day, while in the latter, it is 2020 pair of shoes, indicating room for optimization.

A similar result was found analyzing labor utilization through sensitivity analysis. Hourly productivity of the all production system is decreased by 10% when the number of available operators for the assembly area is reduced from 33 to 27. As expected for this production mix, decreasing the labor availability in the finishing area has no impact on the overall performances.

Some what if analysis were carried on some input parameters managing labor allocation, showing some potential to increase through-put by a fine tuning activity.

8.6 Conclusion

This paper explored the use of simulation to better understand production dynamics as basis for determining an optimization strategy. The real shoe production plant provided a challenging example of a highly flexible production process operating on diverse production mixes.

Through a combination of analyzing simple and complex scenarios, full picture of the production's dynamic was obtained. Simple use-cases were instrumental in identifying basic dynamics and understand the system response of more complex use-cases. The more complex use-cases, although difficult to interpret, had the advantage that they supported the validation of simulation results against real production.

Further research will concentrate on combining detailed modeling such as described in this paper with 'modeling the model' technologies, for overall optimization (testing against realistic use-cases) (Merkureyeva et all, 2008).

We expect a combined approach of a time consuming detailed model and a less detailed but faster model enables to find concrete solutions for optimal sets of process parameters while reducing analysis time.

Authors Biographies

Fred Voorhorst is managing innovation at HUGO BOSS Ticino SA, a department for Product Development and Operation Management for five product groups, one of which is Shoes. He has more than ten years of experience in managing (business) innovation projects, both in industrial as well as academic context.

Antonio Avai is a partner and technical director of Technology Transfer System, an IT company located in Milan. He has managed research programs with focus on methodologies and leading information technologies to support all the life cycle of manufacturing processes. He has more than 15 years of experience in discrete event simulation and its integration with other software tools and technologies, and has authored several papers on these topics.

Contact

Antonio Avai
TTS
Via Vacini 15
20131 Milano
Italy
Antonio.Avai@ttsnetwork.com

Claudio Roberto Boër is Director of the ICIMSI Institute CIM for Sustainable Innovation of the University of Applied Science of Southern Switzerland. He has more than 16 years of industrial experience and research in implementation of computer aided design and manufacturing as well as design and setting up manufacturing and assembly flexible systems. He is author of the book on Mass Customization in the Footwear based on the European funded project EUROShoE that dealt, among several issues, with the complexity and optimization of footwear assembly systems.

References

Baiens, T., Hadfield, L., Mason, S., Ladbrook, J.: Using empirical evidence of variations in worker performance to extend the capabilities of discrete event simulations in manufacturing. In: Proceedings of the 2003 Winter Simulation Conference, pp. 1210–1216 (2003)

Boër, C.R., Avai, A., El-Chaar, J., Imperio, E.: Computer Simulation for the Design and Planning of Flexible Assembly Systems. In: Proceedings of International Workshop on Application and Development of Modelling and Simulation of Manufacturing Systems (1993)

Chung, C.A.: Simulation modelling handbook. Crc Press, Beijing (2004)

Kelton, W.D., Sadowski, R.P., Sturrock, D.T.: Simulation with Arena, 3rd edn. WCB/McGraw-Hill, New York (2003)

Merkureyeva, G.: Metamodelling for simulating applications in production and logistics, http://www.sim-serv.com (accessed June 16, 2008)

Merkureyeva, G., Brezhinska, S., Brezhinskis, J.: Response surface-based simulation metamodelling methods, http://www.simserv.com (accessed June 16, 2008)

Roser, C., Nakano, M., Tanaka, M.: Buffer allocation model based on a single simulation. In: Proceedings of the 2003 Winter Simulation Conference, pp. 1230–1246 (2003)

9 Simulation and Highly Variable Environments: A Case Study in a Natural Roofing Slates Manufacturing Plant

D. Crespo Pereira, D. del Rio Vilas, N. Rego Monteil, and R. Rios Prado

High variability is a harmful factor for manufacturing performance that may be originated from multiple sources and whose effect might appear in different temporary levels. The case study analysed in this chapter constitutes a paradigmatic case of a process whose variability cannot be efficiently controlled and reduced. It also displays a complex behaviour in the generation of intermediate buffers. Simulation is employed as a tool for detailed modelling of elements and variability components capable of reproducing the system behaviour. A multilevel modelling approach to variability is validated and compared to a conventional static model in which process parameters are kept constant and only process cycle dependant variations are introduced. Results show the errors incurred by the simpler static approach and the necessity of incorporating a time series model capable of simulating the autocorrelation structure present in data. A new layout is proposed and analysed by means of the simulation model in order to assess its robustness to the present variability. The new layout removes unnecessary process steps and provides a smoother response to changes in the process parameters.

9.1 Introduction

Variability is an acknowledged driver of inefficiency in manufacturing. Whether it comes in the form of changeable and uncertain demand, product characteristics, resources or processes, it leads to disposing overcapacity, increased work in process and operational risks. State of art process improvement techniques – such as Lean Manufacturing or Just in Time– tackle variability by different mechanisms aimed at reducing it or its impact in production. Manufacturing plants adopt flexible system designs, product and processes standardization, protocols or quality controls among other systems in order to efficiently control and manage variability.

However, there are still sources of variability that cannot be reduced in a profitable way beyond a certain limit. Demand patterns, human resources, machines

D. Crespo Pereira · D. del Rio Vilas · N. Rego Monteil · R. Rios Prado
Integrated Group for Engineering Research - University of A Coruña

failures, natural products or the socio economical context are examples of factors whose variability can only be partially controlled.

This chapter deals with a case study of a manufacturing plant which produces natural slate roofing tiles from irregular blocks of rock extracted from a nearby quarry. The variable characteristics of the input material due to the variable geologic nature of the rock introduce a variable behaviour in the plant.

In this chapter, the definition of a highly variable environment will refer to a subjective circumstance of a manufacturing system that reflects the complexity in the analysis of its variability sources and their impact in performance. We are not aiming at introducing a formal definition of highly variable environments but rather an informal one that a process manager or an analyst might employ to define a system with the characteristics given below. Such a system will exhibit the following features:

- There are sources of variability present that cannot be efficiently controlled.
- These sources of variability are key drivers of process inefficiency and thus design of the production system will be oriented to coping with them in an efficient way.
- The interaction between the sources of variability and the elements on the systems responds to a complex pattern which cannot be immediately determined from the particular behaviour of each element.

Discrete events simulation (DES) is a widely employed tool for manufacturing systems analysis due to its inherent capability for modelling variability. By means of a detailed specification of each element logics and related statistical distributions, the DES model is capable of computing the overall performance even if emergent behaviour may arise.

This chapter covers the analysis of a paradigmatic case of a highly variable environment. The modelling and simulation of a natural roofing slates manufacturing plant will be presented covering the discussion of the appropriate modelling approach plus the analysis of a layout improvement proposal taking into account the high level of variability present.

9.1.1 Sources of Variability in Manufacturing: A PPR Approach

If we consider a manufacturing process as the transformation of a series of input products into output products through a set of processes and given a set of resources, it would be useful to assign the different components of variation to the different elements involved. Thus a product, process, resource (PPR) approach will provide with a useful way of categorizing the variability sources.

Product variability can be originated either by changes in the characteristics of the process inputs or in the outputs of the system. Changes in the output will usually be linked to changes in demand. For example, changes in the quantity or in the mix of demanded products will cause the system to face changes in the throughput rates and occupancy levels. These changes may be linked to seasonal demand patterns, long term trends or random variations in shorter terms like daily or monthly ones.

Changes in the product specifications or design – like those which are typical in make to order environments or mass customization – make process cycle times to vary and consequently generate intermediate product buffers or performance losses due to blocking and starvation.

A special case in which this sort of product variation is strongly evident happens in natural products processing. The variable characteristics of the natural resources – like those extracted in mining, forestry, fishing or agricultural sectors – cause quality, input utilization rates and process cycle times to vary due to the heterogeneity in the source materials [9.1], [9.2].

Process variability might be related to either a lack of standardization in process routines and protocols or to an attempt of active adaptation of the process to the changeable environment. In some manufacturing environments – like small workshops or SME's with low processes standardization – undefined procedures or informal planning and production control schemes lead to a heterogeneous response to similar events and uncertainty. Although this is not necessary a bad feature of a system, since it enhances flexibility, it may often lead to un-optimal responses. Variability in process definition can be intently introduced by management as a means of adapting to different conditions and counteracting the effects of other undesirable forms of variability. Flexible manufacturing is a common approach to improve the robustness of a system to a changeable environment [9.3]. A flexible capacity dimensioning allows for reallocating resources to where they are most needed. However, difficulties may appear in the practical implementation of these practices. Schultz et al. [9.4] show in their work how behaviour related issues may harm the expected benefits from a flexible design of work.

Finally, resources driven variability is a frequent circumstance in manufacturing. Machines tend to feature *quasi*-constant cycle times when performing a single task in uniform conditions, but are subject to stochastic failures that reduce their availability. Human resources introduce several components of variability in a system. Within a process cycle scope, two main effects can be noticed. First, workers use to show larger deviations in cycle times than those of automated devices. Second, human beings display state dependant behaviour that further complicates the analysis of labour intensive processes. Humans are capable of adjusting their work pace depending on the system state and workload [9.5]. The consequence is a form of flexibility in capacity that counteracts some of the drawbacks caused by the larger variability [9.6]. Evidence from just in time (JIT) manufacturing lines shows that lower connection buffer capacities do not necessary produce the loses in performance that would be expected if considering human factors in a mechanistic way [9.7]. Human variations in performance may occur in different time horizons or linked to different process execution levels. Authors such as Arakawa *et al*, Aue *et al* or Baines *et al* [9.8-9.10] have studied hourly variations of human performance along a shift and across different shifts in a day. Baines *et al* have considered as well longer term variations in performance linked to aging, although they claim that further research and results validation are necessary. Another important source of variation is that related to individual differences [9.11], [9.12]. These differences may produce balance loses in serial flow lines [9.13] or more complex effects in parallel arrangements, such as group behaviour and regression to the mean effects when arranged [9.14].

Finally, characterizing variability is also related to the time horizon in which its effects appear. We might find variability between consecutive process cycles, between different days, between different production batches, etc. Accordingly, a reasonable scope and methodology for modelling variability has to be defined depending on different analysis span (yearly, seasonal, monthly, weekly, daily, shift and hourly variation).

9.1.2 Statistical Modelling of Variability

DES models support the high resolution modelling of manufacturing systems via the inclusion of elements' operating logics, sequences of processes and statistical distributions associated to their variability. Common statistical models that are employed span cycle time distributions of machines or workers [9.15], time between failure and time to repair distributions [9.16] or demand stochastic processes [9.17].

Both cycle time and time between failures statistical processes use to be assumed as stationary independent and identically distributed (i.i.d.). Evidence from multiple manufacturing environments justify this assumption [9.18]. Process cycle execution is commonly regarded as the main driver of variability and therefore of longer term variability calculated from it. For instance, Colledani et al.[9.19] calculate buffer capacities with a goal on minimizing weekly overall variance in throughput. He *et al* employ Markov processes for calculating production variations originated in the cycle time distributions [9.20].

However, this assumption is not necessarily valid for all the circumstances. Autocorrelation in stochastic processes and state dependant behaviour are two important deviations from this assumption that could greatly distort simulation results. Autocorrelation patterns are commonly found in demand processes and in the characteristics of natural product inputs [9.21], [9.22], although it might be observed as well in other types of highly variable processes such as semiconductors manufacturing [9.23]. State dependant behaviour also causes important divergences in the simulation results, as noted by Schultz *et al* in the above mentioned work [9.6]. According to them, the overall performance of a flow line may actually improve along with its length when considering a model in which cycle times are positively correlated with the workload.

9.2 Case Study: The Roofing Slates Manufacturing Process

Our case study is based on a Spanish SME company that produces natural roofing slate for institutional and residential buildings. More than 80% of its production is exported to other countries in Europe, especially France, where their slates have been awarded with the NF mark which sets the highest quality requirements in the industry.

The company is mainly devoted to the production of the highest value added roofing slates, that is to say, the thinnest commercial tiles. The thinner the tile is the harder and more wasteful the manufacturing process becomes. On the other hand, there is a quite constant demand of 3.5 mm thick tiles from France which provides a stable market.

In spite of the Spanish slates are the most employed in the world, the sector has scarcely benefited from technological transference from other industries. The level of automation is low as well as the application of lean manufacturing principles. The most arguably reason is perhaps the relative geographic isolation of slate production areas, mainly located in the northwest mountain region of Spain. Besides or as a result, it is labour-intensive and workers are exposed to very hard conditions both environmental and ergonomic. It is indeed difficult to find skilled workers or even convince youngsters to start this career so high salaries have to be paid. Accordingly, labour and operating expenses account for one third each of the total company costs set up.

In this context, the company has started a global improvement project comprising actions in the fields of production, quality, health and safety and environment [9.24], [9.25]. The purpose is to achieve a more efficient process in terms of productivity and the first step is to gain knowledge about the operations involved aiming at reducing uncertainty, defining capacities, and identifying both opportunities and limiting factors for a subsequent process optimization.

9.2.1 Process Description

For the extraction of slate from quarry light explosives are employed. The results are irregular and heavy blocks that are then loaded onto dumpers and transported to the manufacturing plant, located a few kilometres away. These blocks are then

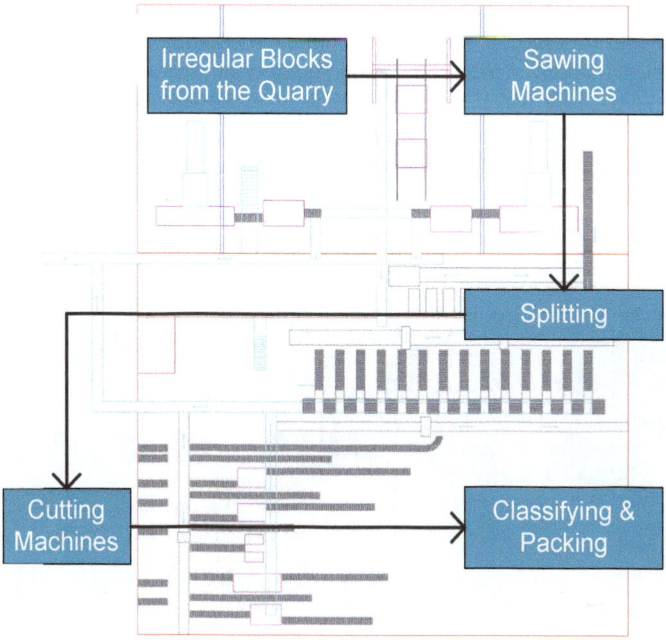

Fig. 9.1 CAD Layout of the Manufacturing Plant.

Fig. 9.2 Slabs Arriving Process from Sawing. Real Process and Simulation Model.

introduced in the Sawing Plant and stocked, so an adequate level of input is always assured. In this plant blocks are first cut into strips by means of circular saws and then a second group of saws cuts the strips into slabs which are then carried to the splitters on an automated conveyor belt.

An operator on an electric rail mounted vehicle receives and distributes slabs among the splitters according to the specified format and their stock level (Figure 9.2).

Slabs are taken by the splitters one by one and cut in several pieces by means of a special type of chisel so they can handle them better and also determine its quality. Then, they change to a smaller chisel for cutting these parts into plates The chisel, placed in position against the edge of the block, is lightly tapped with a mallet; a crack appears in the direction of cleavage, and slight leverage with the chisel serves to split the block into two pieces with smooth and even surfaces. This is repeated until the original block is converted into a variable number of pieces. The resulting number of slates of different formats is variable, depending mostly on the quality of the slate rock from quarry as well as the splitters experience and skill.

A second operator collects the slates lots produced by the splitters on a second electric trolley and takes them to a third one who carries and distributes them

Fig. 9.3 A Splitter (left) and the Resulting Output: The Target Formats (regular lots in the left) and Secondary Less Quality Output Formats (the two series in the right).

amongst the cutting machines. Split stone is then mechanically cut according to the shape and size required. This operation is done both by manual and fully automated cutting machines.

Finally, slate presented is inspected one by one by classifiers with a trained eye prior to being placed in crate pallets. Slate that does not meet with quality requirements is set aside and recycled to be cut again into another shape until it complies with company standards. In case this is not possible, it is rejected. Slate pieces are packed until they are ready for final use. Slates are available under different sizes and grades. Quality is assessed in terms of roughness, colour homogeneity, thickness and presence and position of imperfections –mainly quartzite lines and waving-. Accordingly, the company offers three grades for every commercial size: Superior, First and Standard

Alternatively, the latter operator takes the recycled plates and transports them to their corresponding machines. A third task assigned to this worker is to stock material in buffers previously located to the machines whenever their utilization is full. So a triple flow is shared by one transportation system connecting a push system (lots coming from splitters) and a pull system (lots required by cutting machines). And even more, the assignation rules that the operator follow depend on his criterion, so it is easily comprehensible the complexity of modelling this system.

Fig. 9.4 Distribution of Lots to Cutting Machines.

From the splitting to the packaging 26 transportation and stocking activities take place whilst only 13 value-add operations –mainly transformation and inspection operations- occur. The abundance of these non-value-added operations as well as the presence of feedback lines diminish the overall process performance. It becomes clear the necessity of reducing non value-added activities and rearranging the whole process in terms of layout design.

9.2.2 The PPR Approach to Variability

Natural roofing slate manufacturing is a process perceived by both process managers and workers as highly variable. According to their perception, the system displays the following behaviours:

- The properties of the input slabs to the process are inconstant along time. Some days "good" material enters the process that can be easily split into the target formats and shows good quality in the classification and other days the material is bad and the loss in splitting are large.
- The process bottleneck dynamically moves between the splitters and the classification and packing steps.
- There is a need for large capacity intermediate buffers due to the high variability in products characteristics. Sometimes large work in process accumulates and there is need for space in which to allocate stocks and sometimes queues disappear and material is quickly consumed causing starvation in the last steps of the process. It is this perceived necessity that has configured a layout designed for providing the maximum possible capacity for the connection buffers.

The most relevant source of variability in this process is due to the intrinsic variability of the natural slate. This variability corresponds with the possibility of variations both in mineral composition and morphology so that undesirable visual and structural effects in the final product may appear. It is the geological nature of the specific zone in the quarry that is eventually being exploited which determines this circumstance. Although there is certain knowledge about the quality of rock that is expected to extract in the quarry according to previous experience and/or mineral exploration operations, it is not possible to determine the real continuous mineral profile at a microscopic or visual level.

This uncertainty about the final quality has traditionally configured the whole manufacturing process resulting in a reactive system, that is, a system where there is no previously determined schedule and the assignment of operations to machines or workers is done according to the state of the system [9.26].

In our case, a foreman dynamically decides the formats to be cut as well as the number and identity of splitters, classifiers and machines assigned to each format according to his perception of process performance. Eventually, the functions performed and messages sent are allowed to adapt such that feedback paths in the process occur. It introduces another relevant component of variability related to the process rules and resources capacity. The foreman dynamically adjusts splitters working hours, adds splitters from a nearby plant and reassigns workers to classification and packing. He may also change the target format specifications or the thickness goal for the splitters. All these decisions are taken according to his long experience in the plant.

The labour intensive nature of this process involves another source of variation. Splitting is a task that requires highly skilled workers among which important differences can be observed in performance. Each splitter has their own technique for splitting the slabs, leading to heterogeneous working paces and material utilization. For instance, some of them are able to split high quality slabs in the target thickness 3.5mm and others not. Classification and packing are another two examples of manual tasks presents in which variety of criteria and working procedures can be found. In despite of the quality standards should provide homogenous criteria for tiles classification, different classifiers adopt more or less conservative criteria and thus their decisions may slightly differ. Packing detailed movements are differently performed by workers and the placement of tiles piles and pallets is variable.

The resulting process is complex, reactive and out of statistical control. Then, the overall system may exhibit emergent behaviours that cannot be produced by any simple subset of components alone, defining a complex system [9.27]. When proposing modifications in these systems special care has to be taken since even small changes in deterministic rules (SPT, FIFO, etc.) may result in a chaotic behaviour. Developing DES models of such processes has been proposed as a systematic way for their characterization and analysis [9.26].

9.3 The Model

9.3.1 Conceptual Model

As a first step in the model building phase of the project, a conceptual model was developed in order to identify the key process variables and parameters and to suggest hypothesis of their relations. The notation employed in this model will be introduced below.

Subscripts:

i: Splitter subscript. Its values range from 1 to NS, being NS the number of splitters in the plant. If omitted, the variable represents the sum of all the splitters.

f: Format subscript. The possible values are 32, 30 and 27 for the respective 32x22cm, 30x20cm and 27x18cm formats. Related to the split process, possible formats are *TF* – which stands for target format, frequently 32x22 – and SF – which stands for secondary format, both 30x22 and 27x18 –. If omitted, the variable represents the sum of all the formats.

q: Quality subscript. Its values can be F for first quality, T for traditional quality and STD for standard quality. If omitted, the variable represents the sum of all the qualities.

Th: Thickness subscript. Its values can be 3.5 or 4.5. If omitted, the variable represents the sum of all the thickness values.

t: Time subscript. If used, the variable contains its average value for the day t.

c: Cycle subscript. If used, the variable contains its value for the related process cycle execution c.

Product flow rates:

B: Rate of slabs per unit of time that enter the plant.

B_i: Rate of slabs per unit of time that are consumed by the splitter *i*.

$SL_{f,i}$: Rate of split lots of f format slates per unit of time produced by splitter *i*.

CL_f : Rate of cut lots of 32cm format slates per unit of time.

RL : Rate of recirculated lots of slates per unit of time.

$PL_{f,q,Th}$: Rate of classified and packed lots of format given by subscript *f*, quality *q* and thickness *Th* slates per unit of time.

Size of slates lots:

NSL_f : Number of slates in each split lot by format.

NCL : Number of slates in each cut lot. It is the same for the different formats.

NPL : Number of slates in each classified lot for packing. It is the same for the different formats.

NRL : Number of slates in each recirculated lot.

Lots are formed manually by the splitters upon specific goals on their size. Hence all of them are subject to random variations in content but with well-defined mean values.

Product transformation rates:

NSP: Number of parts generated by each slab in the rough splitting process.

τ_{SU} : Utilization rate of split blocks. It represents the percentage of the blocks' material that can be transformed into split slates.

τ_{TF} : Rate of target format slates produced by the splitters

τ_{rej} : Rejections rate in the classification step.

τ_{32} : Rate of 32 format slates produced in the factory.

α_{30} : Relation between the throughput of 30 format slates and 27 format slates.

α_T : Relation between the throughput of traditional quality slates and standard quality slates.

τ_{recirc} : Rate of slates recirculated after the classification process to lower formats.

τ_{thick} : Rate of slates classified as 4.5mm thickness.

Resources parameters:

γ_i : Relation between the individual throughput rate and the average throughput rate for the splitter *i*.

Figure 9.5 represents the process flow diagram indicating the flows of intermediate products and the transformation and transportation steps. Acronyms for the resources are inserted at the end of each element's name. As it can be noted, the process type corresponds to a disassembly process in which from a single process input different outputs are obtained.

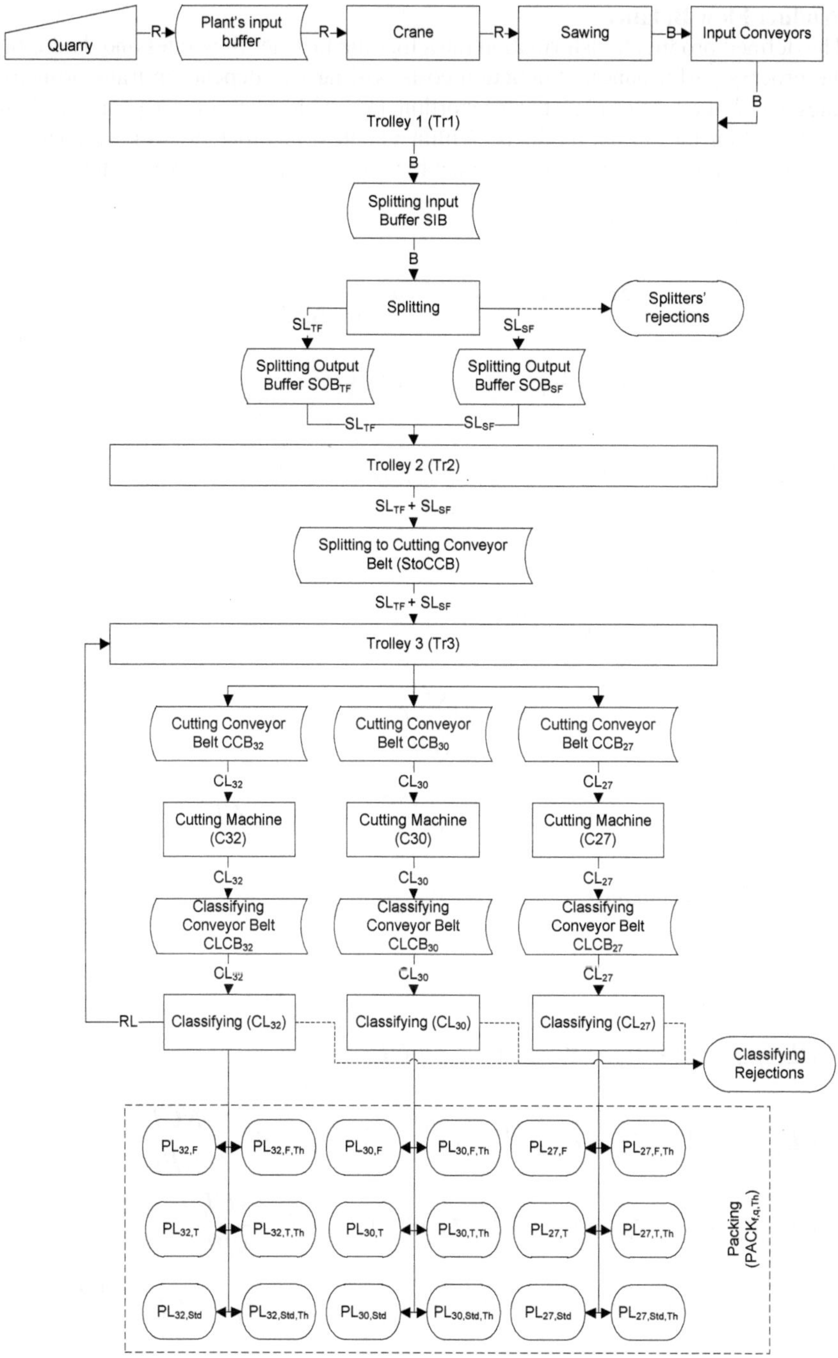

Fig. 9.5 Process flow diagram

Product Flow Balance

The defined product transformation rates link the product flow rates and determine the process performance. Production costs will largely depend on transformation rates since they determine the proportion in which costs of early intermediate products transfer to final products. Within a context in which prices keep constant, economical performance will be subject to variations in process parameters.

Splitting process balance

$$
\begin{aligned}
SL_{TF} &= \tau_{SU} \cdot \tau_{TF} \cdot B \cdot NSP \cdot \frac{w_p}{w_s} \cdot \frac{1}{NSL_{TF}} \\
SL_{SF} &= \tau_{SU} \cdot (1-\tau_{TF}) \cdot B \cdot NSP \cdot \frac{w_p}{w_s} \cdot \frac{1}{NSL_{SF}}
\end{aligned}
\tag{1}
$$

Where w_p is the width of a rough split part of a slab and w_s is the width of a slate.

Cutting process balance

$$
\begin{aligned}
CL_{32} &= SL_{TF} \cdot \frac{NSL_{TF}}{NCL} \\
CL_{30} &= \alpha_{30} \cdot \left(SL_{SF} \cdot \frac{NSL_{SF}}{NCL} + \tau_{recirc} \cdot CL_{32} \right) \\
CL_{27} &= (1-\alpha_{30}) \cdot \left(SL_{SF} \cdot \frac{NSL_{SF}}{NCL} + \tau_{recirc} \cdot CL_{32} \right)
\end{aligned}
\tag{2}
$$

Packing process balance

$$
\begin{aligned}
PL_{f,F,3.5} &= (1-\tau_{rej}-\tau_{recirc}) \cdot \tau_F \cdot (1-\tau_{Thick}) \cdot CL_f \cdot \frac{NCL}{NPL} \\
PL_{f,F,4.5} &= (1-\tau_{rej}-\tau_{recirc}) \cdot \tau_F \cdot \tau_{Thick} \cdot CL_f \cdot \frac{NCL}{NPL} \\
PL_{f,T,3.5} &= (1-\tau_{rej}-\tau_{recirc}) \cdot (1-\tau_F) \cdot \alpha_T \cdot (1-\tau_{Thick}) \cdot CL_f \cdot \frac{NCL}{NPL} \\
PL_{f,T,4.5} &= (1-\tau_{rej}-\tau_{recirc}) \cdot (1-\tau_F) \cdot \alpha_T \cdot \tau_{Thick} \cdot CL_f \cdot \frac{NCL}{NPL} \\
PL_{f,STD,3.5} &= (1-\tau_{rej}-\tau_{recirc}) \cdot (1-\tau_F) \cdot (1-\alpha_T) \cdot (1-\tau_{Thick}) \cdot CL_f \cdot \frac{NCL}{NPL} \\
PL_{f,STD,4.5} &= (1-\tau_{rej}-\tau_{recirc}) \cdot (1-\tau_F) \cdot (1-\alpha_T) \cdot \tau_{Thick} \cdot CL_f \cdot \frac{NCL}{NPL}
\end{aligned}
\tag{3}
$$

9.3.2 Statistical Analysis

Three main sources of information were used for the simulation project: videos, interviews with personnel and production data records. Interviews served us for performing a qualitative analysis of the systems characteristics and behaviours presented before.

Production Data Analysis
The videos provided with observations of cycle time realizations from which to study the statistical distribution of the diverse elements in the system. Statistical distributions were fitted by means of the software utility Statfit. Regression models for the splitters cycle times were fitted in R [9.28]. The following items were considered in this analysis:

- Inter-arrival times for the input slabs process. They were fitted to an exponential distribution in which for a single arrival event several slabs may arrive according to the empirical distribution.
- Loading and unloading times as well as speeds of trolleys.
- Splitters cycle times. Data of joint observations of time per slab, number of produced slates and material utilization rate were collected.
- Cycle times of cutting machines. Cutting time per slate was assumed to be constant due to the low coefficient of variation. Variability is introduced in the loading and unloading times plus the number of slates in a pile.
- Cycle times of classifiers. It was fitted to a triangular distribution with a coefficient of variation of 0.3.
- Cycle times of packing tasks. It was fitted to a triangular distribution with a coefficient of variation of 0.3.

The splitters' cycle time was found to be positively correlated with the width of a slab given by the number of rough splitting parts (NSP) in which it is divided and the utilization rate. Slabs in which large fractions are wasted are processed faster. The coefficients of the model and its equation are given below.

$$ST_c = e^{b_0} \cdot (NSP_c + 1)^{b_{NP}} \cdot \left(\frac{SSP_c}{NSP_c} + 0.5 \right)^{b_{SU}} \cdot e^{\varepsilon_{ST,c}} \quad (4)$$

Where:

ST_c: Splitting time by cycle.

SSP_c: Successful split parts in cycle c.

b_0, b_{NP}, b_{SU} : Model parameters.

$\varepsilon_{ST,c}$: Random error. It follows a normal distribution with zero mean and standard deviation $\sigma_{ST,c}$.

Table 9.1 Coefficients of the splitting cycle time model.

Coefficient	Value	Std. Error	p-value
b_0	2.192	0.156	1.74E-15
b_{NP}	1.367	0.089	1.87E-16
b_{SU}	0.620	0.161	0.00054

Figure 9.6 shows the splitters cycle time and throughput depending on the size of the incoming slabs and the utilization rate of the material. Cycle time graphs show a slight concavity. Hence, for both small and big size slabs the throughout rate is lower. This can be explained by taking into account that processing small slabs increases the proportion of auxiliary tasks such as picking them up or cleaning the workstation. Big slabs are harder to handle, thus reducing productivity as well.

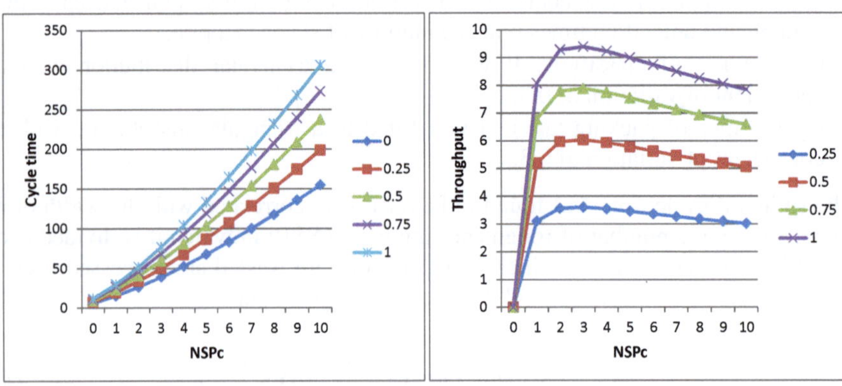

Fig. 9.6 Cycle Time and Split Lots Throughput Rate as a function of the Number of Parts per Slab (NSP_c) and for various levels of Slab Utilization Rate (SSP_c / NSP_c).

Production Data Analysis

Production data were gathered from the company's production daily records a set of 250 days of activity were stored in a relational database implemented in Microsoft Access. Available registries contain the following items:

(Splitters' production records)

- Labour.
- Number of target format lots.
- Number of secondary target lots.

- Average number of pieces in a target format lot.
- Working hours.

(Packing records)

- Number of pallets by format and quality.
- Number of slates in each pallet.
- Price per slate of each pallet
- Cutting machine in which the pallet had been processed.

Two of the process transformation rates were assumed to be constant due to their less relevant role in the process. They are the percentage of 30cm format over the total of secondary formats and the percentage of traditional quality over the sum of traditional plus standard: α_{30}=0.65 and α_{Trad}=0.473.

Although the data sources do not cover all the relevant process parameters defined before, they allowed us to infer those not explicitly contained. Datasets with parameters values for statistical analysis were derived from these sources. Three variables needed to be inferred from the data: the utilization rate of blocks material, the rejections rate in the classification process and the recirculation rate.

According to information provided by the plant managers, the sawing throughput rates stands roughly constant among different days. Thus, assuming B as a fixed value, and taking into account the variations in the splitting process throughput along time, we can estimate the variations in the utilization rate:

$$\tau_{SU,t} = \frac{8 \cdot NSP \cdot B}{NSL \cdot SL_t} \tag{5}$$

Assuming B as a constant, $\tau_{SU,t}$ will follow a stochastic process that will be directly proportional to $\frac{1}{SL_t}$.

The rejections rate in classification could be obtained by the difference between split and packed slates:

$$\tau_{rej,t} = 1 - \frac{NPL \cdot PL_t}{NSL \cdot SL_t} \tag{6}$$

Finally, the recirculation rate can be obtained assuming that the rejections rate is the same for the different produced formats. Then:

$$\tau_{recirc,t} = 1 - \frac{NPL \cdot PL_{32,t}}{\left(1 - \tau_{rej,t}\right) NSL \cdot SL_{TF,t}} \tag{7}$$

Table 9.2 shows the dataset with the seven most relevant process parameters identified before. The statistics summary contains the mean, standard deviation and 1 order autocorrelation of each time series.

Table 9.2 Most relevant process parameters values.

	τ_{SU}	τ_{TF}	$1-\tau_{rej}$	τ_{32}	τ_F	τ_{recirc}	τ_{thick}
AVG	67.72%	87.46%	78.15%	79.37%	41.47%	7.73%	27.93%
DESV	7.37%	3.66%	10.89%	9.57%	12.06%	5.70%	9.09%
Autocor	0.55	0.64	0.54	0.27	0.35	0.93	0.28

Time Series Model of Process Parameters

The first analysis conducted on the process parameters dataset was a principal components analysis (PCA) aimed at identifying the main dimensions of variability present in the data. Data were first standardized and then the PCA analysis performed in R. The first four principal components of variability were selected for further analysis. They account for 80.30% of the total variance. The loadings and standard deviation for each one of these components are given in Table 9.3.

Table 9.3 Loadings and standard deviation of the Principal Components Analysis.

	C_1	C_2	C_3	C_4
τ_{SU}	0.466	-0.254	-	-0.152
τ_{TF}	-	0.566	-0.377	0.236
$1-\tau_{rej}$	-0.537	0.162	0.116	-
τ_{32}	0.451	0.232	0.109	0.151
τ_F	-	-0.343	-0.819	-0.258
τ_{recirc}	-0.492	-	-0.234	-
τ_{thick}	-	-0.397	-0.112	0.897
Standard deviation	1.667	1.344	0.981	0.914

Component 1 is linked to the joint variation in the utilization rate of slabs in splitting, the rate of 32cm format, rejections in classification and oppositely to the recirculation's fractions. Thus component 1 shows a situation in which splitters production is high but there are important losses in classification and the main output is the objective target 32 cm. Component 2 is linked to the joint increase in the target format production in splitting and output, but together with low quality outputs and lower utilization of the slabs. Component 3 is mainly associated to quality, which is a rather independent feature of the output with respect to other process parameters. Component 4 shows the fraction of thick slates to be a fairly independent variable as well. The 1st and 2nd components of variability might interact with process management decisions, since it is possible to alter the priorities with respect to which formats to produce and the incentives in splitting to the different outputs. However, quality and thickness are two variables over which there is no feasible control that can be exerted by the managers. Thus they might be considered as external sources of variation in the process that must be coped with. Then, the principal components time series were fitted first to a multivariate autoregressive process employing the *vars* package in R [9.29]. However, this multivariate model only showed significant 1st order autoregressive effects to be relevant. Cross effects were negligible and they only accounted for a small share of the variance.

The models were simplified and fitted again to independent first order autoregressive models for each variable by means of the R *tseries* package [9.30]. Higher order terms did not improve the accuracy in a significant way so they were rejected. Table 9.4 summarizes the fitted models.

Table 9.4 First order autoregressive models

Component	Lag 1 coefficient	p-value	Std. Error	Model equation
c_1	0.71488	<2e-16	1.167	$c_{1,t}=0.715\cdot c_{1,t-1}+ {}_{1,t}$ where ${}_{1,t}\sim N(0, 1.167)$
c_2	0.61937	<2e-16	1.048	$c_{2,t}=0.619\cdot c_{2,t-1}+ {}_{2,t}$ where ${}_{2,t}\sim N(0, 1.048)$
c_3	0.45178	1.17e-10	0.884	$c_{3,t}=0.452\cdot c_{3,t-1}+ {}_{3,t}$ where ${}_{3,t}\sim N(0, 0.884)$
c_4	0.21142	0.00493	0.899	$c_{4,t}=0.211\cdot c_{4,t-1}+ {}_{4,t}$ where ${}_{4,t}\sim N(0, 0.899)$

Splitters Individual Differences Analysis

As it has been said in the second section, a remarkable component of variability in the process is given by the splitters individual differences. Differences in skill gender differences in cycle times, utilization rates of material and consequently throughput. Differences in throughput were computed from historical data and summarized in a set of individual effect parameters noted as γ_i.

$$\gamma_i = \frac{SL_i}{SL} \tag{8}$$

Time series of $SL_{i,t}$ values were normalized and the significance of two possible hypothesis tested:

- H1: The daily variation of each individual splitter is associated with the daily variations in the average of the rest of the splitters. This hypothesis would be related with a common cause of variability for all the splitters that would be linked to changes in the quality of the material, associated with changes in the mean values of the slabs utilization.

- H2: The daily variation of each individual splitter is associated with the daily variation of the next splitter lying in his visual field. This effect would be related to behavioural issues consisting of regression to the mean phenomena as considered by Schultz et al [9.31]. In this case, due to the spatial arrangement of the splitters –linear-, each one can only see his following workmate. Then, behaviour could only be affected by feedback on the next splitter's work-pace.

The model proposed in order to study the significance of these two possible phenomena is the following:

Let $r_i = \dfrac{SL_{i,t} - \mu(SL_i)}{\sigma(SL_i)}$ be the normalized observation of the splitter i throughput at time t and $r^c{}_i = \dfrac{\sum_{j \neq i} \dfrac{SL_{j,t}}{NS-1} - \mu\left(\sum_{j \neq i} \dfrac{SL_j}{NS-1}\right)}{\sigma\left(\sum_{j \neq i} \dfrac{SL_j}{NS-1}\right)}$ be the normalized observation of the average throughput of all the rest of the splitters but i.

$$\begin{aligned} r_{i,t} &= \beta_{1,i} \cdot r^c{}_{i,t} + \beta_{2,i} \cdot \left(r_{i+1,t} - \operatorname{cov}\left(r_{i+1}, r^c{}_i\right) r^c{}_{i,t}\right) + \\ &+ \varphi_i \cdot \left(r_{i,t-1} - \beta_{1,i} \cdot r^c{}_{i,t-1} - \beta_{2,i} \cdot \left(r_{i+1,t-1} - \operatorname{cov}\left(r_{i+1}, r^c{}_i\right) r^c{}_{i,t-1}\right)\right) + \delta_{i,t} \end{aligned} \tag{9}$$

Where $\beta_{1,i}$ and $\beta_{2,i}$ are the coefficients to be estimated by generalized least squares and φ_i a coefficient for considering autocorrelation in the model's residuals following a 1st order autoregressive process (AR1). $\delta_{i,t} \sim N(0, \varepsilon_i)$ is a white noise error process. The term $r_{i+1,t} - \operatorname{cov}\left(r_{i+1}, r^c{}_i\right) r^c{}_{i,t}$ represents the daily variation of the next splitter to i that is not explained by the first regressor of the model. Thus $\beta_{2,i}$ isolates the effect of possible behavioral phenomena given by the association between the variations of both splitters that are not linked to the variation in the global behavior of the splitters.

Testing H1 and H2 can be performed by checking the significance of the null hypothesis $\beta_{1,i} = 0$ and $\beta_{2,i} = 0$. Table 9.5 shows the coefficients, p-values and errors of the fitted models. As it can be noted, only H1 was found to be significant. The p-value of $\beta_{1,i}$ is near zero for all the analyzed splitters, validating the perceived perception of the existence of good and bad input material conditions that affect to the global performance. However, H2 could not be proved significant. The average positive values suggest that larger dataset were available, it might be found significant; but it has a small effect nevertheless.

Taking into account the cycle time model given by equation (4), the throughput will be depend on both average cycle time and utilization and thus individual differences might be explained by either differences in work-pace, utilization or a combination of both. At this point, the assumption that individual differences would be only explained by differences in cycle times and global splitting variations by differences in utilization rates was adopted. The reasoning behind it is twofold. First, even though there are differences in the utilization rate of slabs by the different splitters, all of them share the goal of maximizing slabs utilization. And second, differences in skill that make it possible higher rates of material utilization are less important than those related to the different work-paces. Partial collected data together with expert judgement supported this assumption.

Table 9.5 Coefficients, p-values and errors of the splitters models.

	Splitter 1	Splitter 2	Splitter 3	Splitter 4	Splitter 5	Splitter 6	Splitter 7
Avg.	762.85	601.73	752.02	863.18	710.66	643.58	864.28
Std. Deviation	127.46	67.34	88.69	99.80	98.34	71.18	102.98
γ_i	1.03	0.81	1.01	1.16	0.96	0.87	1.16
γ_i	0.824	0.744	0.828	0.750	0.508	0.775	0.501
p-value	0	0	0	0	0	0	‘
$\beta_{2,i}$	-	0.206	0.0139	0.0216	0.0992	0.1066	-0.1245
p-value	-	6e-48	0.8681	0.7049	0.2227	0.1755	0.0788
φ_i	0.095	0.244	0.177	0.062	0.212	0.299	0.337
φ_i confidence interval	(-0.029, 0.216)	(0.120, 0.361)	(0.050, 0.299)	(-0.062, 0.186)	(0.083, 0.334)	(0.175, 0.414)	(0.214, 0.449)
Std. Error	0.793	0.756	0.685	0.752	0.865	0.796	0.870

Thus the model for the splitters' cycle time remains as:

$$ST_{i,c} = \gamma_i \cdot e^{b_0} \cdot (NSP_c + 1)^{b_{NP}} \cdot \left(\frac{SSP_c}{NSP_c} + 0.5 \right)^{b_{SU}} \cdot e^{\varepsilon_{ST,c}} \tag{10}$$

Where the splitting utilization rate can be calculated from the principal components time series as given by equation (11). The rest of the variables in the model are calculated according to their statistical distributions.

$$\tau_{SU,t} = \mu(\tau_{SU}) + \sigma(\tau_{SU})(0.466 \cdot c_1 - 0.254 \cdot c_2 - 0.152 \cdot c_4) \tag{11}$$

$$NSP_c \sim Empirical\_distribution \tag{12}$$

$$SSP_c \sim B(NSP_C, \tau_{SU,t}) \tag{13}$$

$$\varepsilon_{ST,c} \sim N(0, \sigma_{ST}) \tag{14}$$

Hence, the proposed model connects the daily variability generated by the principal components time series models with the process cycle variability given by the statistical distributions of the aforementioned variables.

9.3.3 Model Implementation and Validation

The executable model was implemented in the simulation software Delmia Quest V5R20. A parameterized model was developed so that the process parameters can be updated on a daily basis and results stored for further analysis by means of a SCL macro. Geometrical and kinematical similarity in the transportation elements could be achieved thanks to the 3D simulation paradigm built in Quest. This feature provided a good means for visually validating the model and accrediting results in front of the plant managers.

Cycle level variation is introduced in the model by means of the statistical distributions of processing times, lots sizes and the splitting model introduced before. Classified slates are randomly assigned to the different categories of classified products according to product transformation rates. For instance, in the classification of 32 format tiles, each one will be randomly rejected, recirculated or transformed into a packed slate with quality and thickness generated according to the product transformation rates values.

Three variability modelling approaches were considered at this point:

1. A static model in which mean values of process parameters are kept constant along the simulation run. Thus only cycle related variability is introduced.
2. A stationary autoregressive model in which process parameters are generated on a daily basis according to the time series models presented before.
3. The 2nd modelling approach incorporating the individual differences.

The experimentation conducted at this step comprised the simulation of 200 days periods in which statistics are collected. The analysed results span:

- Final product production records in the same fashion as they are recorded in the real plant.
- Splitting production records in the same fashion as they are recorded in the real plant.
- Daily averages of resources utilization and occurrence of blocking.
- Daily averages of buffer levels.

The simulated production records provide with a means for validating the simulation model by comparing the time series autocorrelation structure from the real plant with that generated by the model. As it can be seen in the Table 9.6, the static modelling approach leads to the largest differences with the data from the real plant. Parameters deviations are lower than those present in the plant indicating that variability is being underestimated.

Table 9.6 Average and Standard deviation parameters for the real and simulated systems

Approach	Statistic	1-tDL	tC	1-tDC	t32	tP	tR	tF
Real	Avg.	0.837	0.875	0.781	0.794	0.415	0.077	0.279
	S.d.	0.091	0.037	0.109	0.096	0.121	0.057	0.091
Model 1	Avg.	0.828	0.874	0.803	0.779	0.412	0.087	0.244
	S.d.	0.021	0.005	0.074	0.042	0.050	0.039	0.053
Model 2	Avg.	0.841	0.876	0.788	0.784	0.424	0.085	0.243
	S.d.	0.079	0.034	0.130	0.061	0.130	0.054	0.093
Model 3	Avg.	0.818	0.869	0.822	0.770	0.423	0.095	0.247
	S.d.	0.069	0.032	0.118	0.057	0.135	0.051	0.095

Table 9.7 summarizes the principal components loadings for the real and the simulated time series, their variance and the 1st order autoregressive model coefficient and p-value. Principal components loadings display several further dissimilarities and the autocorrelation coefficients are negative. Modelling approaches 2 and 3 provide with a better modelling of the system variability and display a more similar autocorrelation pattern. However, all the autocorrelation coefficients have lower values than those in the generating time series. This result might be explained taking into account that the cycle level variability generated by the simulation model also affects the daily time series. As model 1 results show, the exclusive consideration of cycle time variability results on negative autocorrelation. Accordingly, the positive autocorrelation structure generated by the temporary series model is slightly counteracted by the cycle random processes.

Results from models 2 and 3 do not present relevant differences. This result can be interpreted as that even though the individual differences are clearly present in data; their impact in the global process performance is not relevant. Thus in the rest of the chapter, model 2 will be adopted for simplicity.

The next step in the validation process consisted of an informal validation in which the behaviour of the models was compared to the manufacturing plant behavior descriptions given by the process managers (Table 9.8). These features can be summarized as:

Table 9.7 Principal components loadings for the real and the simulated time series

Parameter		c1	c2	c3	c4
splitPerformance	System	0.466	-0.254		-0.152
	Model 1			0.564	0.646
	Model 2	0.518	0.239	-0.168	0.226
	Model 3	0.575		0.193	
tauSQ	System		0.566	-0.377	0.236
	Model 1				-0.663
	Model 2	-0.247	-0.472	-0.505	-0.206
	Model 3	-0.452	-0.355		-0.536
tauRej	System	-0.537	0.162	0.116	
	Model 1	-0.149	0.736	-0.135	
	Model 2	-0.536	-0.133	0.234	-0.126
	Model 3	-0.508	0.227	-0.176	0.251
tau32	System	0.451	0.232	0.109	0.151
	Model 1	0.678			
	Model 2	0.229	-0.665	-0.111	
	Model 3		-0.672	-0.167	
tauF	System		-0.343	-0.819	-0.258
	Model 1		0.168	-0.706	0.34
	Model 2	0.121	0.334	-0.58	-0.602
	Model 3	0.31	0.217	-0.349	-0.734
tauRecirc	System	-0.492		-0.234	
	Model 1	-0.686			
	Model 2	-0.502	0.385		
	Model 3	-0.298	0.565		-0.206
tauThick	System		-0.397	-0.112	0.897
	Model 1	0.193	0.642	0.392	-0.145
	Model 2	0.254		0.553	-0.721
	Model 3	0.116		-0.879	0.26
Std. Deviation	System	1.667	1.344	0.981	0.914
	Model 1	1.4654143	1.1581529	1.0732068	1.0105033
	Model 2	1.5718817	1.3638524	0.9884897	0.9638779
	Model 3	1.5440086	1.4516356	1.0101435	0.9127112
AR1 coef.	System	0.71488	0.61937	0.45178	0.21142
	Model 1	-0.53965	-0.4745	-0.41343	0.01391
	Model 2	0.43213	0.28589	0.35404	0.15376
	Model 3	0.548486	0.087541	0.125889	0.224131
AR1 p-value	System	<2e-16	<2e-16	1.17E-10	0.00493
	Model 1	<2e-16	2.44E-14	1.34E-10	0.844
	Model 2	1.88E-09	0.000162	0.00000194	0.049
	Model 3	<2e-16	0.217	0.0735	0.00113

- Feature 1. Splitters workload is highly variable. Under some conditions they are near saturation and under some other conditions their workload is lower and they have idle times.
- Feature 2. Slabs utilization rate is subject to large variations. During some periods the slate quality is optimal and the throughput is high. Under some other periods the throughput is high causing idle resources downstream the production line.

- Feature 3. The connection buffers from splitting to cutting are subject to large variations in occupancy levels.
- Feature 4. The process bottleneck dynamically switches between splitting and classification & packing.

Comparing the graphs of utilization rates of resources and buffers contents generated by models 1 and 2, the model behaviour features can be checked. As we can see in Fig. 9.7 and Fig. 9.8, model 1 displays a much more constant pattern of

Table 9.8 Presence in the Models of the Experts Percepcion of the System

Feature	Model 1	Model 2
1	Partially present	Present
2	Partially present	Present
3	Not Present	Present
4	Not Present	Present

Fig. 9.7 Workload and Blocking Occurrences in model 1 and model 2

variability in buffer contents with some random fluctuations around mean values. On the other hand, model 2 shows much larger variations that better match the system's description.

Content of the StoCCB conveyor presents long periods in which it is fully occupied and long periods in which it is almost empty. The emergence of these periods is a feature that matches system's behaviour although it is not immediate to predict from the individual condition of the other elements. On the contrary, due to the lack of variability inherent to its modelling approach, Model 1 is not capable of displaying such behaviour.

Fig. 9.8 Buffer Loadings for model 1 and model 2

Table 9.9 Main Results of Model 1 and Model 2

	Stationary Model				Time Series Model			
	Avg.	S.d.	Max.	Min.	Avg.	S.d.	Max.	Min.
avgSIB	4,0424	0,7306	6,9085	2,5811	4,3692	1,1037	8,4949	2,3910
ST_Utilization	0,8802	0,0216	0,9805	0,8748	0,8879	0,0340	1,0000	0,8184
Tr2_ Utilization	0,7851	0,0175	0,8247	0,7342	0,7573	0,0690	0,9152	0,5765
StoCCB	0,8087	0,4726	6,7110	0,6789	12,0476	12,3275	27,4666	0,5008
Tr3_ Utilization	0,5975	0,0139	0,6284	0,5602	0,5717	0,0644	0,7174	0,4131
Tr3_Block	0,0036	0,0186	0,1785	0,0000	0,2156	0,2224	0,5869	0,0000
$CCB_{32\ queue}$	0,8783	0,6630	4,9862	0,4016	3,2660	2,8620	9,5937	0,3320
CCB_{30}	0,5363	0,6225	7,4219	0,3201	1,5324	2,5546	11,7239	0,1565
CCB_{27}	0,1392	0,0207	0,2169	0,0964	0,2107	0,1416	0,7010	0,0304
$CL_{32\ queue}$	0,8558	0,0380	0,9622	0,7658	0,8696	0,0708	0,9815	0,6136
CL_{30}	0,4001	0,0343	0,5099	0,3077	0,3867	0,0770	0,5489	0,2064
CL_{27}	0,2378	0,0317	0,3369	0,1726	0,2359	0,0514	0,3832	0,1149
$CLCB_{32\ queue}$	3,4739	1,4201	8,3682	1,3766	5,6296	2,6192	8,9343	0,8763
$CLCB_{30}$	1,8763	2,0006	17,6434	0,3215	3,5704	4,8023	17,3718	0,1461
$CLCB_{27}$	0,2063	0,1293	0,6260	0,0403	0,2606	0,2353	1,6384	0,0219
PL_{32}	0,8117	0,0337	0,8992	0,7267	0,8234	0,0704	0,9330	0,6189
PL_{30}	0,7778	0,0484	0,9032	0,6279	0,7575	0,1460	0,9950	0,3506
PL_{27}	0,4506	0,0400	0,5381	0,3483	0,4366	0,0894	0,7252	0,1997

9.4 Process Improvement

9.4.1 New Layout Description

The original process layout was designed aiming at maximizing the intermediate buffers capacity. The StoCCB conveyor and the length of the conveyors in the classification and packing areas are examples of this traditional plant design concept that was common in the sector. A new layout configuration was proposed by the research team based on the idea that more intermediate steps may be actually amplifying process variability. Thus we compared by simulation the original layout with a new more linear one in which trolley 3 and StoCCB conveyor were removed. Figure 9.9 and Figure 9.10 depict a floor plan of both the old and the new layout.

Two difficulties needed to be overcome with this new layout. First, trolley 2 operation would become more complex. It would have multiple sources and destinations and thus routing logics needed to be defined. A simple nearest pending decision event criterion was adopted as a simple rule for selecting local

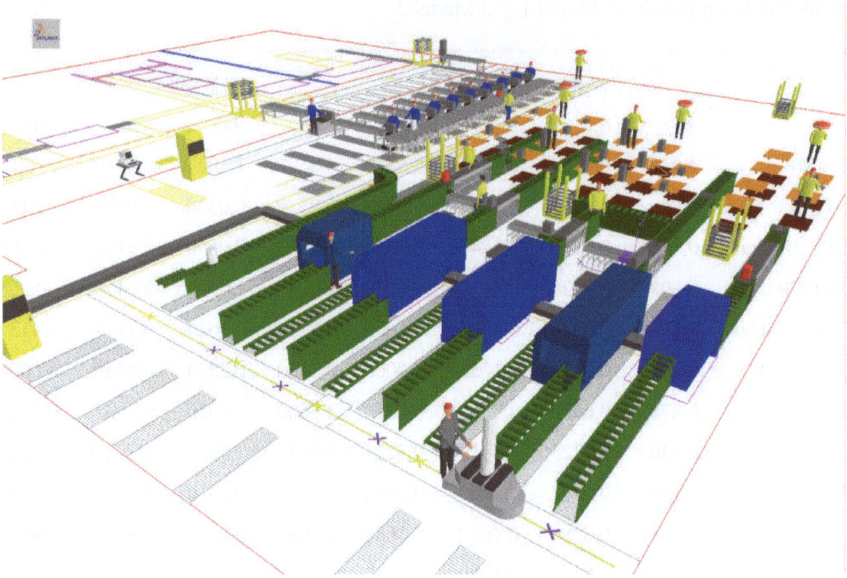

Fig. 9.9 Model of the Old Layout

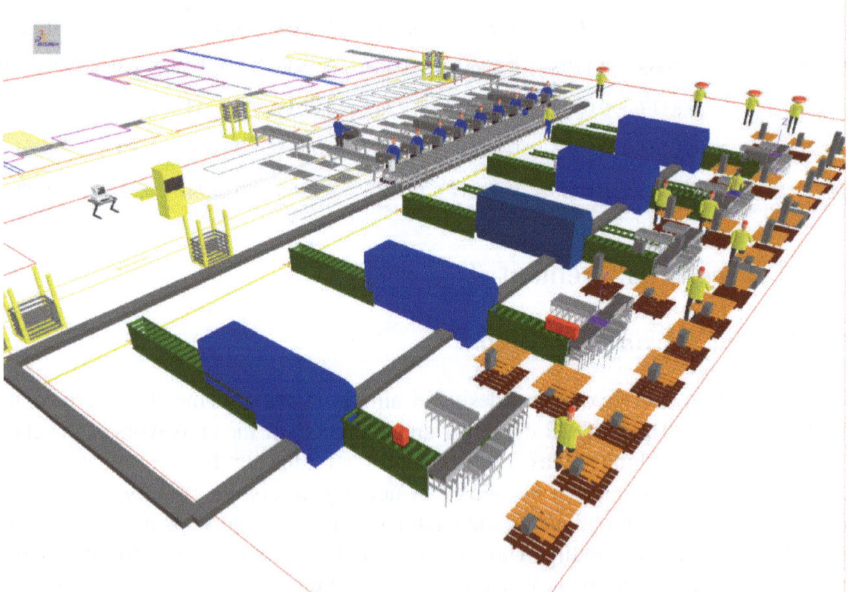

Fig. 9.10 Model of the Proposed Layout

optimal decisions in the routing process. Second, the arrangement of the pallets on the plant needed to be reconfigured. The decision adopted was to locate the pallets of products with the highest throughput rates in the outer positions so that they can be more easily accessed for retrieval. Pallets of first, traditional and standard qualities were located in such a way that the highest throughput qualities are the nearest to the classification roller belts. The cutting and classification lines were also placed so that trolley 2 movements are minimized. Target format cutting machines are the closest to splitting and secondary formats the farthest. In addition, recirculation roller bets in order to transport recirculated lots from the target format lines to the secondary format ones were connected via trolley 2.

9.4.2 New Layout Simulation

A simulation of 150 working days of the new layout was performed and results compared to those of the original layout. Levels of variability, plant saturation, and buffers occupancy were considered. Table 9.10 shows the simulation results for the new layout. Fig. 9.11 shows the evolution of buffers occupancy and resources utilization.

As it can be noticed, the new layout behaviour is different from that of the original one, providing evidence that the intermediate process steps affect system's behaviour by amplifying variability. Thanks to the appropriate placement

Table 9.10 Simulation Results for the New Layout

	Avg.	S.d.	Max.	Min.
$ST_{utilization}$	0,8767	0,0470	1,0000	0,7957
$Tr2_{utilization}$	0,7007	0,0895	0,9284	0,4378
$Tr2_{blocked}$	0,0188	0,0703	0,4496	0,0000
$CCB_{32\ queue}$	0,9372	1,6384	8,9362	0,1046
CCB_{30}	0,3407	0,2703	1,9918	0,0346
CCB_{27}	0,2280	0,2220	1,7115	0,0365
CL_{32}	0,8514	0,0899	1,0000	0,6186
CL_{30}	0,3396	0,0799	0,5674	0,1361
CL_{27}	0,2838	0,0648	0,4506	0,1383
$CLCB_{32\ queue}$	3,0328	2,1494	7,9224	0,5180
$CLCB_{30}$	0,6771	0,7422	4,5029	0,0211
$CLCB_{27}$	0,3093	0,2719	1,4695	0,0251
PL_{32}	0,6744	0,0642	0,8325	0,4343
PL_{30}	0,5550	0,1474	0,9578	0,2021
PL_{27}	0,4614	0,1117	0,7388	0,2042

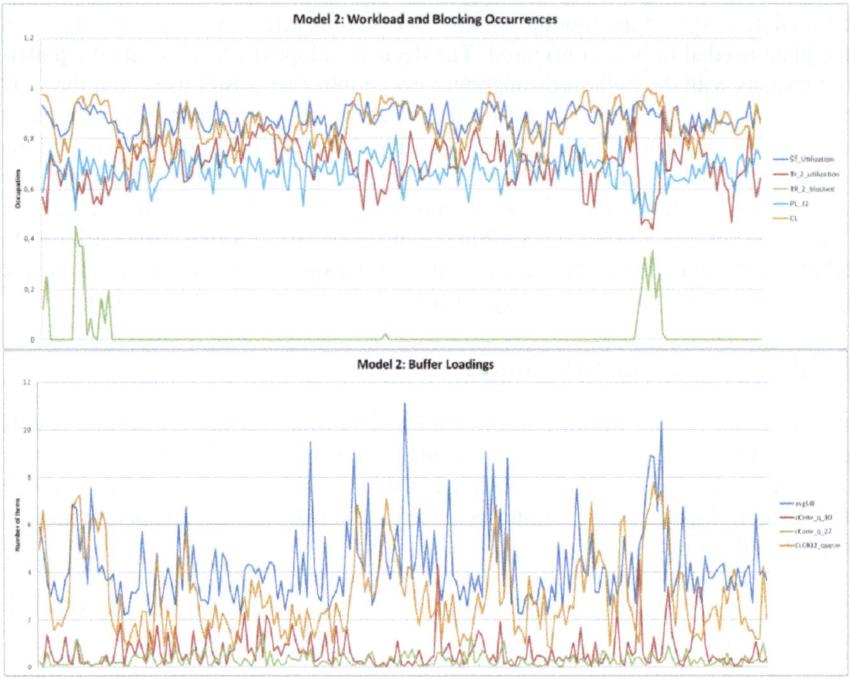

Fig. 9.11 Workload, Blocking Occurences and Buffer Loadings

of cutting lines along the trolley 2 line, trolley 2 utilization rate is similar to that in the original layout. Buffer occupancies are reduced and the periodic saturation of an element like StoCCB does not occur. In general, the model now behaves in a smoother manner.

9.5 Discussion and Conclusions

A paradigmatic case of a highly variable environment manufacturing process has been presented. A DES model has allowed a multilevel characterization of variability, so that improvement proposals have been adequately formulated and evaluated. The heterogeneity in input material properties together with the prevalence of manual operations constituted a process with high levels of variability, large work in process and important performance losses.

Data gathered from the manufacturing plant make it possible the construction of a dataset containing daily time series of performance parameters. PCA and AR models were employed in order to identify principal modes of variation and modelling of autocorrelation patterns. These temporary series models provided with a model for the daily variability in the plant. Video recordings provided with data for fitting models of cycle level variability in the system elements.

Models were validated by comparison of simulation results and actual plant's data along with expert criteria. Two modelling approaches were compared: a classical approach in which only cycle related variability is considered and a multilevel one that combines both cycle and daily levels. A third approach considered consisted of introducing individual differences, although it was rejected due to the low impact in results. Only the multilevel approach was capable of exhibiting a sort of process behaviour such as the large variability present in connection buffers. The classic approach lead to an underestimation of the process variations and negative autocorrelation patterns that unmatched real data.

A new layout aimed at increasing the process linearity was proposed and implemented in the simulation model. The new layout removes unnecessary intermediate transportations and connection buffers and standardises and simplifies the location of pallets in the output area. It also reduces buffers average contents and provides a smoother response to the process inherent variability. The time series model of daily variations ensured a robust design of the new layout that leads to a change of paradigm in the design of slates manufacturing processes. Instead of disposing the maximum capacity feasible for connection buffers, an optimized design in which only the minimum necessary buffers capacity is employed becomes possible.

Integrated Group for Engineering Research - Authors

The University of A Coruña (UDC) is a relatively young (1990) but very active University located on the northwest cost of Spain. It has 22 centers and more than 24,000 students. Additionally, the University manages three technological centers endowed with outstanding instruments and equipment at a European level, and co-participates with other universities in 4 more.

The Integrated Group for Engineering Research (GII, please see www.gii.udc.es) is a multidisciplinary group involved in research in a broad range of fields related to engineering and computing. Currently, GII is made up of a large number of professionals (56) belonging to Computing and Engineering Departments. In the last few years it has participated in more than 100 projects with industry as well as in dozens of publicly funded competitive projects, both national and international. It has collaborations with centers all over the world. The industrial engineering area is mainly devoted to conducting applied research in the following lines: manufacturing and logistics processes optimization, discrete events simulation and human factors engineering.

Diego Crespo Pereira holds an MSc in Industrial Engineering from the UDC and he is currently studying for a PhD. He is Assistant Professor of the Department of Economic Analysis and Company Management of the University of A Coruna. He also works in the GII as a research engineer since 2008. He is mainly involved in the development of R&D projects related to industrial and logistical processes optimization. He has also developed projects in the field of human factors engineering affecting manufacturing processes under a modeling and simulation approach.

David del Rio Vilas holds an MSc in Industrial Engineering and has been studying for a PhD since 2007. He is Adjunct Professor of the Department of Economic Analysis and Company Management of the UDC and research engineer in the GII of the UDC since 2007. Since 2010 he works as a R&D Coordinator for two different privately held companies in the Civil Engineering sector. He is mainly involved in R&D projects development related to industrial and logistical processes optimization.

Rosa Rios Prado works as a research engineer in the GII of the UDC since 2009. She holds an MSc in Industrial Engineering from the UDC and she is currently studying for a PhD. She has previous professional experience as an Industrial Engineer in several installations engineering companies. She is mainly devoted to the development of transportation and logistical models for the assessment of multimodal networks and infrastructures by means of simulation techniques.

Nadia Rego Monteil obtained her MSc in Industrial Engineering in 2010. She works as a research engineer at the Engineering Research Group (GII) of the University of A Coruna (UDC) where she is also studying for a PhD. Her areas of major interest are in the fields of Ergonomics, Process Optimization and Production Planning.

Contact

Mr. Diego Crespo Pereira
Address:
Escuela Politecnica Superior, Mendizabal s/n, Campus de Esteiro
15403, Ferrol, A Coruna (Spain)
Tel: +34981337400 - 1 - 3866 (work)
Mob: +34627598330

References

[9.1] Penker, A., Barbu, M.C., Gronald, M.: Bottleneck analysis in MDF-production by means of discrete event simulation. International Journal of Simulation Modelling 6(1), 49–57 (2007)

[9.2] Mertens, K., Vaesen, I., Löffel, J., Kemps, B., Kamers, B., Zoons, J., Darius, P., Decuypere, E., De Baerdemaeker, J., De Ketelaere, B.: An intelligent control chart for monitoring of autocorrelated egg production process data based on a synergistic control strategy. Computers and Electronics in Agriculture 69(1), 100–111 (2009)

[9.3] Nachtwey, A., Riedel, R., Mueller, E.: Flexibility oriented design of production systems. In: 2009 International Conference on Computers & Industrial Engineering, pp. 720–724 (July 2009)

[9.4] Schultz, K.: Overcoming the dark side of worker flexibility. Journal of Operations Management 21(1), 81–92 (2003)

[9.5] Bendoly, E., Prietula, M.: In 'the zone': The role of evolving skill and transitional workload on motivation and realized performance in operational tasks. International Journal of Operations & Production Management 28(12), 1130–1152 (2008)

[9.6] Powell, S.G., Schultz, K.L.: Throughput in Serial Lines with State-Dependent Behavior. Management Science 50(8), 1095–1105 (2004)

[9.7] Schultz, K.L., Juran, D.C., Boudreau, J.W.: The effects of low inventory on the development of productivity norms. Management Science 45(12), 1664–1678 (1999)

[9.8] Arakawa, K., Ishikawa, T., Saito, Y., Ashikaga, T.: Individual differences on diurnal variations of the task performance. Computers Ind. Engineering 27(1-4), 389–392 (1994)

[9.9] Baines, T., Mason, S., Siebers, P.-O., Ladbrook, J.: Humans: the missing link in manufacturing simulation? Simulation Modelling Practice and Theory 12(7-8), 515–526 (2004)

[9.10] Aue, W.R., Arruda, J.E., Kass, S.J., Stanny, C.J.: Brain and Cognition Cyclic variations in sustained human performance. Brain and Cognition 71(3), 336–344 (2009)

[9.11] Fletcher, S.R., Baines, T.S., Harrison, D.K.: An investigation of production workers' performance variations and the potential impact of attitudes. The International Journal of Advanced Manufacturing Technology 35(11-12), 1113–1123 (2006)

[9.12] Buzacott, J.: The impact of worker differences on production system output. International Journal of Production Economics 78(1), 37–44 (2002)

[9.13] Neumann, W.P., Winkel, J., Medbo, L., Magneberg, R., Mathiassen, S.E.: Production system design elements influencing productivity and ergonomics: A case study of parallel and serial flow strategies. International Journal of Operations & Production Management 26(8), 904–923 (2006)

[9.14] Schultz, K.L., Schoenherr, T., Nembhard, D.: An Example and a Proposal Concerning the Correlation of Worker Processing Times in Parallel Tasks. Management Science 56(1), 176–191 (2009)

[9.15] Mason, S.: Improving the design process for factories: Modeling human performance variation. Journal of Manufacturing Systems 24(1), 47–54 (2005)

[9.16] Shaaban, S., Mcnamara, T.: Unreliable Flow Lines with Jointly Unequal Operation Time Means, Variabilities and Buffer Sizes. In: Proceedings of the World Congress on Engineering and Computer Science, vol. II (2009)

[9.17] D'Angelo, A.: Production variability and shop configuration: An experimental analysis. International Journal of Production Economics 68(1), 43–57 (2000)

[9.18] Inman, R.R.: Empirical Evaluation of Exponential and Independence Assumptions in Queueing Models of Manufacturing Systems *. Production and Operations Management 8(4), 409–432 (1999)

[9.19] Colledani, M., Matta, A., Tolio, T.: Analysis of the production variability in multistage manufacturing systems. CIRP Annals - Manufacturing Technology 59(1), 449–452 (2010)

[9.20] He, X., Wu, S., Li, Q.: Production variability of production lines. International Journal of Production Economics 107(1), 78–87 (2007)

[9.21] Young, T.M., Winistorfer, P.M.: The effects of autocorrelation on real-time statistical process control with solutions for forest products manufacturers. Forest Products Journal 51(11/12), 70–77 (2001)

[9.22] Mertens, K., et al.: An intelligent control chart for monitoring of autocorrelated egg production process data based on a synergistic control strategy. Computers and Electronics in Agriculture 69(1), 100–111 (2009)

[9.23] Mittler, M.: Autocorrelation of Cycle Semiconductor Manufacturing Times in Density. In: Proceedings of the 1995 Winter Simulation Conference, pp. 865–872 (1995)

[9.24] del Rio Vilas, D., Crespo Pereira, D., Crespo Mariño, J.L., Garcia del Valle, A.: Modelling and Simulation of a Natural Roofing Slates Manufacturing Plant. In: Proceedings of The International Workshop on Modelling and Applied Simulation, vol. (c), pp. 232–239 (2009)
[9.25] Rego Monteil, N., del Rio Vilas, D., Crespo Pereira, D., Rios Prado, R.: A Simulation-Based Ergonomic Evaluation for the Operational Improvement of the Slate Splitters Work. In: Proceedings of the 22nd European Modeling & Simulation Symposium, vol. (c), pp. 191–200 (2010)
[9.26] Alfaro, M., Sepulveda, J.: Chaotic behavior in manufacturing systems. International Journal of Production Economics 101(1), 150–158 (2006)
[9.27] Clymer, J.R.: Simulation-based engineering of complex systems, 2nd edn. Wiley, Hoboken (2009)
[9.28] R. D. C. Team, R: A Language and Environment for Statistical Computing, R Foundation for Statistical Computing (2005)
[9.29] Pfaff, B.: VAR, SVAR and SVEC Models: Implementation Within R Package vars. Journal of Statistical Software 27(4), 1–32 (2008)
[9.30] Trapletti, A., Hornik, K.: tseries: Time Series Analysis and Computational Finance (2009), http://cran.r-project.org/package=tseries (accessed 2011)
[9.31] Schultz, K.L., Schoenherr, T., Nembhard, D.: An Example and a Proposal Concerning the Correlation of Worker Processing Times in Parallel Tasks. Management Science 56(1), 176–191 (2009)

10 Validating the Existing Solar Cell Manufacturing Plant Layout and Pro-posing an Alternative Layout Using Simulation

Sanjay V. Kulkarni and Laxmisha Gowda*

Modeling and Simulation techniques are powerful tools for evaluating best layout option by analyzing key performance indicators of a given process. A simulation technique for layout validation has its own unique benefit because the element of risk associated is almost zero. By sensitivity analysis potential process improvement strategies can be identified, evaluated, compared and chosen in a virtual environment much before the actual implementation and this helps in better decision making.

The dissertation work undertaken was on **process improvement** (reconfiguring plant layout in order to achieve effective utilization of resources, cost reduction and throughput improvement) i.e. to identify ways by which the performance could be improved in the system by, simulating the manufacturing process and evaluating the effectiveness of the process in terms of machine, human and system performance to identify bottlenecks and provide means to eliminate these inefficiencies.

Initially relevant data required was collected verified and cleaned using various statistical tools. After building the initial model an "AS-IS" model evolved, as the results were presented and discussed highlighting the pit falls in the current layout which affects the performance of the plant with process owners. At the analysis stage various "WHAT-IF" scenarios were identified and evaluated so as to identify the best alternative depending upon the performance measures which have the most significant improvement.

This would, hence, become a prerequisite for management in arriving at a better decision after evaluation of various alternative results obtained from the simulation.

Sanjay V. Kulkarni
Industrial and Production Engineering Department,
B.V.B College of Engineering and Technology,
Hubli - 580021, Karnataka, India
e-mail: skipbvb@gmail.com

Laxmisha Gowda
Student – Industrial and Production Engineering Department,
B.V.B College of Engineering and Technology,
Hubli - 580021, Karnataka, India

* Co-author.

10.1 Introduction

10.1.1 Problem Statement

Owing to highly fluctuating demand and cut throat competition in the global market, manufacturers are always pressed to implement changes and improve their key activities or processes to cope with challenges like reducing manufacturing costs, improving quality and customer satisfaction. One way to overcome the above and stay competitive is to become more efficient [10.1].

The case study undertaken in a solar photovoltaic (PV) module manufacturing company located in southern India tries to address the above. After interviewing managers who were responsible for design of plant layout, it was noticed that the layout was traditionally designed following conventional methods, which were mostly based on engineering experience and simple calculations of line productivity with constant processing data. Factors pertaining to worker utilization, machine utilization were not taken into consideration. However, the breadth and depth and type of experience "designers" have can vary and conventional methods of design cannot accurately result into an efficient plant layout because of factors such as unevenness of processing data and unpredictability of machine failures and unaccounted resource utilization.

It was clearly evident that the application of simulation study was lacking in this field. Further, no simulation technique was implemented by the organization to validate the proposed layout and to identify the problems faced in the existing layout.

Thus to fill the missing elements of traditional plant layout design; Simulation technique is used along with traditional plant layout design which adds value to the entire process of layout optimization. Here Discrete Event Simulation tool ARENA® is used to systematically examine the key performance variables on performance with more information such as total time in system, waiting time, and utilization. The key performance improvements are in terms of cycle time reduction, productivity increase, reduction in travelling time and resource utilization factor.

10.1.2 Purpose

The purpose of carrying out this project is to find the best layout which will result into optimized resource usage in terms of operators and machines in order to reduce production costs and improve productivity. The result of this study will provide recommendations as well as the validation of the recommendations resulting into the desired improvements. Many "What-If" with different resource combinations will be considered and experimented.

10.1.3 Scope

The scope of this study is limited to the production process of solar PV module manufacturing unit located in southern India.

Other issues that will not be included in the study are as follows:

- Problems about the workers' behavior that may influence the productivity are considered to be out of scope. Morale, learning resistance, behavior and relationship management should be the superintendent's responsibility.
- Management problems are not considered in this study, nor will any changes in management behavior be proposed.

10.1.4 Objective

The main objective of this project work is to identify ways by which the performance could be improved in the system by:

- Validating existing plant layout so as to provide an independent, third party assessment and to confirm that the current layout meets the required key performance indicators during routine production.
- Simulating the manufacturing process and evaluating the effectiveness of the process in terms of machine, human and system performance to identify bottlenecks and provide means to smoothen out which will assist the management in arriving at a better decision after evaluation of various alternative results obtained from the simulation.

The following specific objectives will also be achieved through the process:

- Mapping the layout of the manufacturing system and coming out with improved layout.
- Developing a computer based simulation model.
- Verifying model according to the modeling assumptions.
- Validating the model with actual performance measures.
- Use simulation model to study and analyze different alternatives generated to determine the improvement in performance.

10.1.5 Methodology

Complete literature surveys of manufacturing systems, concepts of modeling and simulation and simulation software's that are currently available which suits the system were studied. The actual factory's manufacturing system was studied, modeled and simulations were performed. Model building requires the following:

- The study of the current layout.
- Gathering the process times and fitting the data to probability distribution. The goodness-of-fit test was used to find out type of probability distribution, the process time taken; Normal, Uniform, Triangular, etc.

Model development, Verification and Validation are the core part of the entire simulation. The verification, that the model is operating the way it should, was determined by series of discussion with the process owners. Finally conclusion and recommendation were made.

10.2 System Background

10.2.1 Plant Layout Details

The plant layout area is shared by crystalline and thin film manufacturing unit. The project work under taken is restricted only to crystalline products being manufactured. The plant layout details were obtained through a detailed study of the process and material movement and a CAD layout drawing, was prepared as shown in fig 10.1.

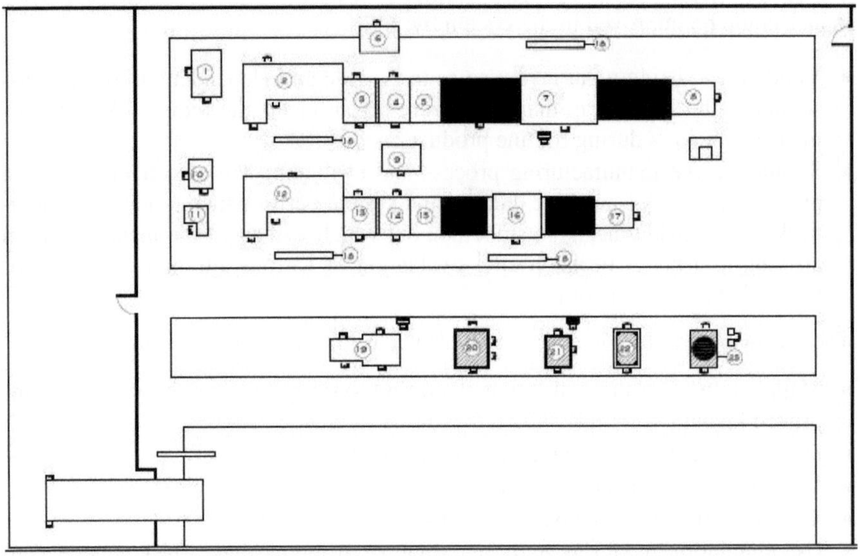

Fig. 10.1 Existing Plant Layout.

(1) Eva/tedlar cutting machine, (2) SPI assembler-1, (3) Bussing station-1, (4) Inspection table-1, (5) layup station-1, (6) rework station-1, (7) Laminator-1, (8) Qc final inspection-1, (9) Rework station-2, (10) Cell testing station (11) Cell cutting station, (12) SPI assembler-2, (13) Bussing station-2, (14) Inspection station-2, (15) Layup station-2, (16) Laminator-2, (17) Qc final inspection-2, (18) Storage rack, (19) Qc final, (20) HIPOT testing station, (21) Sun-simulator testing station, (22) Job fixing staion.

10.2.2 Description of Process

The plant can be divided mainly into pre-lamination stage and post-lamination stage. The assembly line constitutes two production lines in the pre lamination stage and thereafter, both these lines are converged to use the single post lamination stage. The assembly line is a product type where materials or semi-processed components are transferred to subsequent stations manually. After finishing operation on part, the operator places the component on the relevant table where subsequent operation is to take place. If operation is to be undertaken on station not adjacent to the previous, additional crew are used to transfer components.

The basic sequences of assembly line processes are listed in table 10.1. In making PV module, initially the EVA, Tedlar cutting, glass cleaning, cell and TCI cutting operations are carried out. Once these operations are done, the cells are loaded for string assembly process which is carried with the help of assembler. The visual inspection is carried out to find any cell damage; if found, the material will be sent for string rework station. Bussing and layup stages are completed before the modules are sent for the lamination stage.

Table 10.1 Basic sequences of assembly line processes

	Basic process	**Predecessor**
A	String assembly	
B	Layup preparation	
C	Bussing	A and B
D	Layup final	C
E	Dark IV test	
F	Lamination	E and F
G	Trimming	F
H	Quality check	
I	Framing	G and H
J	JB fixing	I
K	SUN SIMULATOR testing	K
L	Labeling	
M	Quality check	
N	Packing	K,L and M

The modules coming out of the laminator are trimmed and inspected for any defects, if found, modules will be sent to laminate rework station. The modules that are given clearance from the quality department are then passed on to the framing station and JB fixing station where the laminated modules are framed and junction boxes are mounted respectively. SUN-SIMULATOR testing and HI-POT testing are carried out to check the performance parameters of the module before the labeling operation is carried out.

10.3 Model Building and Simulation

10.3.1 Assumptions of the Model

A set of modeling assumptions were set according to the system constraints dictated by process type, and the entities movement-sequence between the stations.

The following were the assumptions for simulation:

- Operators are always available during the two shifts (1 shift =8 hours), an hour of lunch break and fifteen minute of tea breaks will be provided.
- There are no significant equipment or station failures.
- The production is continuous.
- Processes for 60 cells and 72 cell modules have independent process time in pre-lamination stage.
- Materials are always available at each assembly station.
- Transfer times between stations are taken as constant values.

10.3.2 Simulation Model

The simulation model is built as per the process flow diagram of the system which is shown in fig 10.2.

The model requires four modules [10.2] to supply all the input data required to perform the simulation experiment. These are:

Capacity Inputs: The information provided in this module indicates the number of machines in each process and the schedule cycles of workers to operate.

Product Specific Data: Data required for processing each product type such as setup, load-unload time, production rates, processing batch size, and flow line.

User Specific Data: The user has ability to customize the simulation experiment by changing certain requirement in the model, such as shifts start time of each process.

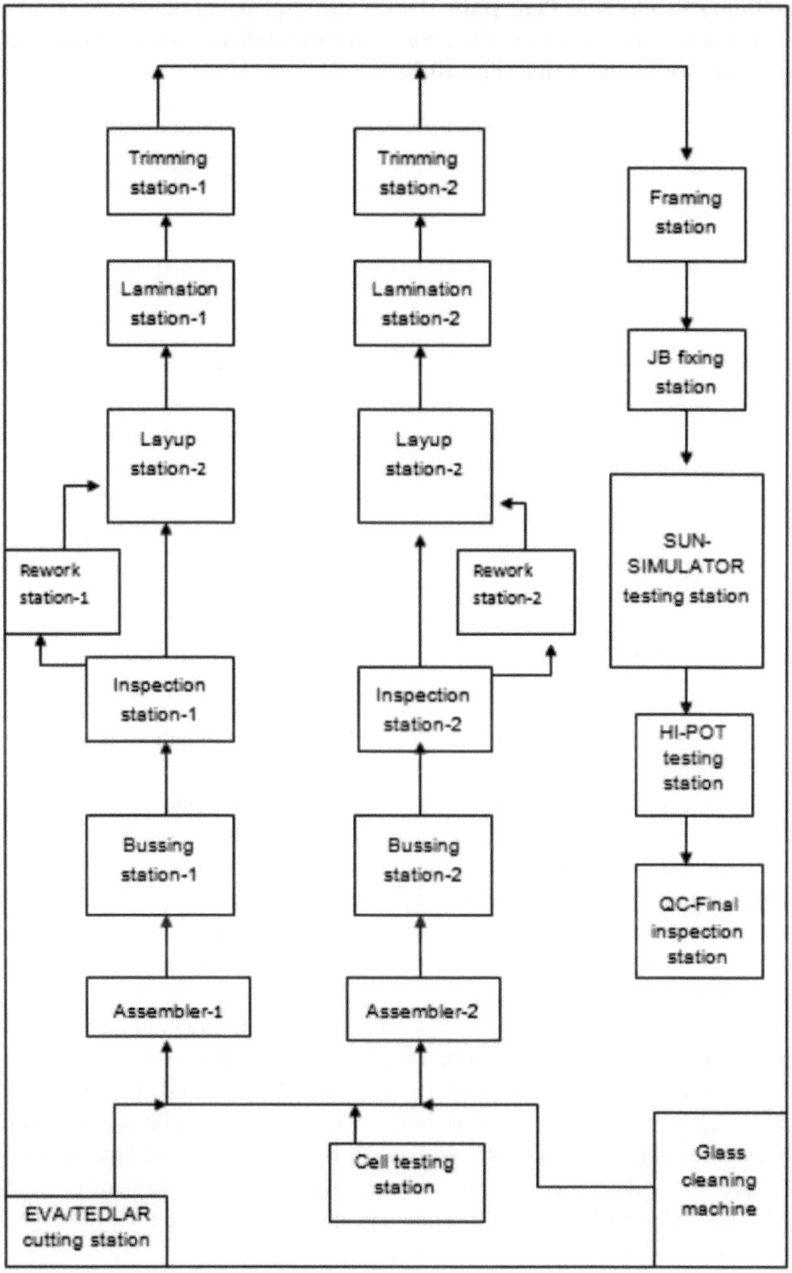

Fig. 10.2 Process flow diagram of the system.

Scheduling Production Plan Data: This is the sequencing time-table of product items for production to follow, this aims to find the ability of the plant to complete the demand within the period (Fig. 10.3).

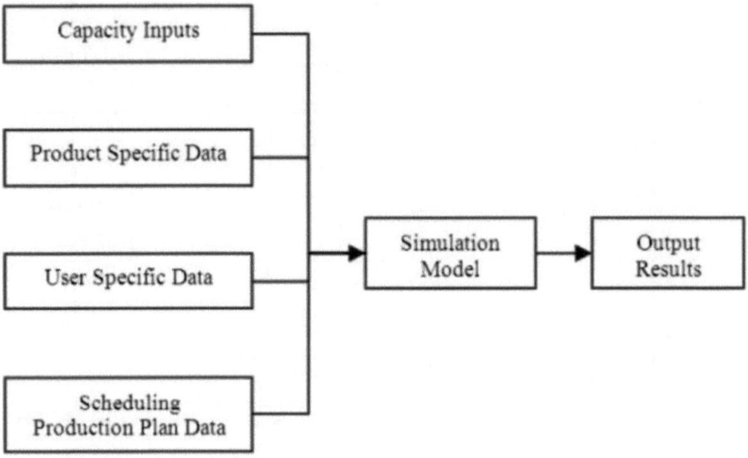

Fig. 10.3 Simulation Model Data Flow.

In order to define machine process times within the simulation, actual process times were collected through time and motion study and were recorded for each & every major event. Individual machine process times were collected from information provided by the shift-in-charge and checked with the production manager. A time study was also conducted on machines that had process variability, either from setup times, or because of the natural variation within the process.

Transport times are also one of the important parameters to be included for building effective models; thus transfer times in between stations are also taken into account while building models.

Machine downtime information was gathered through observations and conversations with the in-charge of shift, quality control super-visor and production manager. Machine downtimes were also collected from records kept in the database for each machine, as well as observations recorded. The use of multiple sources ensured the accuracy of this data. Downtime for the machine in-between each recorded failure was collected from the database from September 2010 to April 2011. Information recorded in the database indicated the total time the machine was down and the number of failures for that specific day.

Table 10.2 Transfer Time to Stations in the process

Station name	Route time(sec)	Station name	Route time(sec)
Route to line1	60	Route to laminator station2	30
Route to line2	60	Route to trimming station	60
Route to rework station1	60	Route to framing station	120
Route to rework station2	60	Route to module rework station	60
Route to bussing station1	30	Route to JB fixing station	120
Route to bussing station2	30	Route to SUN SIMULATOR testing station	120
Route to layup station1	30	Route to HI POT testing station	30
Route to layup station2	30	Route to final inspection station	30
Route to laminator station1	30	Route to exit station	10

Finally the input analyzer tool of ARENA® was used to convert all the time studies, machine break down data, into probability distributions to be used in the simulation model.

Table 10.3 Machine downtime information

Failure name	Failure type	Uptime (hours)	Count	Downtime (minutes)
TCI ribbon change	Count	TRIA(35,40,45)	EXPO(20)	
Flash lamp change	Count	EXPO(100)	EXPO(90)	
Diaphragm change	Count	EXPO(750)	EXPO(120)	
flash lamp of sun simulator change	Count	EXPO(100000)	EXPO(90)	
Shift break	Time	UNIF (430, 435)		UNIF (40, 45)
flux change	Time	EXPO(50)		EXPO(10)
Teflon sheet change	Time	EXPO(50)		EXPO(60)
vacuum oil change	Time	EXPO(500)		EXPO(20)

A summary of the probability distributions for the process times used in the simulation are shown in table 10.3. These probability distributions were selected by the input analyzer as having the best fit of the data, by measuring the square error.

Table 10.3 Probability distributions selected from the Input Analyzer for machine process time studies.

PRE-LAMINATION STAGE		
Machine	**Probability Distribution**	**Expression**
Assembler-1	Log normal	6.38 + LOGN(0.596, 0.457)
Assembler-2	Exponential	5.16 + EXPO(0.481)
Bussing station	Exponential	2 + EXPO(0.894)
Layup station	Normal	NORM(2.25, 0.528)
String rework station	Normal	NORM(9.8, 3.31)
Trimming station	Normal	NORM(51.6, 6.28)
POST LAMINATION STAGE		
Framing station	Beta	1.51 + 1.08 * BETA(1.44,1.25)
JB fixing station	Beta	1.39 + 1.32 * BETA(1.17,0.947)
SUN-SIMULATOR testing station	Triangular	TRIA(1.35, 2.85, 3)
HIPOT testing station	Triangular	TRIA(1.35, 2.85, 3)
QC final inspection station	Exponential	2.12 + EXPO(0.75)
Laminate rework station	Exponential	44.5 + EXPO(33.2)

10.3.3 Model Verification

The model verification stage involved ensuring that the model works as intended and is accurately built and structured in order to replicate the workings of the actual line correctly [10.3]. Most of the verification work was carried out as an iterative process within the overall model building process. The first incarnation of the model was a very simple version which was verified as a proof of concept. Following this verification stage more detail was added into the model. Another verification stage followed. This cycle repeated itself until the model contained all desired detail and functionality.

10.3.4 Model Validation

Validation is the task of demonstrating that the model is a reasonable representation of the actual system and that it reproduces system behavior to satisfy analysis objectives [10.4]. This stage consisted of both face validation tests and statistical validation tests. Face validation was performed by demonstrating the model animation to the process owners. Statistical validation was performed by Historical Data Validation, graphical comparison of data and Event Validity tests [10.5].

10.3.5 Simulation Model Results and Analysis

The analysis is not based on all system variables stated in auto generated report. However system performance variables related to line balancing (bottleneck area) and stations utilization are taken into consideration. These two aspects will assist in focusing future improvement operations. The simulation was done for 400 hours to study the manufacturing system as a whole and 5,483 units were produced in this period. The utilization of resources is shown in table 10.4 below.

Table 10.4 Utilization of resources in the current system.

Process	Station	No. of operator's	Resource utilization	idle time
Stringing operation	Assembler-1	1	85.01	14.99
	Assembler-2	1	83.02	16.98
Bussing operation	Bussing station-1	2	34.80	65.20
	Bussing station-2	2	41.73	58.27
Layup operation	Layup station-1	2	28.03	71.97
	Layup station-2	2	32.77	67.23
Lamination operation	Laminator-1	1	99.72	0.28
	Laminator-2	1	97.05	2.95
Trimming operation	Trimming	2	21.44	78.56
Framing operation	Framing	3	50.04	49.96
JB fixing operation	JB fixing	2	50.90	40.10
SUN SIMULATOR testing	SUN SIMULATOR	4	57.83	42.17
HI POT testing	HIPOT testing	4	56.44	43.56
Final QC inspection and module cleaning	Final QC	5	66.52	33.48
String rework process	String rework-1	1	5.143	94.857
	String rework-2	1	6.662	93.338
Module rework process	Module rework	1	42.05	57.95

With the analysis of the time studies performed, it was established that the lamination stage will be the bottleneck process with the highest utilization factor of the manufacturing line. This was verified by the simulation. The assembler station also consumes a large amount of time due to the fact that the stringing process consists of a number of smaller processes.

The machinery with the highest utilization is the laminator, this is the station were the process time is longer and thus the high utilization rate. The chart shows the utilization of resources in the system.

The cost factor associated with resource utilization is also taken into consideration for analysis part. Here, the cost incurred is categorized as busy cost and idle cost in the system. The results are shown in the pie-chart below (Fig. 10.4).

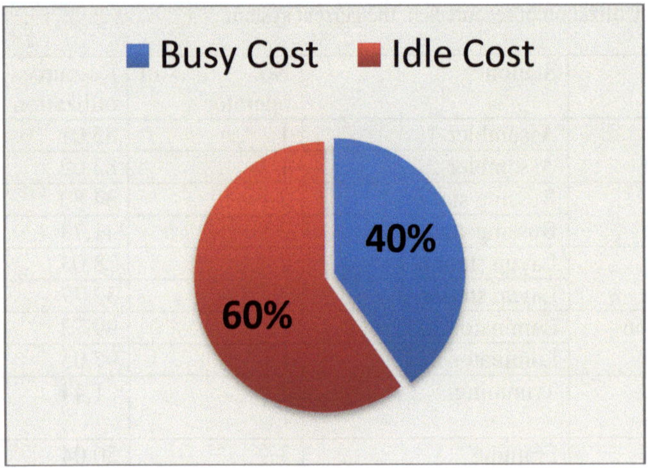

Fig. 10.4 Resources cost in the current system.

As it is clearly seen from the pie-chart, the idle cost of the system overshoots busy cost in this manufacturing policy. Thus in the experimentation stage the experiments are to be designed in such a way that the system idle cost is minimized.

Table 10.5 Summary of current system performance variables.

	EXISTING SYSTEM
Number of hour's simulated	400
System throughput	5483
Work in process	469
Average resource utilization	50.58%
% Reduction in idle cost	60.0%

10.4 Simulation Experiment

The experiments to be carried out are listed below with an objective to increase average resource utilization. Depending on changing level of resources, alternatives that were developed as a proposed system improvements. The possible combinations are:

Scenario 1: with single bussing station
Scenario 2: with single layup station
Scenario 3: with single string-rework station
Scenario 4: with single trimming station.
Scenario 5: with conveyor system for material transfer in pre-lamination stage.
Scenario 6: with additional resource-laminator
Scenario 7: with additional resource-assembler

These changes are a cumulative combination of each other and the effect of these cumulative changes on average resource utilization, WIP, throughput, idle cost, cost for reconfiguring layout and total number of operators saved will be investigated for each scenario.

10.5 Analysis and Discussion

10.5.1 Performance Measures

A variety of measures may be used to evaluate the performance of the plant layout. Traditionally, simulation models use system times or resource utilization levels. For this system, the primary performance measure was the total average resource utilization. This measure was chosen because the objective of the model was to increase the utilization factor of the existing resources.

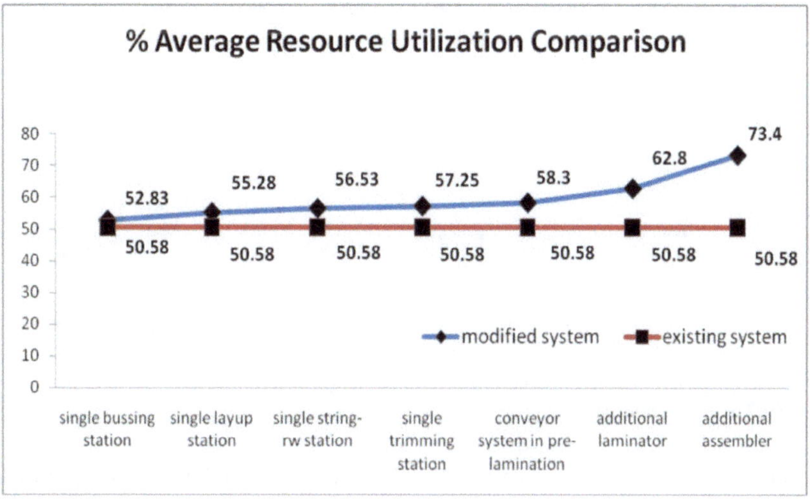

Fig. 10.5 Percentage Average Resource Utilization Comparison.

As seen from the graph (Fig. 10.5), having cumulative combination of all the experimental scenarios resulted in increase of average resource utilization from existing 50.58% to 73.4%.

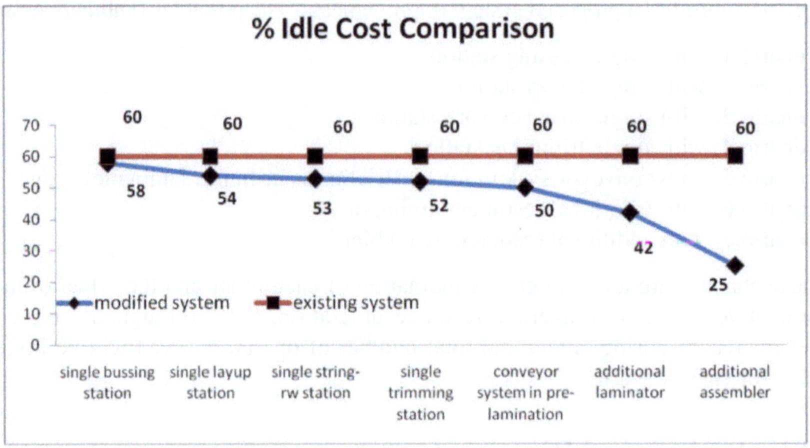

Fig. 10.6 Percentage Idle Cost Comparison.

As seen from the above graph (Fig. 10.6), having cumulative combination of all the experimental scenarios resulted in decrease of idle cost from existing 60% to 25% thus effectively utilizing allocated resources.

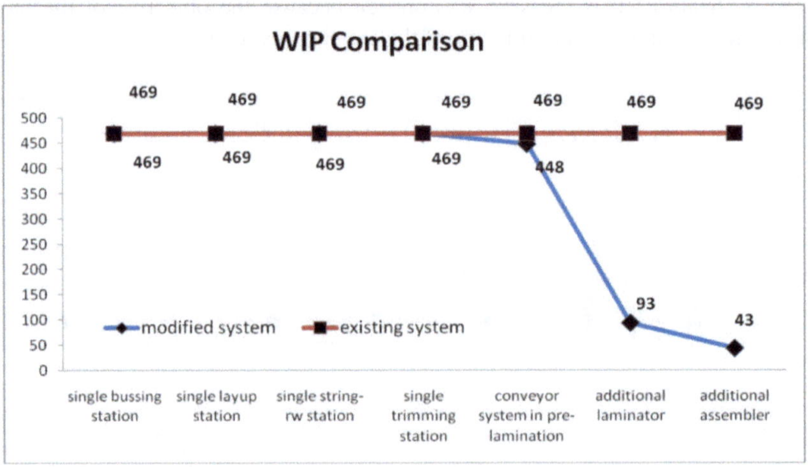

Fig. 10.7 WIP Comparison.

Having cumulative combination of all the experimental Scenarios resulted in decrease of WIP built up in the existing system from 469 numbers to 43 in number (Fig. 10.7).

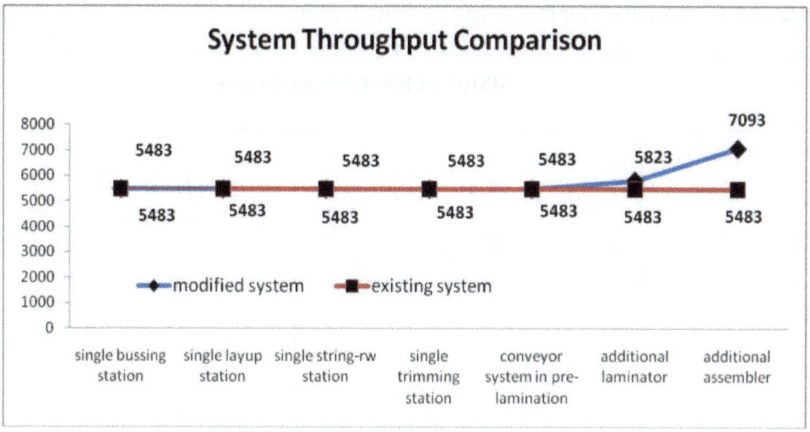

Fig. 10.8 System Throughput Comparison.

As seen from the graph in Fig. 10.8, having cumulative combination of all the experimental scenarios resulted in increasing the system throughput from the existing 5483 units to 7093.

10.5.2 Cost Analysis

Table 10.6 Cost Analysis for Reduction in Manpower.

COST ANALYSIS (Rs)	**SIMULATION EXPERIMENTS**						
	single bussing station	single layup station	single string rework station	Single trimming station	Conveyor system in pre-lamination stage	Additional laminator	Additional assembler
Operator saved	2	4	5	6	6	5	4
Total cost reduced	10,000	20,000	25,000	30,000	30,000	25,000	20,000

The driving factors for the decision of reconfiguring the existing layout are solely dependent on the cost incurred for these changes. The following table summarizes the approximate cost involved for reconfiguring the existing plant layout according to the experimental design.

Table 10.7 Cost Analysis for Reconfiguring Plant Layout.

COST ANALYSIS (Rs)	SIMULATION EXPERIMENTS						
	single bussing station	single layup station	single string rework station	Single trimming station	Conveyor system in pre-lamination	Additional laminator	Additional assembler
	12.5	20.83	28.25	28.25	3,023,80	78,029,75	1,63,060,77.5

The modified system was simulated for a period of 400 hours, and a Pie – Chart graph was plotted for the busy cost v/s idle cost this was compared with the existing system. Pie-chart comparison shows that idle cost has been reduced from 60% to 25% thus effectively in-creasing utilization of the allocated resources.

10.5.3 Summaries of Simulation Experiments

SIMULATION EXPERIMENTS	PERFORMANCE MEASURES							
	Number of hour's simulated	%Average resource utilization	% reduction in idle cost	System throughput	WIP level	Cost for reconfiguring layout	Total operator saved	Total Cost reduced
Existing system	400	50.58	60	5483	469	---	----	---
single bussing station	400	52.83	58	5483	469	12.5	2	20,000
single layup station	400	55.28	54	5483	469	20.83	4	20,000

single string rework station	400	56.53	53	5483	469	28.25	5	25,000
Single trimming station	400	57.25	52	5483	469	28.25	6	30,000
Conveyor system in pre-lamination stage	400	58.30	50	5483	448	3,023,80	6	30,000
Additional laminator	400	62.80	42	5823	93	78,029,75	6	30,000
Additional assembler	400	73.40	25	7093	43	1,63,060,77.5	6	30,000

The above table gives clear comparison in-between the performance measures of the existing system and that of the various other experimental scenarios. It can be seen that modification done in the processes resulted in increasing the average resource utilization and also in bringing down idle cost. The *modification also helped in in-creasing the system throughput with the addition of additional resources*. Thus by comparing the existing production system and the modified system, it can be seen that modified system yields great improvement.

10.6 Conclusions

Current real manufacturing system has been translated into discrete event computer based simulation model using ARENA® simulation package besides being based on the validity test result, the developed model meets validity requirements. It must also be noted that the current manufacturing system has been analyzed promptly and satisfactorily using a valid initial simulation model based upon which it can be concluded that current manufacturing performances are still capable of further improvement. Depending on changing level of resources, alternatives were developed as proposed system improvements and based on the comparison analysis of various scenarios, it can be concluded that cumulative combination of all the changes gives the best system performance improvement rather than individual scenarios.

The throughput was increased from 5483 to 7093 and the percentage average resource utilization increased from 50.58% to 73%. The WIP level was drastically reduced from 469 to 43 numbers. Furthermore, there is still room for improvement as the optimal resource utilization is not achieved.

Since the output and performance parameters for the alternatives are higher than the existing system, it is beneficial for the factory to make use of it.

10.7 Future Scope

Based on the information discovered during the study, the following suggestions on some specific issues have been made.

- After improving resource utilization and bringing down the idle time of the production system along with the analysis pertaining to average waiting time, trying various different scheduling methods for resources can also be considered.
- Reduction in down time of machines and its effect on system throughput can also be considered as another option for further studies.
- Automating the system and its effect on performance measures can also be studied.
- Finally, in order to enhance the features of the simulation animation, 3D graphics could be used. However, the student version of ARENA® does not have this feature.

Authors Biography, Contact

About the College (www.bvb.edu)
The versatile manifestations of engineering have had a profound and lasting impact on our civilization. From the grandeur of the pyramids and mans journey into the space, to the recent information revolution, engineering continues to fascinate and enthrall. The B. V. Bhoomaraddi College of Engineering and Technology (BVBCET) believes in kindling the spirit of this unique and creative discipline in every student who enters its portals. Preparing them for a world in which their contribution truly stands apart.

Established in 1947, BVBCET has achieved an enviable status due to a strong emphasis on academic and technical excellence. From a modest beginning when the college offered only an Undergraduate program in civil engineering, the college has indeed come a long way. Currently college offers 12 UG and 8 PG programs affiliated to Visvesvaraya Technological University, Belgaum and is recognized by AICTE, New Delhi and accredited by NBA. Current annual student intake for Undergraduate & Post Graduate programmes is in excess of 1200. The faculty consists of extremely qualified and dedicated academicians whose commitment to education and scholarly activities has resulted into college gaining Autonomous Status from the University and UGC. The college has adopted Outcome Based Education (OBE) framework to align the curriculum to the needs of the industry and the society. Innovative pedagogical practices in the teaching learning processes form the academic eco system of the institution. The active involvement of faculty in research has led to the recognition of 8 research centers by the University.

Spread over a luxurious 50 acres, the picturesque campus comprises of various buildings with striking architecture. A constant endeavor to keep abreast with technology has resulted in excellent state-of-the-art infrastructure that supplements every engineering discipline. To enable the students to evolve into dynamic professionals with broad range of soft kills, the college offers value addition courses to every student. Good industrial interface and the experienced alumni help the students to become industry ready. The college is a preferred destination for the corporate looking for bright graduates. There is always a sense of vibrancy in the campus and it is perennially bustling with energy through a wide range of extra-curricular activities designed and run by student forums to support the academic experience.

Author: Sanjay Kulkarni

Graduated as a mechanical engineer in the year 1995, Sanjay worked for various engineering industries as a consultant in and around India. He started off as a consultant introducing "Clean Room" concepts to various engineering industries when the technology was very nascent in the Indian region. He had a great opportunity coming across as a software consultant after two years of his first assignment after which he never had to look back. As a software consultant Sanjay had best opportunity to learn various technologies relevant to engineering industry right from Geographical Information Systems, Geographical Positioning Systems, CAD and CAM solutions, Mathematical modeling, Statistical modeling and Process modeling tools to various hardware associated with the above technologies. He spent 14 years serving the engineering industry before he quit and began his second innings with academics.

Presently Sanjay is a professor with one of the oldest and leading engineering colleges of North Karnataka – B V Bhoomaraddi College of Engineering and Technology Hubli, Karnataka, India. He is associated with Industrial and Production department handling subjects like – System Simulation, Supply Chain Management, Organizational Behavior, Marketing Management, and Principles of Management. A rich industry exposure of Sanjay has given an edge while delivering the lectures to the students and it has been a memorable experience to experience both the worlds of engineering profession and engineering academics.

As a consultant he has handled challenging engineering projects in the past for various engineering industries and delivering the results successfully. As a professor he is learning new things every day from his students – actually **learning never ceases**.

Co-Author - Laxmisha

Since childhood, Laxmisha has been interested in mechanical designs. Fascinated by the various combinations and functions in the world of mechanics, he has pursued a career that lets him experiment with and create unique and utilitarian combinations of machines. Wanting to widen his area of expertise in this field, he pursued an M-Tech in Production Management. His interests include concurrent engineering and product life cycle design, computer simulation, and manufacturing systems design and control.

Throughout the last year of his master studies he closely worked with the Process owners of solar cell manufacturing plant in Bangalore, where Simulation technique was tried to incorporate along with the traditional plant layout design so as to add value to the entire process of layout optimization.

In addition to pursuing his academic interests, Laxmisha was active in Scouting Movement and have been given the Rashtrapathi Award(Scout) in 2002 by the then President of India. Laxmisha engages himself in sports like Kayaking, River Rafting and Rock Climbing.

References

[10.1] Roslin, N.H., Seang, O.G., Dawal, S.Z.: A study on facility planning in manufacturing process using witness. In: Proceeding of the 9th Asia Pacific Industrial Engineering & Management Systems Conference, APIEMS 2008 (2008)

[10.2] McLean, C., Kibira, D.: Virtual reality simulation of a mechanical assembly production line. In: Proceeding of the 2002 Winter Simulation Conference, pp. 1130–1137 (2002)

[10.3] Zuhdi, A., Taha, Z.: Simulation Model of Assembly System Design. In: Proceeding Asia Pacific Conference on Management of Technology and Technology Entrepreneurship (2008)

[10.4] Iqbal, M., Hashmi, M.S.J.: Design and analysis of a virtual factory layout. Journal of Materials Processing Technology 118, 403–410 (2001)

APPENDIX

Cost Analysis Calculation for Reconfiguring Plant Layout

1. Cost for reconfiguring plant layout calculation for experimental scenario -1
Cost of manpower per hour=Rs.12.5/hr
Number of manpower required for making changes=4
Time required to makes these changes=15min
Total time required for making changes=4*15=60min
The total cost incurred for making these changes=12.5*1=Rs12.5/-

2. Cost for reconfiguring plant layout calculation for experimental scenario -2
Cost of manpower per hour=Rs.12.5/hr
Number of manpower required for making changes=4
Time required to makes these changes=25min
Total time required for making changes=4*25=100min
The total cost incurred for making these changes=12.5*1.66=Rs20.83/-

3. Cost for reconfiguring plant layout calculation for experimental scenario -3
Cost of manpower per hour=Rs.12.5/hr
Number of manpower required for making changes=4
Time required to makes these changes=35min

Total time required for making changes=4*35=140min
The total cost incurred for making these changes=12.5*2.26=Rs28.25/-

4 Cost for reconfiguring plant layout calculation for experimental scenario -4
Cost of manpower per hour=Rs.12.5/hr
Number of manpower required for making changes=4
Time required to makes these changes=35min
Total time required for making changes=4*35=140min
The total cost incurred for making these changes=12.5*2.26=Rs28.25/-

5. Cost for reconfiguring plant layout calculation for experimental scenario -5
Cost of manpower per hour=Rs.12.5/hr
Number of manpower required for making changes=4
Cost of skilled manpower per hour=Rs30/-
Number of skilled manpower required for making changes=3
Time required to makes these changes=8hours
Total man-hour required for making changes= (4*8) + (3*8) =56hour
The total cost incurred for making these changes= (56*12.5) + (56*30) =Rs2380/-
Cost for procuring conveyor system for pre lamination stage =Rs.3, 00,000/-
Total cost for reconfiguring layout=3, 00,000+2380=Rs3, 02380/-

6. Cost for reconfiguring plant layout calculation for experimental scenario-6
Cost of manpower per hour=Rs.12.5/hr
Number of manpower required for making changes=4
Cost of skilled manpower per hour=Rs30/-
Number of skilled manpower required for making changes=3
Time required to makes these changes=10hours
Total man-hour required for making changes= (4*10) + (3*10) =70hour
The total cost incurred for making these changes= (70*12.5) + (70*30) =Rs2975/-
Cost for procuring conveyor system for pre lamination stage =Rs.3, 00,000/-
Cost for procuring laminator=Rs75, 00,000
Total cost for reconfiguring layout= 75, 00,000+3, 00,000+2380=Rs78, 02975/-

7. Cost for reconfiguring plant layout calculation for experimental scenario -7
Cost of manpower per hour=Rs.12.5/hr
Number of manpower required for making changes=6
Cost of skilled manpower per hour=Rs30/-
Number of skilled manpower required for making changes=5
Time required to makes these changes=13hours
Total man-hour required for making changes= (6*13) + (5*13) =143hour
The total cost incurred for making these changes= (143*12.5) + (143*30) = Rs6, 077.5/-
Cost for procuring conveyor system for pre lamination stage =Rs.3, 00,000/-
Cost for procuring laminator=Rs75, 00,000
Cost for procuring assembler=Rs 85, 00, 000
Total cost for reconfiguring layout= 85, 00, 000+75, 00,000+3, 00,000+ 6, 077.5=Rs **16306077.5**/-

Modified Layout Cad Drawings

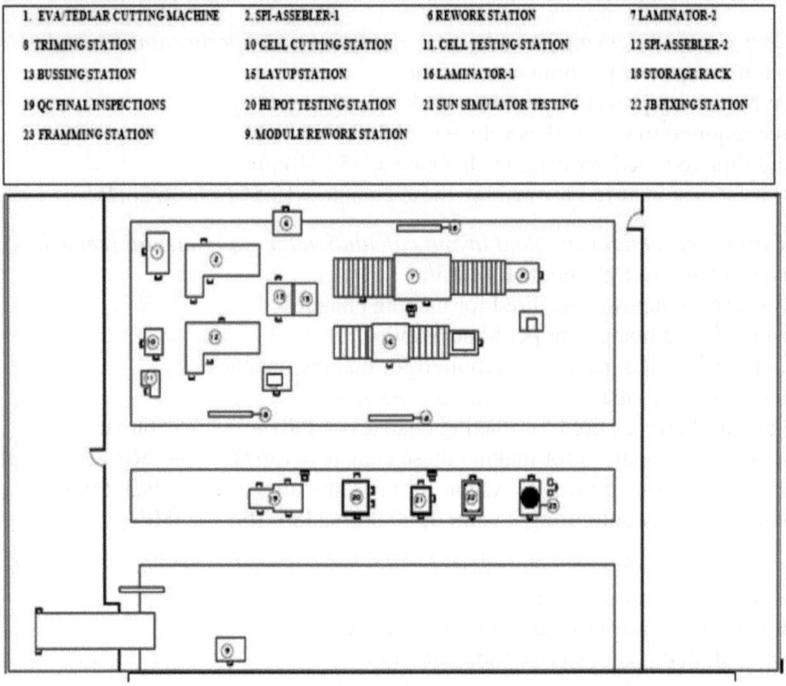

Fig. 10.9 Modified layout with single bussing and layup station.

11 End-to-End Modeling and Simulation of High- Performance Computing Systems

Cyriel Minkenberg, Wolfgang Denzel, German Rodriguez, and Robert Birke

Designing large-scale High-Performance Computing (HPC) systems, including architecture design space exploration and performance prediction, is a daunting task that can benefit enormously from discrete event simulation techniques, as the interactions between the various components of such a system generally render analytic approaches intractable. The work described in this chapter specifically deals with end-to-end, full-system simulation, as opposed to simulation of individual components or nodes. The tools described here can be used in the design phase of a new HPC system to optimize system design for a given set of workloads, or to create performance forecasts for new workloads on existing systems.

We have taken a network-centric approach, as the scale of current high-end HPC systems is in the range of hundreds of thousands of processing cores, so that the impact of the communication among so many cores will be a key factor in determining overall system performance. To this end, we developed an Omnest-based simulation environment that enables studying the impact of an HPC machine's communication subsystem on the overall system's performance for specific workloads.

Full system simulation at an abstraction level that still maintains a reasonably high level of detail is infeasible without resorting to parallel simulation, the main limiting factors being simulation run time and memory footprint. By applying Parallel Discrete Event Simulation techniques, the power of modern parallel computers can be exploited to great effect to perform these kinds of simulations at large scales.

11.1 Introduction

High-performance computing (HPC), commonly referred to as "supercomputing" in popular parlance, has become a pervasive tool for product development in many industries, e.g., in the design of automobiles and airplanes, in the development of pharmaceutical products, for reservoir discovery in the oil and gas industry, for

Cyriel Minkenberg · Wolfgang Denzel · German Rodriguez · Robert Birke
IBM Research — Zurich, Säumerstrasse 4, 8803 Rüschlikon, Switzerland
e-mail: `sil@zurich.ibm.com`

producing films with computer-generated images in the entertainment industry, or for economic forecasting in the financial industry.

Moreover, HPC is an indispensable means for cutting-edge academic research in many disciplines of science, including weather and climate research (e.g. weather prediction, global warming), geology (e.g., earthquake prediction), life sciences (human genome, DNA sequencing), chemistry and physics (for modeling of various processes at the molecular level, e.g., protein folding), and astronomy.

Another important field of application for HPC is that of *analytics*, which deals with analyzing *big data*, i.e., massive, complex, constantly changing, and often *unstructured* sets of data, for instance in the context of business optimization or healthcare. An example of such a system is IBM's Watson, which competed in and won a special challenge in the US television quiz *Jeopardy!* in February 2011.

Twice a year, namely at the International Supercomputing Conference (ISC) and at the International Conference for High Performance Computing (SC), a list of the 500 most performant supercomputers (www.top500.org) is compiled and published. They are ranked by their performance according to the LINPACK benchmark in solving a large, dense system of linear equations. Performance is typically expressed in billions of floating point operations per second (gigaflops); the top machines of today achieve peak rates of several millions of gigaflops, elevating them to the rank of *petascale* machines.

The next major challenge in the field of HPC is to build a so-called *exascale* machine capable of executing at least one quintillion (10^{18}) floating-point operations per second, or, equivalently, 1 billion gigaflops. Designing such a system is in itself a task with a complexity worthy of such computing power. In this chapter, we will present an approach to the design of HPC systems that is based on discrete event simulation to model crucial parts of the system and their interactions, with the ultimate objective of helping guide system design by determining the impact design decisions pertaining to the communication subsystem have on the application-level performance.

11.2 Design of HPC Systems

11.2.1 The Age of Ubiquitous Parallelism

The performance of HPC systems, as demonstrated by the data collected for the Top 500 lists since 1993, has grown at an exponential rate with remarkable regularity. For a long time, HPC systems could simply exploit CMOS scaling according to Moore's Law, taking advantage of the ever-decreasing feature sizes of subsequent CMOS technology generations. Roughly every two years, the scaling of each technology generation resulted in a) doubling of the integration density, b) an increase in clock rates by about 40% (mostly because smaller dimensions imply lower latency), and c) roughly a halving of the active power consumption.

In addition to the gains purely due to pure technology scaling, Pollack's Rule states that the increase in microprocessor performance due to micro-architecture advances is roughly proportional to the square root of the increase in complexity, i.e., the processor logic (area). In contrast, the increase in power consumption is roughly linearly proportional to the increase in complexity.

However, two main trends have caused a significant slowdown in the advancement of single-thread performance in recent years.

First and foremost, the half-century trend postulated by Moore's Law in 1965 is coming to an end. The physical limits of silicon CMOS scaling are drawing nigh – the gate oxide thickness of a transistor in the 22-nm CMOS process is in the range of 0.5-0.8 nm, which is equivalent to the diameter of just a few atoms, implying that the gate thickness cannot be scaled down further. This in turn implies that voltage cannot be scaled down further significantly, which means that the power density of CMOS circuits (in Watts per area), which thus far has remained basically constant, will increase dramatically, exacerbated by increasing passive power leakage. This power density increase not only induces massive challenges in terms of dissipation (cooling), but also constrains the clock frequency, because the active power scales linearly with frequency. As a result, processor clock rates have barely increased for a number of years now.

Second, realizing higher single-thread performance by means of micro-architectural innovations has become increasingly difficult. Instead, the vast majority of the additional transistors becoming available with each technology generation have been invested in "simply" replicating the CPU multiple times onto the same die, giving rise to the now ubiquitous *multi-core* processor.

As the single-core computational performance has not budged much, HPC performance scaling in recent years has mainly been achieved through massive parallelism. Processor chips nowadays have 6, 8, 12, or even 16 cores. With tens of thousands of such multi-core units, the largest machines feature hundreds of thousands of cores—with the #1 system as of June 2011 having no fewer than half a million cores.

This truly massive level of parallelism implies that the means by which all of these processors are connected has become a crucial factor in the overall system performance. At the intra-node level, communication is still largely performed via busses, but between different computing nodes, packet-switched networks are widely being used. The design of these networks is the key to unlocking the full potential of parallel computers at the peta- and, in a number of years, the exascale.

11.3 End-to-End Modeling Approach

Traditional performance evaluation of parallel computers most often focuses either on small-scale compute-centric models including a high level of detail of compute nodes and workloads (sometimes even resorting to execution-driven simulation), or on large-scale communication-centric models including detailed network models, but highly simplified representations of nodes and workloads. Our approach attempts to strike a balance between these two extremes.

By *end-to-end* modeling we mean that the full system is represented, from workloads through communication libraries and protocols down to networking hardware, but always with a focus on the end-to-end performance experienced by the workloads.

11.3.1 Traditional Approach

The design of HPC systems relies to a large extent on simulations to optimize the various components of such a complex system. The three key hardware components that determine system performance are the following:

- Processor (CPU): instruction set architecture, number of CPU cores, CPU clock rate, bus rate, number of threads per core, arithmetic logic units, floating point units, other execution units (superscalar, SIMD), pipelining, out-of-order execution, etc.
- Memory hierarchy: memory size, memory bandwidth, number of cache levels and cache sizes, cache coherence protocol, integration on CPU die (caches, memory controllers), etc.
- Interconnection network: network interface bandwidth, link bandwidth, network topology, network technology, routing algorithm, deadlock prevention, etc.

Examples of tools to evaluate processor performance are MAMBO (Peterson et al. 2006) or SIMICS (Magnusson et al. 2002). These tools can perform cycle-accurate simulations of entire applications at the instruction level. They are very valuable for evaluating the impact of micro-architectural changes on the performance of specific applications. However, such a level of detail prevents the scaling of this type of simulation to large systems. Similar tools are available to evaluate the performance of the memory hierarchy.

The interconnection network of an HPC system is usually modeled at a higher level of abstraction, resorting to discrete-event simulation to enable scaling to systems with many thousands of network ports. The purpose of interconnection network simulation is to optimize the network topology, the switch and adapter architectures and parameters, scheduling and routing policies, the link-level flow control mechanism, and the end-to-end congestion control mechanism. This type of simulation is commonly of the "Monte Carlo" variety, i.e., the workload applied is of stochastic nature, generated synthetically with random destination and inter-arrival-time distributions rather than by real applications.

11.3.2 Taking the Application View

Although such simulations are useful in determining load-throughput and load-delay characteristics, they are not necessarily a reliable performance indicator for the communication phases of specific applications. Therefore, the instruction-level processor simulation, although accurate, does not scale to the desired system sizes, whereas the interconnect simulation does so, but suffers from unrealistic stimuli. Bridging this gap is the key to enabling true end-to-end full-system simulation.

HPC workloads exhibit two basic characteristics that are fundamentally different from the synthetic workloads generally used in performance studies of interconnection networks:

- Synthetic workload generators are not reactive; their injection behavior does not depend on current or previous network conditions. HPC workloads, on the other hand, typically are highly reactive. As communications are often blocking, a communication that is delayed because of network contention will cause the corresponding thread(s) to block until it has completed. This *self-throttling* behavior implies that the traffic injection rate will, to some extent, automatically adapt to network conditions. In addition, HPC applications typically proceed in a highly synchronized fashion, with alternating communication and computation phases, where the next computation phase needs to wait until all communications from the preceding communication phase have completed. This synchronization is usually achieved by means of so-called *barrier* operations, and these barriers also act as a global means for self-throttling.
- Synthetic workload generators are *competitive*. Each traffic generator is considered an isolated entity that is primarily interested in its own benefit, possibly at the expense of others. HPC workloads, on the contrary, are *collaborative* in nature, at least when considering tasks that belong to the same job. The main HPC performance metric is neither network throughput nor mean message latency, but the execution time, i.e., the time from start until the slowest thread has finished. This means that the individual threads have an interest to balance their computation and communication needs such that they all proceed at roughly the same pace.

To some extent, this discrepancy is due to historical circumstances, as network design and modeling originated for a large part in long-range telecommunications and wide-area data networks, where a high degree of statistical multiplexing is assumed, so that the impact of the individual contributor (at the network's edge) is negligibly small.

One class of approaches to capture the behavior of HPC traffic injection and bridge this gap employs *trace-driven* simulation. Rather than by an exact model of its behavior, an application is represented by a *post-mortem* trace, collected during its execution on a real parallel computer. Such a trace contains two basic kinds of records: computation and communication records. Computations are represented only by the amount of CPU time they consumed, not by the kind of operation(s) actually performed. Communications are represented by their key parameters, including source and destination thread, message size, start and end times, communication operation and communication mode.

During simulation, a playback engine replays the trace, taking into account the semantics of the communication operations for a given parallel programming model. Computation records are transformed into delays between subsequent communications. Communication records are transformed into data messages that are fed to a model of the interconnection network. To ensure accurate results, the simulation should preserve causal dependencies between records, e.g., when a particular computation depends on data to be delivered by a preceding communication, the start of that computation must wait for the communication to complete. As many scientific HPC applications are based on the Message Passing Interface (MPI), tracing MPI calls is a suitable method for characterizing the communication patterns of an important class of HPC workloads. This approach is adopted in the two projects presented in Sections 1.5 and 1.6 of this chapter.

11.3.3 Model Components

The simulators developed for these two projects share a common high-level structure. The main components are i) workload, ii) compute node, iii) interconnection network, and iv) simulation statistics and control.

Workloads are represented in four different ways:

i. Stochastic traffic generators, which generate random traffic patterns according to configurable probability distributions for inter-arrival times, traffic destinations, and message or burst sizes.
ii. Deterministic traffic generators, which generate traffic according to predetermined temporal and spatial patterns. Examples are permutation patterns, TDM patterns, hotspot patterns, etc.
iii. Workload models, which model the behavior of applications or their underlying communication libraries, typically focusing on specific parts that are especially communication-intensive.
iv. Application traces, which are collected from real executions and played back, using an integrated or separate trace-replay engine.

The compute node is represented by abstract resource models, which represent node resources such as CPUs (cores), memory (banks, bandwidth, latency), buses (number, bandwidth), and IO (interfaces, bandwidth).

As the focus of our study is on the effect of the system-wide interconnect, the latter is modeled at a suitably fine level of detail. We chose the abstract level of the so-called *flow-control digit*, or *flit* for short. In essence, this is the atomic unit of transfer across a communication link, which may vary across different networking technologies. In cell-switched networks, each cell can be considered a flit, whereas in packet-switched networks, the packets may or may not be further subdivided into smaller units. For instance, in Ethernet networks, frames (of variable lengths of up to 9 KB) cannot be subdivided and are therefore uninterruptible, whereas in so-called *wormhole-switched* networks, which are often used in HPC, variable-length messages are subdivided into smaller units, allowing transmission of a message across a link to be interrupted in mid-flight.

This abstraction level allows the modeling of all important networking aspects, including flow control, routing, buffering, contention resolution, and scheduling policies, without having to resort to even lower abstraction levels (byte or bit level) that would significantly increase the simulator complexity and simulation runtimes, without resulting in deeper insights. The interconnect model comprises two basic module types, namely *adapters*, which form the interface between the compute nodes and the network, and *switches*, which form the network itself.

11.3.4 Tools: Omnest

Our simulators are based on a third-party discrete event simulation package called *Omnest*, which is the commercial version of the academic OMNeT++ package (www.omnest.com and www.omnetpp.org, respectively). In addition to a rich API to create discrete event models, Omnest includes an Eclipse-based integrated development environment featuring rich functionality for post-processing and visualizing simulation results. Omnest contains a model topology specification language (expressed through network description `ned` files that can be loaded dynamically) that, by means of polymorphism, enables highly versatile models. Of particular importance is Omnest's built-in support for parallel distributed simulation, which enables relatively easy execution of an Omnest model on top of the MPI, see Sec. 11.7.

11.4 Computer Networks

As we use discrete event simulation as a tool for research on interconnection networks in an HPC context, this section provides a brief introduction to such networks. HPC networks must be able to interconnect thousands to hundreds of thousands of processing elements in a way that is feasible, achieves high performance, and, ideally, minimizes cost compared with other alternatives providing similar performance.

11.4.1 Network Topologies

The ideal way to connect any number of CPUs is to provide a single link from each CPU to each other CPU in the system. This topology is typically called a *fully-connected mesh* network or, more simply, *full mesh*, and its graph representation is called a *complete graph*. For instance, to connect four nodes, six links would be required (see Figure 11.1). To connect N nodes, we need $N\times(N-1)/2$ links (cables).

To connect a modest machine with 1,024 nodes, we would need 523,776 links, and each node would need 1,023 network interfaces. The cost and difficulty of wiring this network quickly exceed the cost of the processing elements themselves.

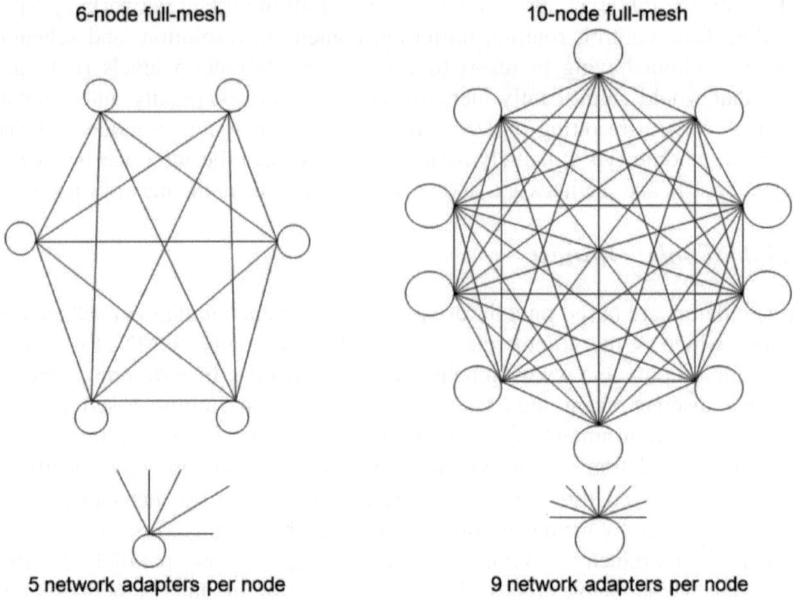

Fig. 11.1 Top: small and big full-mesh; bottom: small and big crossbar.

If we assume that the CPUs will only send data to a single destination at a given time, we could connect each CPU to a *switch,* a device with several inputs and outputs, similar in concept to telephone exchange boards, which will establish the connections across the communicating CPUs. This approach greatly reduces the complexity of the network compared with a full mesh, but still is challenging to design. Connecting N nodes without any possibility of blocking still incurs a complexity proportional to N^2, but only one network interface is needed per node. If blocking (i.e., waiting for a connection to be finished before another one is established) is allowed, the complexity can be reduced. It is common practice to use non-blocking switches (called *crossbars*) having a moderate number of ports as building blocks for other network topologies. In current CMOS technology, typical single-switch port counts are in the range of 16 to 64 ports.

As illustrated in the bottom panel of Figure 11.1, in a crossbar the nodes are no longer directly connected (as in Figure 11.1 top), but *indirectly* connected through one or several intermediate switching stage(s). This distinction brings us to one of the main classifications of network topologies: direct and indirect networks. In *indirect* networks, the terminal computing nodes act exclusively as sources and sinks of packets but do not participate into the forwarding of packets, i.e., they do not act as a router, whereas in *direct* networks they do both (Dally and Towles 2004). Because in direct networks, switches and compute nodes are often integrated on the same chip, the hardware complexity of each individual switch is necessarily rather limited. This implies that direct networks typically feature low-radix switches, whereas indirect networks, in which each switch is a discrete component (or even box), are usually built using high-radix switches. The recent

Dragonfly topology (Sec. 11.4.4) is a one of the first proposals for a high-radix direct network.

Finally, there are two important practical properties regarding network design, i.e., that networks are regular and partitionable. The first property is important because if the basic components are identical, they can be mass-produced, reducing the cost. The second property is useful for scalability (deploying a small network initially, but being able to scale it up in the future) and to be able to share a single large machine among many workloads, assigning a sub-network with similar topological and performance properties as the entire network to each workload.

For a network, performance comparisons are not possible without referring to the actual kind of traffic it has to deliver. To facilitate comparison, a metric that is generally used and employs only topological properties is used: the bisection bandwidth. The bisection bandwidth of a network is the bandwidth between two equal parts of the network. It is a useful metric assuming that each node sends data to some other node in a uniformly distributed fashion, i.e., the destinations are uniformly distributed. For this kind of traffic, it is a very important estimator of the performance of the network. The bisection bandwidth of a full mesh with N nodes and links with a bandwidth of R bits/second equals $R \times (N^2/4)$ if N is even or $R \times (N^2 - 1)/4$ if N is odd. The bisection bandwidth of a crossbar with N nodes is $R \times (N/2)$.

11.4.2 Indirect Networks: Fat Trees

From the family of indirect networks, fat-tree-like networks are the most popular ones in current supercomputers. A fat tree is a multi-stage tree-like topology in which the widths (bandwidths) of the connections increase towards the roots of the tree, much like the branches of biological trees become thicker from top to bottom. An ideal fat-tree topology is shown in Figure 11.2. Square boxes represent switches, whereas circles represent computing nodes.

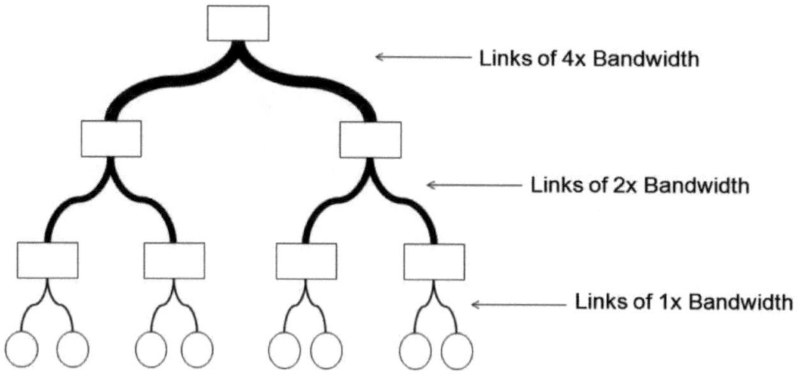

Fig. 11.2 Idealized fat tree.

For large clusters, however, idealized fat-tree networks are not realizable, as the bandwidth requirements per link are multiplied with the trees radix at each level towards the root, which would lead to an exponential increase in either the number of ports per switch or the bandwidth per link. The CM-5 (Leiserson et al. 1992) was the first machine to implement a modification of the idealized fat-tree network. The main advantage of a fat tree is that any half of the network can communicate with the other half of the network without experiencing any contention (blocking). A communication between two nodes is performed by sending the message from the source node up to the nearest common ancestor (NCA) of the two nodes and then down to the destination node.

Figure 11.3 shows how a tree-like network with the same bandwidth as a fat tree could be constructed using identical switches, i.e., without increasing the number of ports per switch or their bandwidth at each level. Note that in Figure 11.4 there are now two overlapped trees from each top-level switch.

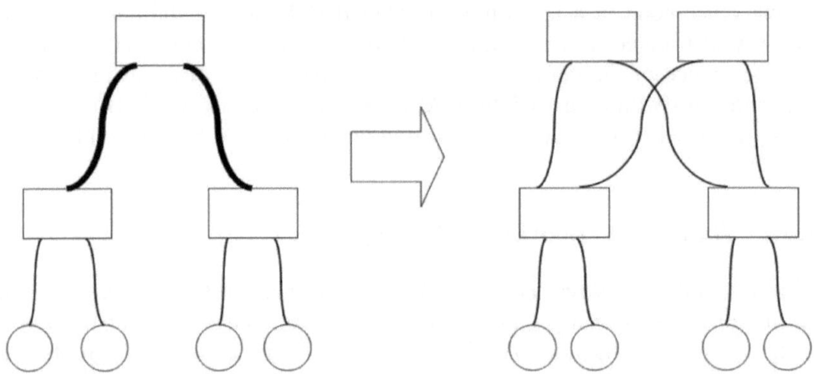

Fig. 11.3 Idealized to realizable fat tree.

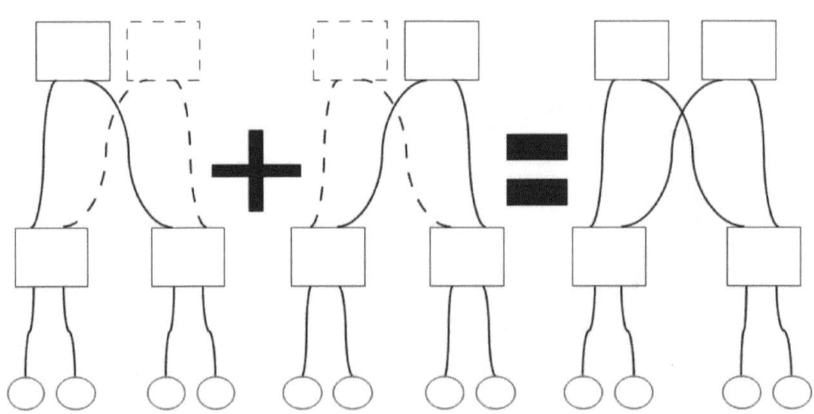

Fig. 11.4 *k*-ary *n*-trees as a composite of two trees.

The idea shown in Figure 11.3 can be applied recursively to all levels of a fat tree. If the connections are arranged in a certain manner, all crossbar switches have the same number of ports, and the bisection bandwidth is retained, then the resulting network belongs to the popular class of *k*-ary *n*-trees (Petrini and Vanneschi 1997).

A formal representation of the broadest class of such multi-tree networks resembling fat trees is the class of Least Common Ancestor Networks (LCANs, Scherson and Chien 1993)

Another formalization of multi-tree networks, less general than LCANs, but still very broad, is the class *extended generalized fat tree* (XGFT) topologies (Öhring et al. 1995). This class covers most tree variations proposed in the literature with a very compact notation.

The property of full bisectional bandwidth provided by *k*-ary *n*-trees generally ensures good performance, but incurs significant cost in terms of switch hardware and cabling. As these costs represent an increasing fraction of the overall system cost, the prospect of trading a modest reduction in performance for a significant *slimming* of the topology is quite attractive (Desai et al. 2008). Perfectly suited to this task are XGFTs with any kind of slimmed (or fattened) tree topology. Slimming (also known as oversubscription) implies that the bisection bandwidth decreases towards the roots, which is achieved by providing more downward ports (towards the leaves) than upward ports (towards the roots). In principle, XGFTs can also describe *fattened* (or overprovisioned) networks, which provide increasing bandwidth towards the roots. A *k*-ary *n*-tree is a particular case of an XGFT with constant bisection bandwidth.

Advantages of *k*-ary *n*-trees are that they can be built using same-radix switches while providing the bisection bandwidth of an idealized fat tree. Keeping a constant bisection bandwidth at each level implies that any permutation traffic pattern, in which the destinations nodes are a permutation of the source nodes, can in principle be routed in such a way that any half of the network can communicate with the other half without contention. There is always a *routing*, i.e., an assignment of paths to communicating pairs, such that there is no contention among any of the communicating pairs in the permutation. In other words, all of the assigned paths are completely edge-disjoint.

Several works have suggested that full-bisection *k*-ary *n*-trees provision more bandwidth than required for certain common HPC traffic patterns (Kamil et al. 2005, Desai et al. 2008), implying that network cost could be reduced without incurring significant performance reductions. Consequently, "slimming" *k*-ary *n*-trees has been proposed to design variations on this topology with fewer switches. One example of employing discrete event simulation in this context is to study the effect of such slimming, using for instance XGFT topologies, on the performance of HPC workloads of interest, see Sec. 11.6.4.

The Myrinet[1] interconnect of the Mare Nostrum supercomputer is an example of a fat-tree network. Similarly, the IBM Roadrunner machine at the Los Alamos National Laboratory features an InfiniBand-based fat-tree topology.

11.4.3 Meshes and Tori

A mesh is a multi-dimensional grid-like network topology in which each node has 2*N* neighbors per dimension. Figure 11.5 shows several meshes with increasing number of dimensions. Meshes can be "square" or "rectangular", depending on whether all dimensions are of the same length. In meshes, a particular node does not always have the same average distance to the rest of nodes. Nodes situated at the borders have a higher average distance and a different "view" from their perspective of the whole network than the interior nodes.

In a torus, "wraparound" links connect the edge nodes at opposing ends of each dimension, creating rings of nodes. A torus is a node-symmetric network: Each node has the exact same view of the network. Average distances from any node to the other nodes are identical. A one-dimensional torus is a ring.

Meshes and tori, particularly low-dimensional ones (2D and 3D), require low-radix switches (the switch radix equals twice the number of dimensions), are easy to scale (no or only very few long links are needed), and are partitionable: any mesh or torus of $N + 1$ dimensions can be partitioned into several meshes or tori of N dimensions.

An example of an HPC architecture featuring a torus interconnect is the IBM Blue Gene[1] class of machines. The Fujitsu "K" computer at the RIKEN institute is based on a six-dimensional torus.

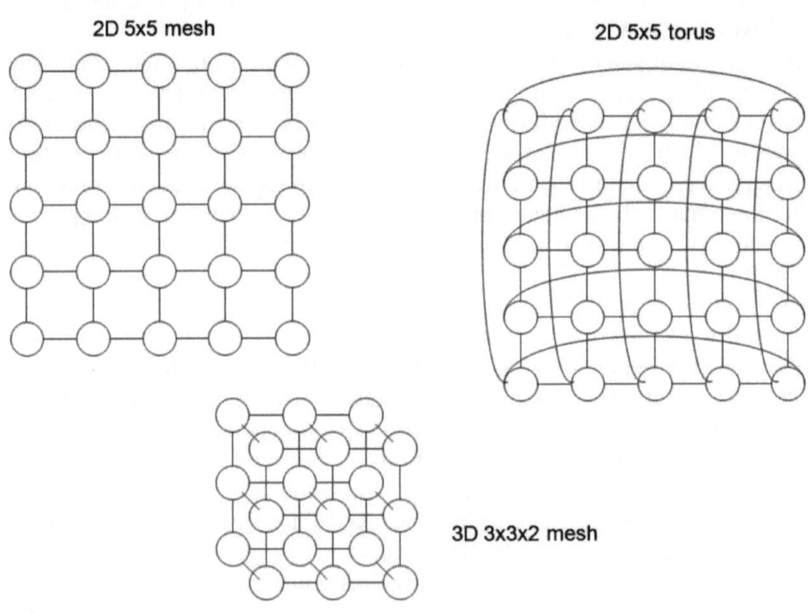

Fig. 11.5 Two- and three-dimensional mesh and torus topologies.

11.4.4 Dragonflies

A dragonfly (Kim et al. 2008) is a hierarchical network that has a fully meshed (complete graph) connection pattern at each level. The network as a whole is not a full mesh, but groups at any particular level of the hierarchy are connected as a full mesh, as shown in Figure 11.6. In a dragonfly, each switch has three kinds of links: i) links connecting to the end nodes, ii) local links, connected to all other switches in the local group, and iii) global links, which connect the local group to all other groups. When all the global links connecting the switches of a group are considered together, they connect all groups in a fully-meshed fashion. A dragonfly can be described by the following three parameters relative to the switch: the number of node ports p, the number of switches per group a, and the number of groups h. Each switch requires $(p + (a - 1) + h)$ ports. Each local group has a switches, which altogether connect to $a \times h$ other groups. The total number of nodes therefore equals $(a \times h + 1)\, a \times p$.

The Dragonfly topology has been adopted by the IBM PERCS class of machines, see Sec. 11.5.

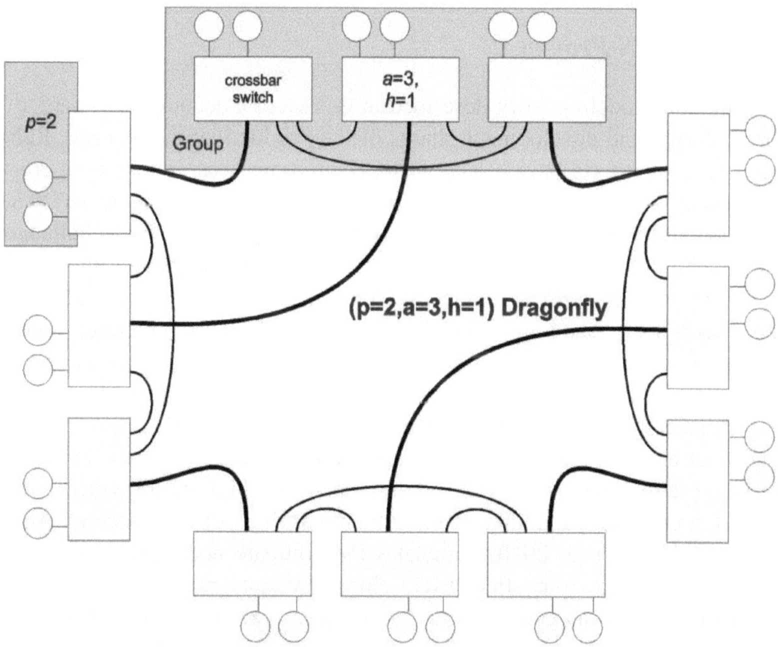

Fig. 11.6 Dragon topology with $p = 2$, $a = 3$, $h = 1$.

11.4.5 Deadlock

Depending on the paths chosen to route the messages from their sources to their destinations, resource-dependency cycles can occur in the network. If the network by itself has no cycles, and the routing does not introduce cycles, deadlocks are not possible or easy to avoid, as is the case for shortest-path routing in fat-tree networks.

If the network topology itself contains cycles, as in a torus, deadlock avoidance techniques are necessary. If a deadlock occurs, a part of the network will no longer be able to forward messages, which generally quickly leads to a network-wide standstill, which may require a reboot of the entire machine. Avoiding deadlocks is therefore an absolute must.

Dragonflies and torus networks have physical link cycles that can easily lead to a deadlock situation. These cyclic dependencies can be broken by means of adding *virtual channels* in conjunction with appropriate routing policies. Examples of deadlock-avoidance mechanisms are *dateline routing* and *bubble injection.*

11.5 Case Study 1: PERCS Simulator

11.5.1 PERCS Project

The simulation modeling work described in this section accompanied parts of the research, design and development phases of the Productive, Easy-to-use, Reliable Computing System (PERCS). The PERCS program was initiated in 2001 as IBM's response to the Defense Advanced Research Projects Agency (DARPA) High-Productivity Computing Systems (HPCSs) program. The challenge of building a highly productive HPC system forced IBM to pursue an integrated approach to system construction by tightly integrating all aspects of an HPC system, such as packaging, power delivery, cooling, architecture, topology, interconnect, communications software, compilers, operating system, and even programming language.

The PERCS system (Rajamony et al. 2011) uses POWER7[1] microprocessors (Sinharoy et al. 2011) arranged on a quad-chip compute node that runs a single operating-system image managing 32 homogeneous high-performance compute cores with a compute capability of more than 900 gigaflops per second. An IBM hub chip (Arimilli et al. 2010) completes the compute node, providing network connectivity to the four POWER7 chips by integrated network adapter functionality and an integrated switch with a peak switching bandwidth of more than 1.1 TB/s. This switch serves not only as a gateway into the network for the

[1] IBM, Blue Gene, POWER7 are trademarks of International Business Machines Corporation, registered in many jurisdictions worldwide. Myrinet is a registered trademark of Myricom, Inc. Intel and Intel Xeon are trademarks or registered trademarks of Intel Corporation or its subsidiaries in the United States and other countries. Other product and service names might be trademarks of IBM or other companies.

processors, but also as a router for traffic between other compute nodes in a direct interconnect topology. Therefore, without requiring costly external switches, a full PERCS system of up to 16,384 compute nodes with more than half a million compute cores can be constructed by using a two-level direct interconnect topology of the dragonfly type (Sec. 11.4.4) that fully connects every element in each of the two levels.

During the early studies for the PERCS system, it became clear that, in addition to various established component simulation efforts, end-to-end full-system performance modeling by means of event-driven simulation with a strong focus on the interconnection network and in conjunction with a realistic workload model is indispensable to evaluate system design options and to help optimize the performance of the compute nodes, the interconnection network, and eventually the entire system, including system software and HPC applications (Denzel et al. 2010). The wide range of ideas and options that needed to be considered during the project required a very flexible simulator. Moreover, the unprecedented system scale required a highly efficient simulator with built-in support for distributed parallel simulation. We selected the Omnest framework as a suitable basis for our simulator.

11.5.2 PERCS Compute Node Model and Interconnect

Figure 11.7 shows a simplified overview of the PERCS HPC system model used for our Omnest implementation. The model reflects in a one-to-one manner the module structure and module nesting of the real hardware including simulation modules for the *Compute Nodes* containing the four *Processor* and the *Hub* submodules. At the next higher nesting level, eight compute nodes are incorporated in *Drawer* modules, and another nesting level higher groups of four drawer modules constitute *Supernode* modules. The interconnection network of a PERCS system with up to 512 supernodes is formed by the direct link wiring between all compute nodes. There are two logical-link hierarchy levels using three different link technologies, offering the best compromise between required link length, bandwidth and cost, respectively. The first logical hierarchy level consists of the so-called *L* (*local*) links, which fully interconnect all 32 compute nodes of a supernode via seven electrical intra-drawer links and 24 optical inter-drawer links. The second logical link hierarchy consists of the so-called *D* (*distant*) links, which fully interconnect all supernodes via a total of 512 optical links per supernode, i.e., 16 links per compute node. The largest system topology of 512 supernodes has only one D link between any pair of supernodes, whereas in smaller systems multiple D links may be used between any pair of supernodes. An end-to-end shortest path between two compute nodes may require up to three routing hops (L-D-L). Indirect paths with up to five hops (L-D-L-D-L) are also allowed.

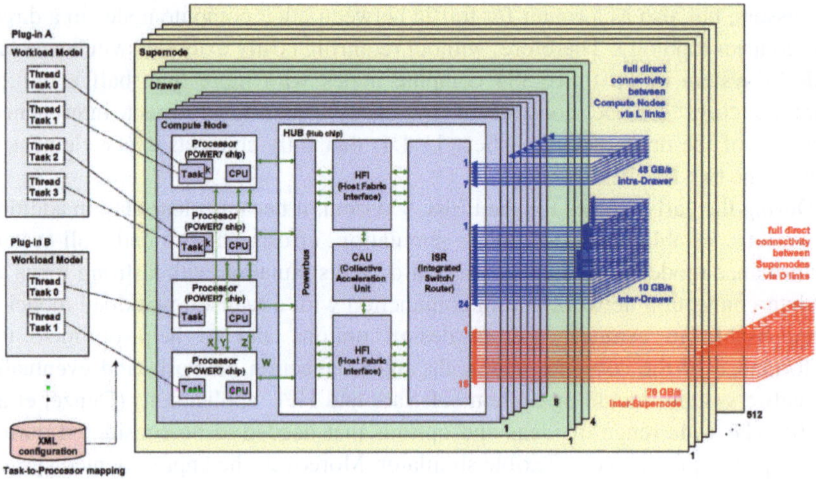

Fig. 11.7 Overview of the PERCS simulation model.

The *Hub* module serves as network adapter for the four processor chips of the compute node. This functionality is shared by two host fabric interface (*HFI*) submodules, which support a total of eight data ramps into and out of the network. The key modeled function of these HFIs is the sending and receiving of packets to/from the compute node's memory in segments of 128-byte flits. The hub chip also contains a collective acceleration unit (*CAU*), not covered in detail here, with special support for accelerating frequently used collective communication (CC) operations such as barrier synchronization or global sum. The key network component is the integrated switch/router (*ISR*) that routes flits between its 55 switch ports, namely, eight HFI ports, seven intra-drawer L link ports, 24 inter-drawer L link ports and 16 D link ports. The ISR is a packet switch module with input and output FIFO buffers at each port, whereby the L port buffers are logically split into three separate virtual channel (VC) partitions and the D port buffers into two. This VC configuration is required for proper deadlock-free operation in the dragonfly topology for routes of up to five hops. The arbitration of the crossbar fabric between the input and output buffers is relatively complex, having to take into account the desired route, the availability of flow-control credits, buffer occupancies, rules for changing VCs for deadlock prevention, the requirement that flits of the same packet must not be interleaved within a given VC, etc. As all these details can have significant impact on network throughput and latency, we had to select a modeling abstraction level that includes all of them. Furthermore, as flits are the smallest transmission units queued, arbitrated, moved, or transmitted contiguously and non-interleaved, we chose to model at flit-level granularity to produce sufficiently realistic results on the one hand and to avoid the unnecessary simulation overhead associated with lower-level byte-wise handling on the other hand.

For the compute node model, the large system scale precludes the use of a very detailed model of the processor chips with their memory. Nevertheless, the throughput and delay performance of node-internal communication between tasks running on the same processor or on different processors of the same compute node as well as the communication across the interconnection network are determined by the inherent performance of the shared processing, memory, and intra-node connectivity resources (busses) and by the queuing and service disciplines (FIFO or processor sharing) used by these resources. Hence we opted for a simple resource-based model represented by a *CPU* module that models these resources and their utilization by the computation or communication activities of parallel application tasks running on the considered processor.

A parallel application is typically specified in terms of a job comprising parallel tasks that perform computations and communicate with each other via a communication protocol such as MPI. Hence our processor modules also contain a *Task* module for each workload task executing on a particular processor. These task modules are created dynamically at simulation initialization time as required by the workload to be simulated. Its task-to-processor mapping is specified in an XML configuration file for each particular simulation run. External workload models are represented by one or several code *plug-ins*. During simulation, each task module requests a next step to execute from its plug-in. The plug-ins respond with the next step, which the task modules then have to handle. While handling a next step, a task module may need to request, possibly compete for, and use computing resources from processor cores as well as transmission bandwidth resources from the intra-node busses and memory bandwidth from the memory, which are all modeled in our CPU resource module.

Workload jobs are modeled by dynamically loadable and exchangeable code plug-ins, i.e., pieces of code that act like a parallel application job with multiple tasks running in individual execution threads. Through the job and task mapping specified in the XML configuration file, the task modules inside the processor modules know which plug-in to load initially and from which plug-in thread to request the next action to handle during the simulation run phase.

11.5.3 Plug-In Concept

Like any computer program, a task of a parallel application can be viewed as a sequence of steps at some lower level of abstraction. An MPI program specifies both communication steps for sending/receiving messages to/from other tasks as well as computation steps, i.e., the gaps between the communication steps. A reasonable workload model should represent a semantically correct sequence of the steps and gaps at the MPI call level. This sequence could be specified to the simulator by MPI trace files captured from an application run on a similar real system. For certain investigations, we have used trace files and corresponding trace-reader plug-ins that simply execute the calls for a next step as calls for a next record from a trace file.

However, trace-driven simulation has its limitations. Using traces does not scale well, becoming unwieldy at larger scales. A drawback is that a set of recorded trace files can only represent a specific run on a specific platform with a specific task count, a specific task to processor placement, and a specific underlying MPI library implementation. A real algorithm, in an MPI code or MPI library, exhibits adaptations or variations coming from the current environment, from the current task placement, and the currently used MPI library, whereas a trace file only reflects what happened on the environment where it was captured. For example, the implementation of a collective communication algorithm may be topology-aware and result in a different trace depending on where the pairs of communicating tasks are placed in the topology. To study software implementation options exposed or prone to such effects, we use an alternative way of specifying workload steps to the simulator, one that does not depend on trace files and one that adapts itself to the scale and to the environment, much like tasks of a real application do.

This alternative way of specifying workload steps to the simulator is via plug-in code that is an abstracted form of the real MPI application code or a fragment of interest, such as the implementation of a collective communication algorithm. This code is abstracted down to a level that includes only point-to-point communication steps and computation steps. For example, *MPI_Send*, *MPI_Recv*, *MPI_Isend*, *MPI_Irecv* and *MPI_Wait* calls in a real application are represented by *SIM_Send*, *SIM_Recv*, *SIM_Isend*, *SIM_Irecv* and *SIM_Wait* calls in plug-in code, and a *SIM_Pause* call is used to model the computing time of a computation activity in a real application. Plug-in code is multi-threaded, just like the parallel application it models. Thereby each execution thread represents a task of the application modeled. Plug-in code can be written by application developers or developers of MPI library components much in the way they are used to writing MPI code, taking into account simple plug-in author guidelines, without any need to understand either the simulator or how to interface to the simulator. This is accomplished by using a small set of specific *SIM* calls and by using predefined infrastructure code that deals with the set-up of the multi-threading infrastructure (based on POSIX pthreads) of the plug-in and its interfacing with the simulator. Plug-ins can be exchanged easily, and several identical or different plug-ins can be plugged into the simulator concurrently.

The simulator only needs to support the semantic actions for the small set of defined plug-in operations issued by a plug-in on the request for the next step. A semantic action might, for example, be an eager blocking send action that sends the message body and waits for an acknowledgement from the destination before requesting the next step, or it might be a rendezvous blocking send action that first sends a message envelope and waits for an ok-to-send from the destination before proceeding like an eager send. We chose to focus on making the models of this small number of elementary MPI actions as accurate as practical and on providing a means to build complex patterns on top of these few primitives. A collective communication algorithm involves highly organized patterns of point-to-point communication, so using plug-ins to model such algorithms allows better models of specific algorithms and opens the door to using simulation for tuning them. The detailed implementation of a collective communication algorithm is thus modeled in a plug-in rather than in the simulator. Different implementations would simply be represented by different plug-ins or different library functions used by a plug-in.

Because of the flexibility of the plug-in concept, we are not restricted with respect to the plug-in contents. As mentioned above, a plug-in could also contain code that represents an MPI trace reader. Alternatively, a plug-in could represent a set of statistical traffic generators, which might be sufficient for throughput studies of network topologies. In this case, we use a *SIM_Inject* function, for example, that can inject a message into the simulator without MPI semantic, i.e., without requiring a matching receive call at the destination.

11.5.4 Sample Results

In the early phase of the PERCS project, we used the simulator to assess different possible network topologies ranging from fat-tree variations to dragonfly topologies with various switch module architectures and link bandwidth assumptions (Denzel et al. 2010). The modular modeling approach supported by the Omnest simulation framework was very helpful for quickly and efficiently accomplishing the numerous modifications required for the topological changes or adaptations in module functionality. Topology changes were limited to rewriting network description files in the Omnest NED language, whereas functional adaptations, e.g. in switch arbitration, could be limited to modifying C++ code in the corresponding Omnest module. The parallel simulation support of Omnest allowed us to gradually grow our models up to the maximum size of the current PERCS system. Over time, the simulator code itself stabilized, and modifications were essentially limited to parameter changes and writing new plug-ins.

For the very early simulations, we used workload models for statistical uniformly distributed all-to-all traffic of fixed-length short messages modeling the GUPS (Giga updates per second) benchmark from the HPC Challenge suite (Luszczek et al. 2006), an important performance measure for HPC systems. Without our having to struggle with a huge number of trace files, the statistical workload model allowed simulations up to the full PERCS system size of 512 supernodes arranged in 32 parallel Omnest partitions each covering 16 supernodes. In this way, we could not only verify the simulator and its scalability, but were also able to obtain and validate analytically predicted GUPS performance results for various fat-tree and dragonfly system configurations. We could also validate a predicted discontinuity effect at a very large system size above which the first-level links became the system bottleneck, whereas below that system size the second-level links were the system bottleneck.

To apply realistic workloads, we began to run trace-driven simulations early on by using MPI traces from scientific applications such as HYCOM, LAMMPS, UMT2K and others. Because of the relatively limited size of the available traces (in terms of number of tasks), we could not really exploit the system and simulator scale, but obtained a number of useful results nevertheless. For example, we could show that a slimmed fat tree with hierarchy levels that have only 50% of the bandwidth of the preceding level would have been a good cost/performance compromise, but more slimming should be avoided. In another study, we investigated indirect versus direct routing in dragonfly topologies when using application traces. We evaluated the performance impact of using indirect (five-hop) routes between supernodes, i.e., when each route visits an intermediate supernode rather than taking the only direct route between the source and destination supernode. This study revealed that the concurrent use of a modest number of indirect routes (e.g., eight) is beneficial, i.e., using only one or two indirect routes is not yet beneficial, and many indirect routes are not worth the effort.

In the late phase of the system development, our focus shifted towards studying the behavior of frequently used MPI communication patterns, such as ping-pong communication, neighbor message exchanges in matrices or MPI collective algorithms (e.g. MPI_Alltoall), and their dependency on task placement and communication concurrency, and the impact of algorithmic improvements.

To study the task placement dependency, we considered an application plug-in of 1,024 tasks forming a 32 by 32 matrix, whereby each of the 1,024 tasks concurrently exchanges 256 times four 20,000 Byte messages with its four neighbor tasks in the matrix. In this example, we predominantly have near-neighbor communication. For comparison, we considered an all-to-all pattern as typically performed by an MPI_Alltoall collective call. In this case, each of the 1,024 tasks concurrently exchanges a 20,000 Byte message with each of the other 1,023 tasks. This results in the same total message count as in the previous example, but with predominantly distant communication. Both applications were placed across 1,024 POWER7 processors on eight PERCS supernodes, once using linear task-to-processor placement and once randomly shuffled placement.

As could be expected, the runtime of the 2D neighbor exchange application is much higher for random placement (52.2 ms) than for linear placement (19.5 ms). On the other hand, the runtime of the all-to-all application behaves in the opposite way: It is much lower for random placement (49.2 ms) than for linear placement (273.1 ms). This can be explained by the fact that in the case of linearly placed tasks at any point in time clusters of neighboring tasks are sending concurrently to the same distant clusters of neighboring tasks over the same network links, which appear as temporary bottlenecks. On the other hand, in the case of randomized task placement, all messages of the same cluster of tasks, which simultaneously squeeze through these bottlenecks, are spatially spread across the system, and hence at any spatial point they appear spread also temporarily.

Another experiment demonstrates the potential of algorithmic optimizations. Rather than running one all-to-all process across all 1,024 tasks, we considered running 32 concurrent all-to-all communication processes, i.e., one process per row of the matrix across all columns. The straightforward implementation of the all-to-all algorithm, as it can be found in existing MPI_Alltoall library implementations, revealed a relatively high runtime of 6.78 ms for linearly placed tasks, which is only half the time it would have taken to run the 32 processes sequentially rather than concurrently. In contrast, with random task placement, we obtained a much better runtime of 1.54 ms, because in this case all concurrent processes start with the same starting task, causing a spatially moving contention during the entire time. With a small modification, we were able to improve the algorithm and spread out the contention by shifting the starting tasks of the processes. This resulted in a drastically improved runtime of 1.57 ms, which is not much different from that for random task placement. In this way, we could optimize the algorithm and make it robust with respect to task placement. We could show that this also holds in the presence of background noise traffic generated by another plug-in, although the absolute runtime will roughly double when, for example, a second similar application is running on different cores of the same processors.

11.6 Case Study 2: Venus

The PERCS simulator described in the preceding section was designed specifically for the purpose of evaluating one particular class of systems. To enable performance studies of a much broader range of systems, we developed a more generic full-system simulator, dubbed “Venus”. In this section, we will describe the Venus simulation environment and its uses.

This simulation framework originated from a joint project between the Barcelona Supercomputer Center (BSC) and IBM, in which a follow-on machine to the currently installed *Mare Nostrum* system is being designed under the working title of *Mare Incognito*. The Venus environment was originally created to aid in the design of the interconnection network of this system, but has evolved to a generic interconnection network simulator capable of simulating many different kinds of networks.

Venus was designed to interoperate closely with two tools developed by BSC, namely, *Dimemas* and *Paraver*, which enable a detailed performance analysis of parallel programs.

11.6.1 Tool Chain

This section describes the Venus tool chain and how its constituent tools are integrated.

11.6.1.1 Dimemas

Dimemas is a tool for parametric simulation studies of the performance of message-passing programs. It is an event-driven simulator that reconstructs the time behavior of message-passing applications on a machine model characterized by a set of performance-related parameters.

The input to Dimemas is an execution trace containing a sequence of operations for each thread of each task. Each operation can be classified as either computation or communication. Such traces are usually generated by instrumenting an MPI application, although they can also be generated synthetically. During instrumentation, each computation is translated into a trace record indicating a “busy time” for a specific CPU, whereas the actual computation performed is not recorded.

Communications are recorded as send, receive, or collective operation records, including the sender, receiver, message size, and type of operation.

Dimemas replays such a trace using an architectural machine model consisting of a network of symmetric multi-processing (SMP) nodes. The model has many configurable parameters, allowing the specification of the number of nodes, the number of processors per node, the relative CPU speed, memory bandwidth, memory latency, the number of communication buses, communication bus latency, etc. Dimemas outputs various statistics as well as a Paraver trace file.

11.6.1.2 Paraver

Paraver is a tool to create visual representations of the behavior of parallel programs. Each Dimemas simulation outputs a Paraver trace representing the state of each thread at every time in the simulation, as well as communications between threads, and occurrences of punctual events.

A Paraver trace is a series of records, each one being associated with a specific thread. There are three basic kinds of records: A *state record* specifies the state of a particular thread for a particular time interval. Typically, the state indicates the type of operation (e.g., computation, MPI send, MPI receive, MPI wait, MPI collective, etc.) the thread was engaged in at a specific time. A *communication record* specifies a point-to-point communication between two threads, including physical and logical start and end times, size of the communication, and a tag. An *event record* specifies the occurrence of particular event at a particular thread, including the type of event, the time of occurrence, and an associated value.

Although Paraver was developed to analyze the performance of parallel programs, its input trace format is highly generic and can easily be adopted for other uses. Specifically, we adopted it to visualize the state of queue backlogs throughout the network.

11.6.1.3 Integration

System design is tightly coupled to the workload that will be executed on the machine. Accurately simulating entire parallel applications with detailed hardware models is a complicated task, mainly because of the difficulty of writing a single simulator combining the capability of simulating the software and hardware stacks in sufficient detail. Therefore, one common approach is to simulate the behavior of an application with drastically simplified hardware models, estimating the parameters of such simplified models either from measured characteristics of real components or through more detailed cycle-accurate unit simulations of the components. An example is the bus-based network model used by Dimemas, which is a highly abstracted representation of real interconnection networks.

A complementary trend is to employ drastically simplified application models, feeding detailed hardware simulators with synthetic (often stochastic) traffic, and drawing conclusions about the hardware design under the assumption that they also apply to the applications.

Our approach lies halfway between these two extremes: we believe that, to optimize the design of the interconnection network of a new massively parallel computer, reasonable abstractions of applications, compute nodes, and interconnection network are both necessary and sufficient, in the sense that too much detail limits simulation scalability, whereas too little detail compromises simulation accuracy.

Existing tools did not meet these requirements: Dimemas has the right abstraction layer at the application and node level, but its bus-based interconnect model does not capture important network-related aspects, such as topology, routing policies, flow control, traffic contention & congestion, deadlock prevention, and anything relating to switch and adapter hardware implementations.

Although the PERCS simulator would provide the necessary, highly detailed, network abstraction level, its trace in- and output capabilities are not compatible with Dimemas and Paraver. Moreover, as it was designed to simulate one specific system (PERCS), it does not provide sufficient flexibility for design space exploration in terms of network topologies, routing schemes, switch architectures, etc. Therefore, we provided the following capabilities in our Venus environment:

- Detailed models of switch and adapter hardware corresponding to different networking technologies, including Ethernet, InfiniBand, Myrinet, and more generic input-, output-, or combined input- and output-queued switch architectures.
- Support for various regular direct and indirect network topologies.
- A *server mode* to support co-simulation with Dimemas via a standard Unix socket interface.
- Output of Paraver-compatible trace files to enable detailed observation of network behavior.
- Support for irregular (as well as regular) topologies by means of a translation tool to convert generic topology descriptions (conforming to the `map` format used by Myrinet) to Omnest `ned` topology description files.
- Import facility to load generic source routing descriptions (conforming to the `routes` format used by Myrinet) at simulation runtime.
- Tools for topology generation, route generation, and route optimization given a specific application traffic pattern.
- Support for multi-rail networks.
- A flexible mechanism to map Dimemas tasks to network nodes.

Figure 11.8 depicts the complete tool chain of our simulation environment. The following subsections describe each of the above features in some detail.

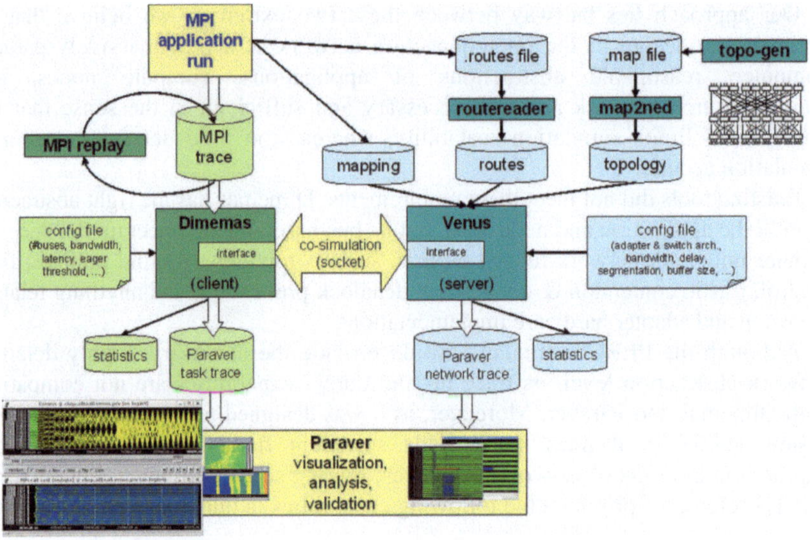

Fig. 11.8 Integrated tool chain.

11.6.1.4 Server Mode

To achieve interoperability between Venus and Dimemas, we implemented a hybrid approach combining *Parallel Discrete Event Simulation* (PDES) with a client/server model. The PDES approach enables distributed simulation of a single system model involving multiple independent simulation engines. The natural boundary of the co-simulation lies between the detailed simulation of the network (Venus) and the replaying of an application's trace (Dimemas). We extended the PDES framework to make Venus act as server and Dimemas as client. We defined a communication interface between the two sides that allows one or more Dimemas instances to be plugged into one or more instances of Venus.

The main challenge of such an approach is to synchronize the event schedulers of the simulators, so that global temporal ordering and causality of events are guaranteed. As Dimemas uses a proprietary event-scheduling engine, whereas Venus is based on Omnest, porting one simulator to adopt the engine of the other would have been too invasive. Therefore, we adopted a conservative version of the "Null Message Algorithm" (Bagrodia and Takai 2000, Varga et al. 2003). We assume that the *earliest input time* of each of the simulators is 0, so that each one of the parallel simulators can expect an event from the other one at the current timestamp. The *earliest output time* is set to the next event in the local event queue. Although the algorithm is borrowed from a PDES technique, the actual lookahead settings make it run in a serial way: the simulations take turns, performing at least one action at a time, so that they always progress. To reduce the communication overhead due to the "null messages" between the simulators, they are only exchanged when Venus has work to do (i.e., some simulated nodes are communicating). Otherwise, Dimemas runs without synchronizing with Venus until some communication event is reached. On the other side, Venus runs without synchronizing as long as the time stamp of the next event in the Dimemas queue is strictly greater than or equal to Venus's current simulation time, unless an event is processed that could change the state of Dimemas, in particular the completion of a message transfer, i.e., when a message has arrived in its entirety at an output of the network.

Venus has been extended with a module that acts as a server receiving commands from Dimemas. Upon initialization, a listening TCP socket is opened, and Venus awaits incoming connections. Once a client connects to Venus, it can send new-line separated commands in plain text. Venus understands several types of commands, including `STOP` and `SEND`. `STOP` is the actual "null message" exchange: it only serves to inform Venus of the timestamp of the next relevant event in the Dimemas queue. The `SEND` command will force the server module to send a message through the network simulated by Venus. When a message has arrived at a network output, Venus passes it back to the server module, which in turn sends a corresponding `COMPLETED SEND` message to Dimemas.

11.6.1.5 Paraver Trace Output

Paraver was originally intended to represent the state of and communication between MPI threads. However, owing to its generic trace format and high level of configurability (association of semantic labels with threads, states, and events) and the myriad ways in which data contained in trace records can be translated to numbers and colors for visual rendering, Paraver is also highly suitable to represent the state of the interconnection network.

We chose the following, natural mapping from network entities to Paraver tasks and threads: Each adapter and each switch is represented by one task. Each port in an adapter and switch is represented as a thread belonging to the corresponding task.

Table 11.1 shows the structure of all Paraver records, including their semantics, both as originally intended at the MPI level and as newly assigned in the context of the network. The main difference in the communication records is that the logical and the physical send time now correspond to the first and the last flit of a message, respectively. At the network level, there will be one communication record for each link traversal of a given message. In each record, the sending entity corresponds to the transmitting port of the switch, whereas the receiving entity corresponds to the peer port of the receiving switch. The size in both cases corresponds to the size of the MPI message in bytes, whereas the tag uniquely identifies the message, enabling easy tracing of the progress of a specific message through the network.

Table 11.1 Structure of state (fields S0-4), event (fields E0-4), and communication records (fields C0-8) in a Paraver trace. GID = global identifier. Global thread identifiers comprise application, task, and thread; Global port identifiers comprise switch or adapter ID and local port ID.

Fld	Content	MPI-level meaning	Network-level meaning
S0	'1'	State record type	*same*
S1	Entity	Sending thread GID	Port GID
S2	Begin time	Starting time of state record	*same*
S3	End time	Ending time of state record	*same*
S4	State	Activity carried out by thread	Quantized buffer backlog at port
E0	'2'	Event record type	*same*
E1	Entity	Sending thread GID	Port GID
E2	Time	Time at which event occurred	*same*
E3	Event type	Type of event	*same* (but different set of events)
E4	Event value	Value associated with event	*same* (but different semantics)
C0	'3'	Communication record type	*Same*
C1	Sending entity	Sending thread GID	Sending port GID
C2	Logical send time	Time at which send is posted	Arrival time of first flit of message
C3	Physical send time	Actual sending time of message	Sending time of first flit of message
C4	Receiving entity	Receiving thread GID	Receiving port GID
C5	Logical receive time	Time at which receive is posted	Reception time of first flit of message
C6	Physical receive time	Actual message reception time	Reception time of last flit of message
C7	Size	Message size in bytes	*Same*
C8	Tag	Message type (MPI operation)	Unique message identifier

In each state record, the entity identifies the specific switch and port to which the record applies. The state value indicates the state of the entity from begin to end time. The main difference between state records at the MPI and at the network level is that at the MPI level, the states correspond to certain MPI thread (in)activities (idle, running, waiting, blocked, send, receive, etc.), whereas at the network level the state represents a buffer-filling level. The actual state value is quantized with respect to a configurable buffer quantum. The backlog can be traced either per input port or per output port.

An event record marks the occurrence of a punctual event. At the network level, we implemented events to flag the issuance of stop and go flow-control signals, the start and end of head-of-line blocking, and the start and end of transmission of individual message segments, all at the port level. The semantics of the value depend on the specific type of event.

This tracing capability enables "debugging" of the interconnection network. For instance, hot spots due to overloaded links can easily be identified. Imbalances caused by late message arrivals can be tracked down by inspecting the ports those messages traversed. Moreover, inefficiencies such as head-of-line (HOL) blocking can be exposed, and also *underutilization* can be diagnosed, which can be used to reduce network over-dimensioning and save cost.

11.6.2 Workload Models

Different ways of applying workloads to the system under study were outlined in Sec. 11.3.3. Venus supports the following three main workload models:

- Random traffic generators. These are generally used to determine throughput vs. load and delay vs. load characteristics for different stochastic temporal and spatial traffic distributions. Traffic scenario files can be provided to change the spatial distributions during the simulation to create, for instance, transient hot-spot scenarios.
- Application traces: As described in detail in Sec. 11.6.1, one of the main objectives of this work is to study the behavior of specific applications, represented by traces of representative phases of their execution. This is currently limited to message-passing (MPI) applications. Support for playing back traces of *partitioned global address space* (PGAS) applications, in particular Unified Parallel C (UPC) and SHMEM, has also been implemented, but still is at an experimental stage.
- Workload models: Venus also provides support for modeling workloads by mimicking the behavior of typical communication patterns encountered in HPC programs, in particular those found in *collective operations*, such as barriers, reductions, all-to-all exchanges, scatter-gather, which usually involve all nodes.

The tasks belonging to a given workload need to be mapped to available compute nodes. In a real many-user environment, the job scheduler takes care of this task. Depending on the overall system load, job sizes, job durations, and scheduling policies, this can lead to significant *fragmentation*, meaning that tasks end up being allocated to nodes that are not topological neighbors. To enable the evaluation of the effect of fragmentation and of different task-mapping policies, our environment allows arbitrary mappings of tasks to compute nodes in Venus. This is accomplished by means of a simple configuration file that contains one hostname (as known to Venus) per line; task n is mapped to the host corresponding to the hostname specified on line n. In principle, multiple tasks can be mapped to the same host.

11.6.3 Network Models

11.6.3.1 Topologies

Venus supports a variety of topologies, either directly by means of a `ned` topology description, or indirectly via an additional topology generation tool that outputs a `map` file, which is then converted to a `ned` file through the `map2ned` conversion utility (see Sec. 11.6.3.2). This indirect way is somewhat less convenient, as for each different topology configuration a new `ned` file must be generated. However, it provides more flexibility than the native `ned` format. The following regular topologies are supported:

- Fat tree (k-ary n-tree)
- Extended generalized fat tree (XGFT)
- Omega network (unidirectional)

- Full mesh (complete graph)
- Hypercube (binary *n*-cube)
- Multi-dimensional mesh (*k*-ary *n*-mesh)
- Multi-dimensional torus (*k*-ary *n*-cube)
- Flattened butterfly
- Dragonfly
- Hierarchical mesh

11.6.3.2 map2ned

In addition to various regular direct and indirect topologies, Venus also supports arbitrary irregular topologies by means of a topology specification adopted from the Myrinet interconnect used in Mare Nostrum. Such a specification consists of a simple, but very generic, ASCII-based topology file format referred to as a `map` file, which describes an arbitrary topology comprising hosts and switches.

We implemented a translation tool to convert such a `map` file to an Omnest `ned` file corresponding to the specified topology and a matching initialization file (`ini`) containing network address and host/switch labels. This `map2ned` tool assumes generic base module definitions for both host and switch, taking advantage of the polymorphism mechanism provided by the `ned` format, such that the generated `ned` files can be used with all kinds of network technologies implemented in Venus.

The Omnest `ned` file format is not very well suited for the specification of topologies that require a vector of values rather than a single value that is the same for all levels/dimensions. Examples of such topologies are XGFTs, multi-dimensional (non-square) meshes and tori, hierarchical meshes, and others. Therefore, we also adopted the `map2ned` conversion utility to provide support for such topologies and implemented an additional tool that takes the topology specification as a parameter to generate the corresponding `map` files for these topologies, which are then converted to the `ned` format by `map2ned`. Furthermore, `map2ned` enabled us to *exactly* model the topology of the Mare Nostrum machine by obtaining the `map` description from the real machine's Myrinet network, converting this to the `ned` format, and loading the result into Venus.

11.6.3.3 Routing

A crucial aspect of communication performance is how data is routed through the network. To this end, we have implemented very generic support for routing algorithms. Three basic types of routing are supported:

- Algorithmic routing: This refers to routing algorithms that, at each hop, use a mathematical expression or algorithm to infer the next hop from the current position and the destination and/or source address. This method works very well for the regular topologies encountered in HPC and data center environments.

- Pre-configured routing: With pre-configured routing, the routes are loaded into the simulator from routing table files provided by the user. These files may specify either *source routing* tables (i.e., for each source-destination pair, the file specifies the exact sequence of hops to take), or *distributed routing* tables (i.e., for each destination and each switch, the file specifies which hop to take next). This method was mainly provided to enable the use of routes as collected from real machines.
- Self-learning routing: With this routing method, each switch automatically programs its routing tables by learning the addresses of incoming packets, i.e., associating a packet's source address with the port it arrived on. Packets for as yet unknown addresses are flooded to all ports except the one it arrived on. This is modeled after the way in which traditional Ethernet networks operate, and is especially useful for irregular topologies. The main drawback is that it leads to broadcast storms if the network is not loop-free.

With respect to the pre-configured routing method, we adopted the Myrinet `routes` format, which specifies the routes between any pair of hosts. Myrinet networks use turn-based source routing, meaning that each sender specifies the full route and each message embeds the route in its header. The route consists of one *relative* port index ("turn") for every switch on the path. In each switch, the corresponding turn is added to the index of the port on which the message arrived to obtain the output port index. Turns can therefore also be negative numbers. There may be multiple routes between any pair of hosts to support multi-path routing. To be useful, a `routes` file must match a given `map` file, in terms of both topology and host naming. We implemented a library to import a `routes` file into the simulator at runtime to exploit the routes corresponding to given `map` file. We used this to import the real routes programmed in Mare Nostrum, thereby faithfully capturing contention issues caused by routing conflicts.

11.6.3.4 Networking Technologies

So far, we have considered the network as consisting of generic network interfaces (adapters) and switches. From the modeling point of view, this is indeed our approach. To enable rapid prototyping of different network implementations, we created substrate (abstract) switch and adapter models upon which specific implementations can be created. The abstract base models take care of generic functionality required in any implementation, such as efficient management of transmission events, ensuring a consistent timing model, the interface to Paraver tracing facilities, statistics collection, and interfaces to the generic routing, mapping and topology services.

Venus provides adapter and switch models for Ethernet, InfiniBand, and Myrinet (Boden et al. 1995) networks. In addition, it provides a very generic combined input- and output-queued switch implementation that provides full support for virtual channels, making is especially suitable for networks requiring multiple virtual channels for deadlock prevention, such as tori and Dragonflies.

The abstraction level of these models is the somewhat oddly named "flow control digit" *flit*, which is the atomic unit of data transfer across a link. Basically, a flit corresponds to a packet, cell, or frame, depending on the type of network being modeled. In essence, these models are queuing models: Their core components are input and/or output queues, schedulers (arbiters, allocators), routing algorithms, flow-control policies, congestion management schemes, and service differentiation policies (including priorities, virtual lanes, virtual channels, etc.). For the specific Ethernet, InfiniBand, and Myrinet models, most of these aspects are fixed (according to the respective standard or proprietary implementation), whereas for the generic switch and adapter model, they are entirely configurable in a plug-and-play fashion. Moreover, they are easily extensible through cleanly defined interfaces.

11.6.4 Sample Results

To demonstrate the potential of coupling the two discrete event simulators, we will summarize a research study done to evaluate the impact of different configurations and parameters for XGFT topologies on the application performance (Rodriguez et al. 2009).

The co-simulation structure allowed us to perform this study with the appropriate level of detail at each level (from the application's MPI activity, with events occurring at intervals on the order of tens of microseconds) to the transmission of the data at the physical link layer (with events occurring at intervals on the order of tens of nanoseconds). Without the proper granularity at each level, the simulation of a significant part of the application would not have been feasible.

Our simulations revealed that network contention induced by how messages are routed through the network (i.e., by the method of path selection between a given source and destination) was the most significant factor affecting the communication performance of the workload. Contention occurs when multiple communications (partially) share a path through network, causing increased delays for all contending communications.

We studied several existing routing algorithms and also developed new ones that target the specific needs of the regular exchange patterns present in HPC applications. In particular, we performed offline route optimization, taking into account the communication matrix of a given workload on a given topology, and minimizing contention by assigning concurrent communications to disjoint paths as much as possible.

The highly flexible support for different topologies provided by Venus made it possible to study the network cost-performance trade-offs under different routing schemes. We determined to which extent network slimming (i.e., the reduction of the network's bisection bandwidth, and, hence, cost) impacts the performance of a particular HPC application. As an example, Figure 11.9 shows the relative slowdown suffered by a very common computational kernel in many parallel HPC applications, namely, the Conjugate Gradient (CG, class D) using 128 compute nodes. The graph plots the execution slowdown on the *y*-axis (the slowdown is relative to the execution time on an ideal single-stage crossbar network) as a function of the number of second-level switches in a two-level XGFT on the *x*-axis. Reducing the number of switches means lower cost, but also less bisection bandwidth and fewer alternative paths and therefore more contention and higher delays. The *y*-axis shows the slowdown experienced for that specific XGFT configuration and for different routing schemes. The higher the slowdown, the worse the performance.

Two main conclusions can be drawn:

- The routing scheme has a significant impact on the performance of this communication pattern. A pattern-aware routing scheme ("colored') significantly outperforms oblivious schemes.
- Reducing the number of middle switches from 15 to just 9 has little or no impact on the performance, implying that—for this workload—a significant savings opportunity exists in terms of network cost.

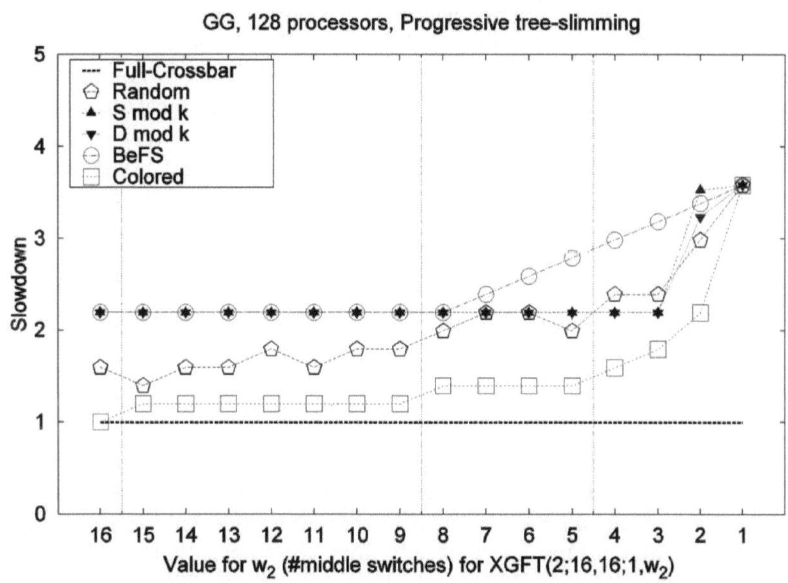

Fig. 11.9 Slowdown of CG for various XGFT configurations and routing schemes.

11.7 Scalability

As the demand for computational power grows and technology advances, HPC systems and their interconnection networks are becoming larger and more complex. To study the performance of such systems, discrete event simulation is an important tool. Nevertheless, the need to simulate ever larger and more complex models puts new emphasis on the scalability of such tools.

The main factors affecting scalability of discrete event simulators are twofold. First, the increased number of events to simulate and their complexity might lead to unacceptably long simulation times. Second, the larger size of the models directly affects the resource usage footprint of the simulators, especially in terms of allocated memory. A suitable solution to both problems can be *parallel discrete event simulation* (PDES). In this section, we discuss our experience in parallelizing the Venus simulator, and how this affected the simulation time for different use cases.

11.7.1 Parallel Discrete Event Simulation

Parallel discrete event simulation splits a model into partitions called *logical processes* (LPs). Each LP is executed on a different processor or host. As the partitions are only a subpart of the original model, they will, in general, have fewer events to simulate and a smaller memory footprint. However, it is difficult to achieve ideal (linear) speedups and memory reductions. Memory reductions are hampered by the overhead due to the need for replicating some common information over all partitions, whereas execution time suffers from a) the overhead introduced by the necessity for synchronization and communication across the LPs and b) intrinsic limitations of the code, such as non-parallelizable code sections. As each LP has its own *future event set* and local *simulation time*, synchronization is needed to prevent violations of the causality of events. Without synchronization, an LP could send an event to another LP with a timestamp that is in the past of the receiving LP, therefore breaking the causality of events in the receiving LP. The overhead itself is due both to the additional messages the LPs need to exchange and process and to an intrinsic overhead of the synchronization algorithm itself.

Synchronization is one of the main issues in PDES, and three broad categories of synchronization algorithms exist:

1. **Conservative algorithms** prevent incausalities from happening by waiting until all events having smaller time stamps have been processed before they execute an event. A critical model property for the speedups obtainable through this category of algorithms is the so-called *lookahead*, i.e., the future time interval for which an LP knows it will not receive any event from any other LP.
2. **Optimistic algorithms** allow incausalities to happen, but detect them and revert to using rollbacks. If rollbacks happen too frequently, their computation cost will prevent the simulation from achieving good speedup. Furthermore, rollbacks are quite complex and require extra functionalities from both the simulation kernel and the model.
3. **Statistical methods** are based on the exchange of statistical properties of the message flow between LPs rather than on their individual messages. Even though these methods can achieve good speedups, their application area is limited.

More information about the three categories can be found in (Fujimoto 1989, Lencse 2002).

11.7.2 Parallel Simulation Support in Omnest

Omnest supports parallel simulation based on conservative synchronization via the Chandy–Misra–Brandy (Chandy and Misra 1979) algorithm (Varga 2010), also known as the *null message algorithm* (NMA).

The NMA maintains two sets of variables. The first set stores the earliest input times (EITs) at which the LP may receive an event for each input LP. The second set stores the earliest output times (EOT) at which the LP may send an event to a given output LP, i.e., the local simulation time plus the lookahead. LPs can safely process events until the minimum of the EITs. If the LP reaches any EIT, it has to block until the EIT is updated by a null message, which contains the EOT of the sender. If null messages are sent sufficiently often, i.e., at least once before the EOT of the last message expires, deadlocks are avoided.

Almost any Omnest model can be run in parallel. Apart from some programming-dependent issues, the main model constraints hereby are the use of static topologies and the presence of lookaheads in the form of link delays. In the case of the PERCS simulator and Venus (and network simulators in general), this latter requirement is easy to fulfill because propagation delays and/or minimum transmission times are natural sources for lookaheads. Omnest offers support for different communication libraries, including MPI, making it therefore easy to run on multi-core shared-memory machines as well as on clusters.

The performance of NMA is impeded by two main factors: the frequency of null messages and the time spent by any LP in the waiting state. Based on these observations, Varga et al. (2003) propose a simple criterion to predict NMA performance and therefore the gain to be expected from parallel simulations. The criterion is based on four variables, the number of processed events per second P, the number of events per simulated second E, the lookahead time L, and the communication latency τ. The parameters L and E are model-dependent, whereas P depends both on the hardware and the computational complexity of the model's events. L is readily available from the model, whereas P and E can be estimated from a serial execution of the model. Finally, τ depends on the communication library and the hardware used. It can be inferred from simple benchmarks; typical values are in the range of microseconds. From these values, the coupling factor λ can be computed as:

$$\lambda = (L \times E)/(\tau \times P).$$

If $\lambda >> 1$ then good performance can be expected from the parallel simulation, whereas $\lambda < 1$ will result in poor performance. The rationale behind this equation is that the model should have a sufficiently large number of events in the lookahead (given by $L \times E$) to keep the CPU busy during the communication time, i.e., the events processed during the communication time (given by $\tau \times P$). The number of partitions n mainly affects the event density E. The partitioning does not influence the model, meaning that the total number of events over all partitions remains constant, so that each partition gets fewer events to simulate. Hence, if the partitions are of the same size, λ_n will also decrease: $\lambda_n = \lambda/n$.

Partitioning is done by assigning a partition identifier to each module in the model (via Omnest's configuration file). The partitions should be as homogeneous as possible, so that the simulation load is evenly balanced across LPs because the overall speedup is gated by the slowest LP. The number of partitions should not be too high so that the value of λ_n is not too small. Moreover, the partitioning should maximize the lookahead between LPs to increase λ and minimize the number of events crossing the LP boundaries, thus reducing the communication overhead.

11.7.3 Venus

Venus includes support for parallel simulations based on the Omnest framework. As Venus is an interconnection network simulator for HPC systems, it is easy to satisfy both parallel simulation model constraints. Most HPC topologies are regular and static, and lookaheads can easily be set to the minimum packet delay on a link.

Venus supports different network topologies. Here, we focus on three regular topologies: the mesh, the hierarchical full mesh (referred to as h-mesh), and the fat tree. To facilitate comparison, we use a configuration connecting 4,096 end nodes for each network type. Specifically, we consider a 2D 64x64 mesh, a 2-level 8-cluster 16-switch h-mesh, and a 4-level 8-radix fat tree. All links delays are the same and the lookaheads are set to the minimum packet transmission time. The traffic pattern is random uniform (Bernoulli) traffic.

Figure 11.10 presents an example of how to partition each of these topologies into LPs. For clarity, the examples are given for a 16-node configuration rather than for the full 4,096-node configuration. The mesh and h-mesh nodes include the switch and any host directly connected to it, whereas in the fat-tree representation hosts and switches are separated. The rationale behind the partitioning is, first, to have equal-sized partitions of equal complexity. This avoids having one slow LP dragging down the performance of all other LPs. Second, the partitioning tries to minimize the number of links crossing LP boundaries, so that as much traffic as possible remains local to a single LP; in other words, traffic with source and destination in the same LP should not be forced to cross an LP boundary, to avoid the additional LP communication overhead. Finally, the partitioning should try to maximize the lookahead values between LPs to reduce the synchronization overhead. This last criterion is less important in this example as all lookaheads are equal. For different links and/or different traffic patterns, other partitions than the ones presented here may have to be considered.

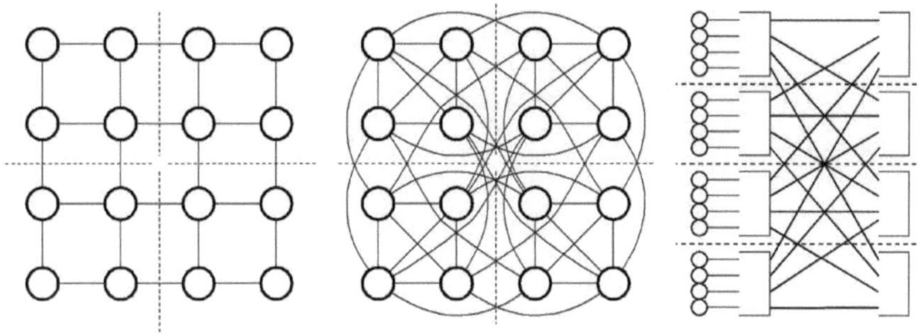

Fig. 11.10 Partitioning example for a 16-node mesh (left), h-mesh (middle) and fat tree (right) topology into 4 LPs. The dashed lines delimit each LP.

The three models were simulated using Omnest v4.1 and OpenMPI v1.4 on one high-end server equipped with 4 Intel[1] Xeon[1] X7560@2.27 GHz CPUs (32 cores total).

Figure 11.11 shows the results of simulation runs using $n \in [1, 2, 4, 8, 16, 32]$ LPs. The upper and middle panels show the absolute and relative speedups achieved, respectively, whereas the bottom panel presents the corresponding λ values measured. All simulations behave similarly, and all have high λ values, indicating that the models are sufficiently large and complex for parallel simulations. As predicted, all simulations gained from parallel simulation. However, we observed some differences between the models. As expected, the relative speedup with increasing numbers of LPs reaches a peak and then decreases. The peak indicates where the model achieves the best tradeoff between the gain from parallel simulation and the overhead incurred. The mesh topology attains the peak earlier than the fat tree and h-mesh do. Hence, fat trees and h-meshes can achieve better speedups with a high number of LPs. Surprisingly, the models (especially the mesh) achieve super-linear speedups for certain values of n, i.e., relative speedups greater than one. One reason can be the reduced overhead in the simulator itself because of the lower event density. In particular, the cost of operations on the future event set (inserting/removing events) is directly related to its size. Another reason is faster memory accesses because more cache memory is available, as each core has its own local cache.

Speedups are not the only benefit of parallel simulations. Resource constraints are another one. Figure 11.12 shows the memory footprint of different h-meshes with increasing numbers of nodes run on a 64-node cluster. Each node is equipped with two Intel® Xeon® X5670@2.93 GHz CPUs and an InfiniBand interconnect. As the number of hosts increases, the total peak memory footprint rapidly grows to hundreds of GBytes. In these simulations, the scale of the simulated system was clearly limited by the available RAM. However, the maximum memory footprint per partition is much more reasonable, which enabled simulation of up to 128K nodes.

In conclusion, we showed that network simulators such as Venus can achieve very good speedup values using PDES techniques. Moreover, parallel simulations can be used to overcome hardware resource constraints, especially memory requirements.

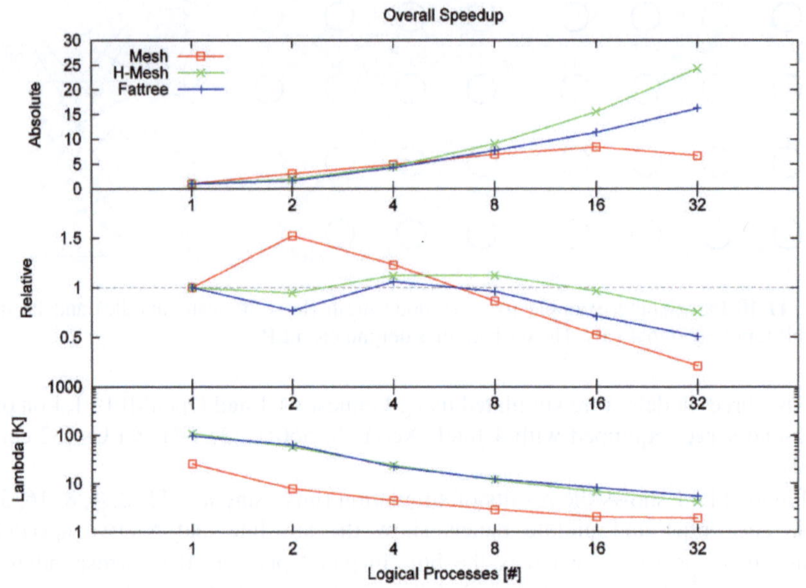

Fig. 11.11 Absolute speedups (top), relative speedups (middle) and corresponding λ values achieved for the three reference topologies.

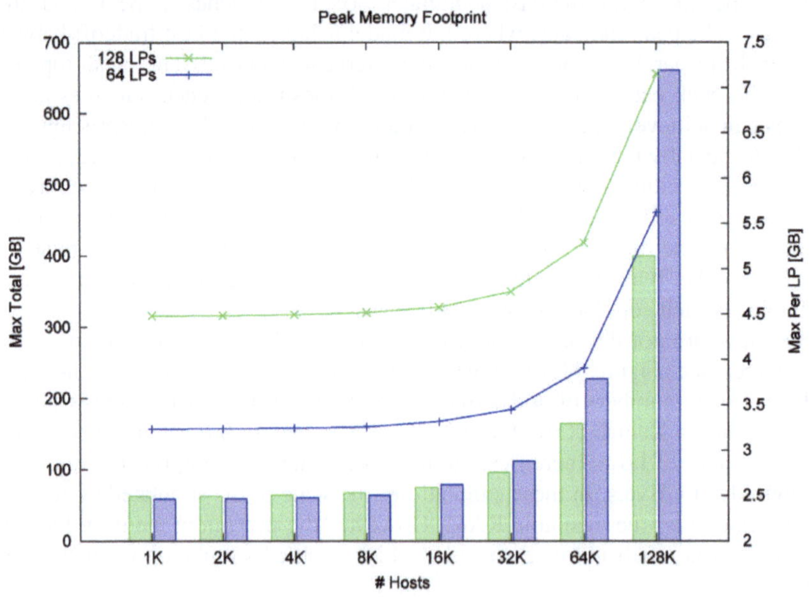

Fig. 11.12 Maximum total and per-LP h-mesh memory footprint with increasing number of hosts.

11.8 Conclusion

Designing large-scale HPC systems is a daunting task that can benefit enormously from discrete event simulation techniques, as the interactions between the various components of such a system generally render analytic approaches intractable. The work described in this chapter specifically deals with end-to-end, full-system simulation, as opposed to simulation of individual components or nodes. To overcome the intrinsic complexity of simulating such large systems, choosing reasonable levels of abstraction is essential; workloads are represented either by stochastic patterns, by execution/communication traces of real workloads, or by so-called "plug-in" code modules that model communication-intensive workload phases.

We have taken a network-centric approach, as the levels of parallelism (up to hundreds of thousands of cores) imply that the impact of the communication between all these cores will be a key factor in determining overall system performance. The network is essentially represented by a huge queuing model that models the network traversal of each communication at the level of individual data units.

Using this approach, we identified and solved unexpected interactions between the various system layers, ranging from the application to the communication library (e.g. MPI), the network layer (e.g. routing) and the hardware (adapter and switch implementation details) that without this holistic approach would not have been discovered.

The tools described here can be used in the design phase of a new HPC system to optimize system design for a given set of workloads, or to create performance forecasts for new workloads on existing systems. We have shown that the power of modern parallel computers can be exploited to great effect to perform these kinds of discrete event simulations at large scales, obtaining linear speed-up factors with up to 16 cores for simulations of 4,096 end nodes, and enabling simulations of more than 100,000 nodes by overcoming the memory footprint bottleneck.

In closing, we would like to remark that this approach is by no means limited to HPC environments. Our current efforts are directed towards applying the same methodology to the optimization of networks for commercial datacenters, which are subject to workloads of an entirely different nature.

Acknowledgments. This material is based on the work supported in part by the Defense Advanced Research Projects Agency under its Agreement HR0011-07-9-0002. Any opinions, findings, and conclusions or recommendations expressed in this material are those of the author(s) and do not necessarily reflect the views of the funding agencies.

This work was funded in part by the U.S. Department of Defense and used elements at the Extreme Scale Systems Center, located at Oak Ridge National Laboratory and funded by the U.S. Department of Defense.

This work was supported in part also by the European Union FP7-ICT project TEXT under contract no. 261580.

Authors Biography, Contact

IBM Research – Zurich is one of IBM's nine research laboratories around the globe. This network of some 3000 scientists is the largest industrial IT research organization in the world. The Zurich laboratory was established in 1956 and is home to world-class scientists representing more than 30 nationalities. Cutting-edge research and outstanding scientific achievements—most remarkably two Nobel Prizes—are associated with this lab. The spectrum of research activities range from exploratory research in nanoscience and –technology for future chips, to advanced server and storage technologies, to high-performance computing, to software and services in areas such as security and privacy, risk and compliance, analytics and information management, or business optimization. Very recently, the Binnig and Rohrer Nanotechnology Center - a facility for world-class nanoscale research recently opened on the campus of IBM Research – Zurich. The building is the centerpiece of a 10-year strategic partnership in nanoscience between IBM and ETH Zurich where scientists will research novel nanoscale structures and devices to advance energy and information technologies. Also co-located at IBM Research - Zurich is the Industry Solutions Lab. This think-tank and briefing center is an integral part of the European IBM Forum Center network and provides a unique place in Europe to gain insights from IBM researchers, industry and trend experts, in order to tackle today's and tomorrow's challenges. URL: www.zurich.ibm.com

Cyriel Minkenberg obtained MSc and PhD degrees in electrical engineering from the Eindhoven University of Technology, the Netherlands, in 1996, and 2001, respectively. He is currently a research staff member at IBM Research - Zurich and manages the System Fabrics group, which is concerned with interconnection networks for high-performance computing and data center networks. Previously, he participated in the IEEE 802.1Qau Working Group to standardize congestion management in Convergence Enhanced Ethernet networks, was responsible for the architecture and performance evaluation of a crossbar scheduler for a 2.5 Tb/s optical switch (OSMOSIS), and contributed to the design and testing of several generations of the IBM PowerPRS switching chips. His research interests include interconnection networks, switch architectures, networking protocols, performance modeling, and simulation. Minkenberg has co-authored over 45 publications in international journals and conferences proceedings. He received the 2001 IEEE Fred W. Ellersick Award for the best paper published in an IEEE Communications Society magazine in 2000, the Hot Interconnects 2005 Best Paper Award, and the IPDPS 2007 Architectures Track Best Paper Award.

Wolfgang Denzel received M.S. and Ph.D. degrees in Electrical Engineering from Stuttgart University, Germany, in 1979 and 1986, respectively. Since 1985 he is a researcher at IBM Research - Zurich in Rüschlikon, Switzerland. He was responsible for architectural design and performance evaluation of IBM's PRIZMA switch. He worked on system aspects of ATM-based corporate networks and corporate optical networks. In these fields he participated in several European RACE projects and coordinated the ACTS COBNET project. His recent interests are in server interconnection networks and end-to-end simulation techniques for large-scale high-performance computing systems. In this context, he created the full-system simulator for the US DARPA PERCS project.

German Rodriguez earned his Ph.D. in Computer Architecture in April, 2011 with his dissertation "Understanding and Reducing Contention in Generalized Fat Tree Networks for High Performance Computing" issued by the Technical University of Catalonia, Spain. He has done research on network performance and routing for High-Performance Computing Systems at the Barcelona Supercomputing Centre (Spain) during his Ph.D, and currently as a Post-Doc at the IBM Research - Zurich. His main research interests focus on the simulation and optimization of network performance of supercomputing clusters for High Performance Computing applications.

Robert Birke holds a double master degree in information engineering from the Politecnico di Torino, Italy and the University of Illinois at Chicago, US and acquired his PhD title in February 2009 from the Politecnico di Torino. In the past he participated in various international research projects, both Italian (Bora-Bora, Mimosa and Recipe) and European (Napa-Wine), as well as networks of excellence (Euro-NGI and Euro-NF). He is currently a post-doctoral researcher at IBM Research - Zurich, Switzerland. His research interests include high speed switching architectures, software routers, and traffic analysis.

References

Arimilli, B., Arimilli, R., Chung, V., Clark, S., Denzel, W., Drerup, B., Hoefler, T., Joyner, J., Lewis, J., Li, J., Ni, N., Rajamony, R.: The PERCS high-performance interconnect. In: 2010 IEEE 18th Annual Symposium on High-Performance Interconnects on Proc. High Performance Interconnects (HOTI), August 18-20, pp. 75–82 (2010)

Bagrodia, R., Takai, M.: Performance evaluation of conservative algorithms in parallel simulation languages. IEEE Transactions Parallel Distributed Systems 11(4), 395–411 (2000)

Boden, N.J., Cohen, D., Felderman, R.E., Kulawik, A.E., Seitz, C.L., Seizovic, J.N., Su, W.K.: Myrinet: A gigabit-per-second local area network. IEEE Micro. 15(1), 29–36 (1995)

Chandy, M., Misra, J.: Distributed simulation: A case study in design and verification of distributed programs. IEEE Transactions on Software Engineering 5, 440–452 (1979)

Dally, W.J., Towles, B.: Principles and practices of interconnection networks, 1st edn. Morgan Kaufmann (2004)

Denzel, W., Li, J., Walker, P., Jin, Y.: A framework for end-to-end simulation of high-performance computing systems. SIMULATION - Transactions of The Society for Modeling and Simulation International 86(5-6), 331–350 (2010)

Desai, N., Balaji, P., Sadayappan, P., Islam, M.: Are nonblocking networks really needed for high-end-computing workloads. In: Proc. 2008 IEEE International Conference on Cluster Computing (Cluster 2008), Tsukuba, Japan, September 29-October 1, pp. 152–159 (2008)

Fujimoto, R.M.: Parallel discrete event simulation. In: Proceedings of the 21st Conference on Winter Simulation, pp. 19–28 (1989)

Geoffray, P., Hoefler, T.: Adaptive routing strategies for modern high performance networks. In: Proc. 16th IEEE Symposium on High Performance Interconnects (HOTI 2008), Stanford, CA, August 27-28, pp. 165–172 (2008)

Kamil, S., Shalf, J., Oliker, L., Skinner, D.: Understanding ultra-scale application communication requirements. In: Proc. Workload Characterization Symposium, October 2005, pp. 178–187 (2005)

Kim, J., Dally, W.J., Scott, S., Abts, D.: Technology-driven, highly-scalable dragonfly network. In: Proc. International Symposium on Computer Architecture (ISCA), Beijing, China, pp. 77–88 (2008)

Leiserson, C.E., Abuhamdeh, Z.S., Douglas, D.C., Feynman, C.R., Ganmukhi, M.N., Hill, J.V., Hillis, W.D., Kuszmaul, B.C., St. Pierre, M.A., Wells, D.S., Wong, M.C., Yang, S.W., Zak, R.: The network architecture of the Connection Machine CM-5. In: Proc. 4th Annual ACM Symposium on Parallel Algorithms and Architectures (SPAA), San Diego, CA, pp. 272–285 (June 1992)

Lencse, G.: Parallel simulation with OMNeT++ using the statistical synchronization method. In: Proceedings of the 2nd International OMNeT++ Workshop, pp. 24–32 (2002)

Luszczek, P., Bailey, D., Dongarra, J., et al.: The HPC challenge (HPCC) benchmark suite. In: Proc. 2006 ACM/IEEE Conference on Supercomputing, SC 2006, Tampa, FL, USA (2006)

Magnusson, P.S., Christensson, M., Eskilson, J., Forsgren, D., Hallberg, G., Hogberg, J., Larsson, F., Moestedt, A., Werner, B.: Simics: A full system simulation platform. IEEE Computer 35(2), 50–58 (2002)

Minkenberg, C., Rodriguez, G.: Trace-driven co-simulation of high-performance computing systems using OMNeT++. In: Proc. SIMUTools 2nd International Workshop on OMNeT++ (OMNeT++ 2009), Rome, Italy, March 6 (2009)

Öhring, S., Ibel, M., Das, S.K., Kumar, M.J.: On generalized fat trees. In: Proc. 9th International Symposium on Parallel Processing (IPPS 1995), Santa Barbara, CA, April 25-28, pp. 37–44 (1995)

Peterson, J.L., et al.: Application of full-system simulation in exploratory system design and development. IBM Journal of Research and Development 50(2/3), 321–332 (2006)

Petrini, F., Vanneschi, M.: k-ary n-trees: High-performance networks for massively parallel architectures. In: Proc. 11th International Symposium on Parallel Processing (IPPS 1997), Geneva, Switzerland, April 1-5, pp. 87–93 (1997)

Rajamony, R., Arimilli, L.B., Gildea, K.: PERCS: The IBM POWER7-IH high-performance computing system. IBM Journal of Research and Development 55(3), 3:1–3:12 (2011)

Rodriguez, G., Beivide, R., Minkenberg, C., Labarta, J., Valero, M.: Exploring pattern-aware routing in generalized fat tree networks for HPC. In: Proc. 23rd International Conference on Supercomputing (ICS 2009), New York, NY, June 9-11 (2009)

Scherson, I.D., Chien, C.K.: Least common ancestor networks. In: Proc. 7th International Parallel Processing Symposium (IPPS), pp. 507–513 (1993)

Sinharoy, B., Kalla, R., Starke, W.J., Le, H.Q., Cargnoni, R., Van Norstrand, J.A., Ronchetti, B.J., Stuecheli, J., Leenstra, J., Guthrie, G.L., Nguyen, D.Q., Blaner, B., Marino, C.F., Retter, E., Williams, P.: IBM POWER7 multicore server processor. IBM Journal of Research and Development 55(3) 1, 1:1–1:29 (2011)

Varga, A.: The OMNeT++ discrete event simulation system. In: Proc. European Simulation Multiconference (ESM 2001), Prague, Czech Republic (June 2001)

Varga, A.: OMNet++ User Manual (2010), `http://www.omnetpp.org/doc/omnetpp41/Manual.pdf` (accessed October 27, 2011)

Varga, A., Sekercioglu, Y.A., Egan, G.K.: A practical efficiency criterion for the null message algorithm. In: Proc. European Simulation Symposium (ESS 2003), Delft, The Netherlands, October 26–29 (2003)

12 Working with the Modular Library Automotive

Jiří Hloska

This chapter deals with the modular library 'Automotive' (in original VDA Automotive Bausteinkasten) of the software Plant Simulation with the focus on point-oriented elements from this library. First, a general introduction to specific modular libraries in Plant Simulation, their purpose, way of use and limits is presented. A brief description of the library 'Automotive', its historical as well as current development, structure and field of use follows. The core of this chapter presents two sample models which show the use of the library 'Automotive'. The aim is to give the reader insight into the variety of the modules and objects of the library 'Automotive' which enable the user to efficiently simulate various processes we can encounter in the automotive industry.

12.1 Creating and Managing User-Defined Libraries in Plant Simulation

In Plant Simulation it is possible to create simulation models of material flow which can reflect number of logistic and production systems running on various principles. For this reason Class Library provides the user with a range of active and passive material flow objects (built-in objects). By default the Class Library contains the following hierarchically structured folders: MaterialFlow, Resources, InformationFlow, UserInterface, MUs, Tools and Models (see Fig. 12.1).

Jiří Hloska
Institute of Automotive Engineering
Faculty of Mechanical Engineering
Brno University of Technology
Technická 2896/2
616 69 Brno
Czech Republic
e-mail: `yhlosk00@stud.fme.vutbr.cz`

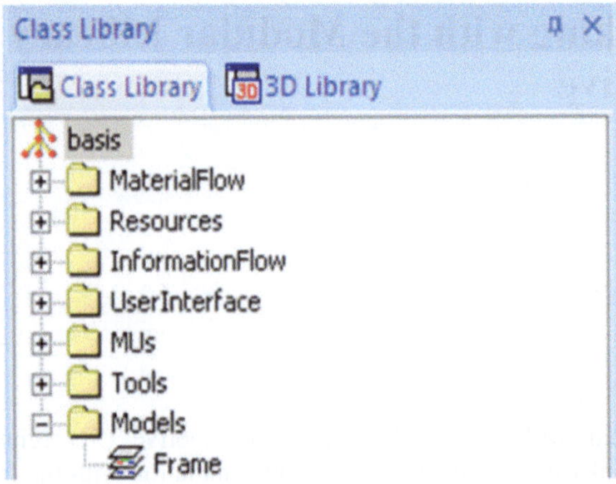

Fig. 12.1 Structure of the class library

However, when modeling specific real processes built-in objects contained in the folders shown in Fig. 12.1 might fail to meet the requirements for the desired functionality. For this reason, it is possible to create user-defined objects with custom functionality. User-defined objects should be organized in toolboxes for transparency reasons (each toolbox should than be dedicated for a set of objects representing the same field of application). Therefore, the very first step should be creating a new folder in the class library (by clicking the right mouse button at the basis or any folder and selecting New – Folder). In this folder, you can create the new toolbox. Basically, there are two ways how to accomplish this:

1. By selecting the basis in the class library, clicking the right mouse button and then selecting New – Toolbar (see Fig. 12.2, left part). A new toolbar will be created on the basis level in the class library. Additionally, in the toolbox window a new tab Toolbar will emerge (highlighted in Fig. 12.2, right part).

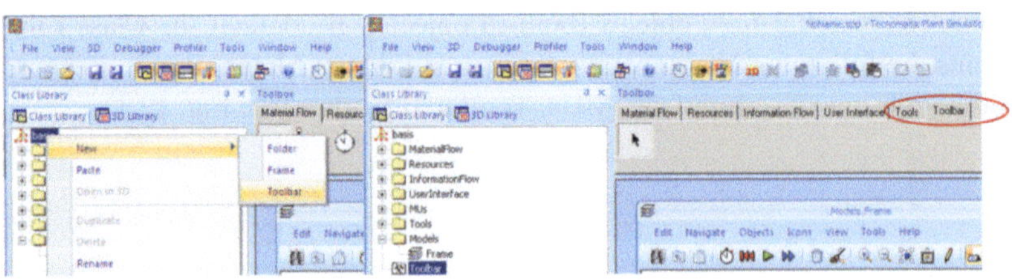

Fig. 12.2 Creation of a new user-defined toolbox in the basis

In the same way the toolbar can be created in any folder of the library (instead of the basis) when selecting the particular folder (optimally a newly created folder designed for user-defined objects). This procedure is depicted in Fig. 12.3. Again a new tab Toolbar in the toolbox will be created. It is possible to rename the toolbox so that its name matches the functionality of the intended objects the toolbox will contain.

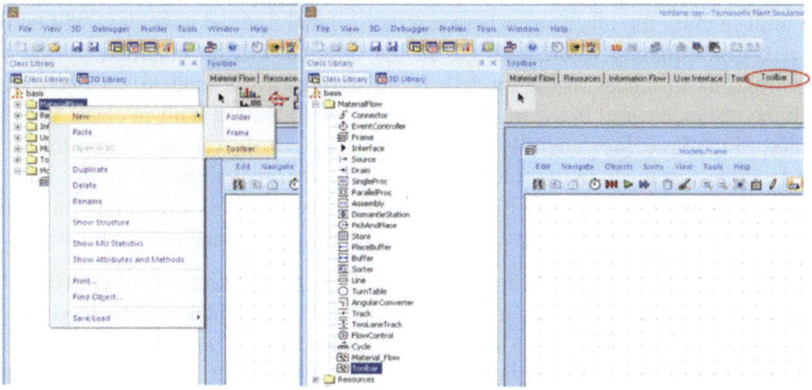

Fig. 12.3 Creation of a new user-defined toolbox in the folder MaterialFlow

2. Alternatively user-defined toolboxes can be created by a method. The command .createFolder; creates a new folder on the basis level. For creating the folder or (if you like) a subfolder in one of the already existing folders, the command has to be specified like this: <path>.createFolder; – let us say we would like to create a folder in the MaterialFlow folder, the command will then be .MaterialFlow.createFolder; In the same way the command .createToolbar; (eventually <path>.createToolbar;, e.g. .Resources.createToolbar;) will create a new toolbar. If the user's aim was creating a new folder and a new user-defined toolbox in that folder, the method (which will additionally rename both the folder and the contained toolbox) could look like this:

```
is
      foldername, toolbarname : string;
do
      .createFolder;
      .newFolder.createToolbar;
      toolbarname := "UserObjects";
      foldername := "UserToolbars";
      .newFolder.Toolbar.name := toolbarname;
      .newFolder.name := foldername;
end;
```

Once the toolbox is created it can be filled with user-defined objects. Depending on the purpose of those objects, one of the built-in classes in the class library can be either derived or duplicated and its features can be modified.

Example: Let us suppose we intend to repeatedly use the object SingleProc always with the processing time 5:00 in our model. Instead of inserting the instance of the class SingleProc from the folder .MaterialFlow into the frame with our model and changing the processing time (since the default value is 1:00) in every single dialog window of those instances it is possible to duplicate (or derive) the SingleProc first and then use the derived (duplicated) and appropriately modified class. If we derive the built-in class the inheritance between the original and the new class will be preserved. Consequently future changes in the original will be inherited by the new class (and its instances in the model). If we wish to avoid inadvertent interference with the instances used in the model we need to duplicate the built-in class. The duplicated SingleProc (or generally any other duplicated object) can be arbitrarily altered and then dragged in the folder containing newly created toolbox.

It is also possible to create classes of more complex tools which can be used to simulate mechanisms or equipments typical for specific branches of industry.

Example: Let us suppose we intend to simulate the function of a jack (or a lift) which lifts objects from lower to upper storey. The single jack will be modeled using several built-in objects, so that it is useful to create a user-defined tool which can then be repeatedly placed in the model (from toolbox or a user-defined folder). To build such a complex tool a new frame (preferably in the folder created by the method above) will be created. This frame will contain all built-in objects needed for modeling the correct function of the jack. To ensure the connections between an instance of the jack inserted into the model with other objects in the model we use the built-in object Interface. The structure of the frame (named Jack) is shown in Fig. 12.4. Here a variable v_Vehicle (type object) refers to the subclass of the movable unit Platform (with a reedited icon) which was duplicated from the base class Transporter. After initialization a reference of the variable v_JackVehicle to an instance of the subclass Platform will be created. This variable is also referred by methods of the objects Jack and LowerStorey. Additionally, these methods define the value of the variable v_Status used for indicating the state of the Jack according to its position. They are responsible for moving the content of the LowerStorey onto the Platform (i.e. into the movable unit) and from the Platform to the UpperStorey as well as for sending the Platform upwards, downwards or stopping the platform in its (default) down position to wait for another entity.

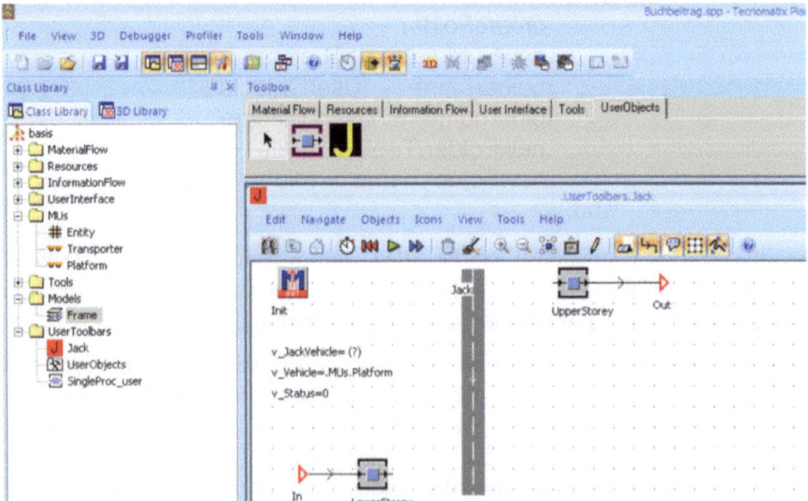

Fig. 12.4 Creation of a user-defined tool for simulation of a function of a jack

To achieve an illustrative animation of the function of the jack the icon of the frame Jack has been edited as shown in Fig. 12.5.

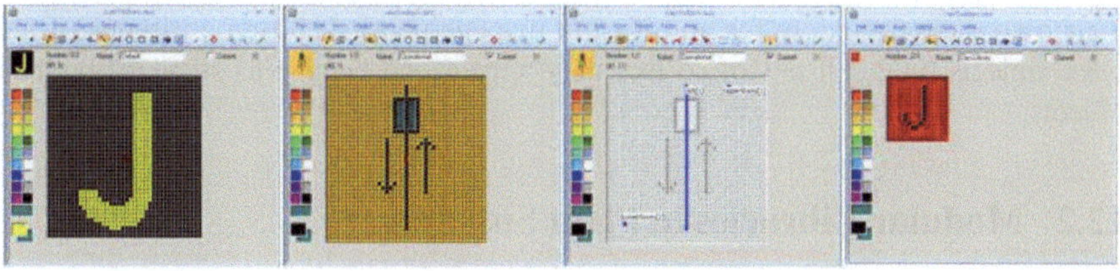

Fig. 12.5 Default and operational icons, animation setting of the user-defined tool Jack and its Classlibrary icon (from left to right)

The tool can then be repeatedly used in the model by inserting the newly created class Jack in the model and connecting it with other objects in the model. This is shown in Fig. 12.6 where the toolbar UserObject of the toolbox is depicted, too. Adding an object to the toolbar can be accomplished by dragging it from its folder in the class library to the desired toolbar (here from .UserToolbars into .UserToolbars.UserObjects). To remove the button representing the object from the toolbar, right-click it and select Delete. To change the order of the buttons representing particular objects in the toolbar click the button which is to be moved, and drag it to a different position on the toolbar. In the model, two instances of a newly created class SingleProc_user have been used. They inherited the processing time 5:00 from the subclass.

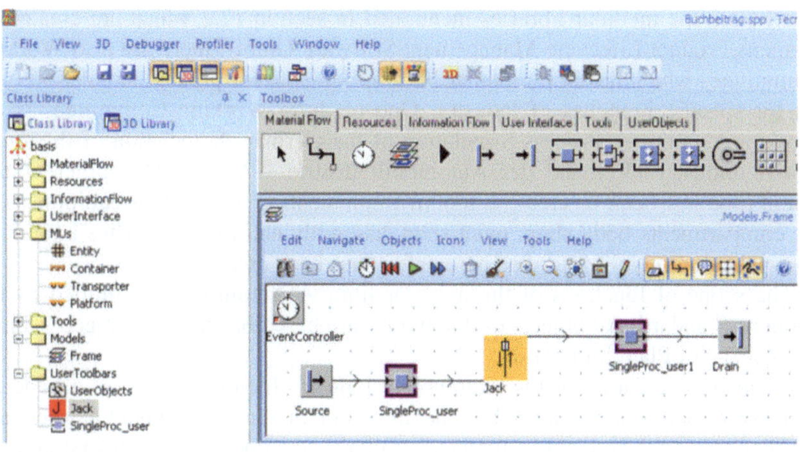

Fig. 12.6 Instance of the user-defined tool Jack

By deriving from built-in objects it is possible to create a range of user defined interactive objects with their own graphics and functionality. These objects can be sorted in modular libraries for an efficient modeling. The modular libraries can contain objects, networks, part of models or whole models. These objects can have their own dialog window for an efficient management of chosen attributes of the basic objects they consists of so that they appear and behave as standard built-in objects.

12.2 Modular Libraries in Plant Simulation

Plant Simulation can be supplemented with a wide range of modular libraries which consist of specific (non-standard) objects. Thanks to these modular libraries the process of creating simulation models suitable in particular branches of industry is very efficient. Moreover, objects in these modular libraries are open for changes so that the user can modify them to meet the requirements of the respective simulation.

Usually, objects from modular libraries provide a user interface in the form of dialog windows, pop-up menus or list for data entry.

As an example some of the modular libraries provided by Siemens Product Lifecycle Management Software GmbH can be mentioned, e.g. AGVS (Automated Guided Vehicle System), Assembly, Carbody, Conveyor 3D, EOM 3D (Electro Overhead Monorail), HBW (High Bay Warehouse), Kanban, Personnel, Shop etc.

12.3 German Association of the Automotive Industry and the Modular Library 'Automotive'

Apart from the modular libraries mentioned in section 12.2 which are created by Siemens Product Lifecycle Management Software GmbH, there are also Working Committees which develop modular libraries suitable for process simulation typical for specific branches of industry. One of those Working Committees is the Working Committee Process Simulation of the German Association of the Automotive Industry (Verband der Automobilindustrie – VDA). The target of this Committee has been to create a toolkit implementing automotive equipment for the compartments body shop, paint shop, assembly and logistics. Thus a modular library 'Automotive' (described hereinafter) has been created.

The scope of functions of the modular library 'Automotive' is limited by the Siemens-Standard. All objects of this library are based on standard objects – built-in objects – of Plant Simulation.

German Association of the Automotive Industry (denoted VDA) is a trade association which groups automobile manufacturers, their development partners and suppliers, as well as manufacturers of trailers, containers, body superstructures, vehicle parts and accessories. The VDA nationally and internationally promotes the interests of the entire German automotive industry in all fields of the motor transport sector. [12.1]

The Process Simulation Working Committee (Arbeitsgruppe Ablaufsimulation) of VDA was founded in March 2005. The founder members were AUDI AG, BMW Group, Daimler Chrysler and Volkswagen AG. Currently the members of the Working Committee are AUDI AG, BMW Group, Daimler AG, Ford of Europe, Adam Opel AG, Volkswagen AG and ZF Friedrichshafen AG. The Working Committee is a subsidiary committee of the working group 'Digital Factory'. Since 2011 it became one of the subsidiary committees (from now on denoted VDA AG Ablaufsimulation) of the working group 'Digital Factory'. Together with service providers the Process Simulation Working Committee manages and develops the modular library 'Automotive' which is based on the simulation software Plant Simulation. [12.2], [12.3]

The aim is to normalize and optimize the use of process simulation in the automotive industry and cooperate in developing the library. With the aid of this library automotive companies and their suppliers generate simulation studies during the planning process and for the support of running operation.

The modular library 'Automotive' is continuously being extended and updated. It is no commercial product but it is used by the OEMs (Original Equipment Manufacturer) who place emphasis on the following requirements to be met by the library [12.4]:

- applicability for models of various level of abstraction,
- possibility to extend the library by new objects from other libraries,
- extensibility for new objects which an OEM whishes not to make accessible for other OEMs (members of the Process Simulation Working Committee),
- all methods must not be encoded,
- the modular library and its objects have to be updatable,
- the modular library has to be compatible with new versions of Plant Simulation

Moreover, to provide certain place for individual creation of models, the particular objects of the library have a user interface, are encapsulated and provide the possibility to create user-defined objects while all objects have to enable their modification.

12.3.1 Structure of the Modular Library 'Automotive'

The basic structure of the library is based on standard built-in objects as well as central frames (which also contain built-in objects) for statistic evaluation and model management. Further elements of the library are parts of one or more modules.

For loading the modules or just some folders into the class library there is a frame .LoadModules in the class library. The frame consists of several methods and a table of objects, nevertheless the user operates it through a dialog window. The dialog window has three tabs for loading modules of the library, folders or OEM-specific elements. In Fig. 12.7 these dialog window for loading the modules (left), folders (centre) and OEM-specific elements (right) are depicted.

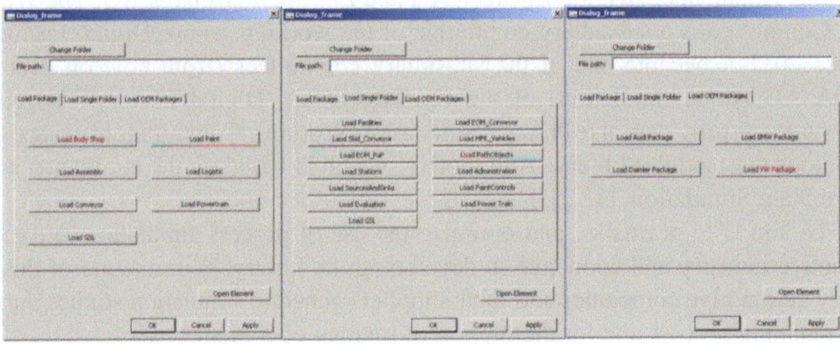

Fig. 12.7 Modules of the library 'Automotive'

Each module consists of several folders and toolbars (only the GSL module – generic standard solution (Generische Standardlösung) - for automatic modeling of the conveyor system on the grounds of layout dates contains no toolbars). Thus, the use of the frame .LoadModule is another way how to extend the toolbox by additional toolbars.

The structure of the class library is shown in Fig. 12.8. In the experimenting level there are frames for conducting simulation experiments, i.e. each frame contains an event controller. All data are stored in this level. In the modeling levels there are models or sub-models and objects for creating user-defined classes of models (usually without event controller). Modified objects or objects with changed parameters can be stored in this level. In the folder ApplicationObjects there are objects which should not be used or modified as they are important for the right function of the up-date mechanism. Therefore, the very same objects are derived and stored in the folder User_Appli_Objects. These objects (which can be used in models) can be found in the toolbox, too. Additional objects in the folder Tools are special material flow objects, methods and variables. The last folder contains all objects which are necessary for updating. The update mechanism guarantees that previously created models always correspond to the up-to-date library. [12.4], [12.5]

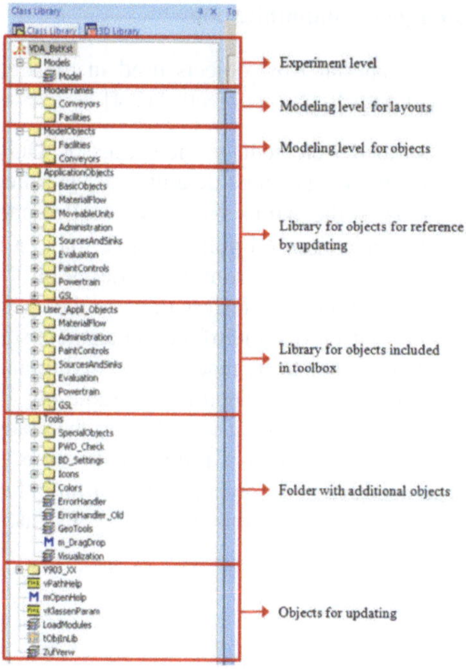

Fig. 12.8 Structure of the class library (taken over from [12.4])

The subfolder MovableUnits of the folder ApplicationObjects consists of twelve different movable units (hereinafter denoted MUs) which have been derived from the basic elements placed in .BasicObjects.MUs. According to their attributes and icons they represent car body or its part, conveyor, carrier etc. and can be generated by sources or jacks (lifts).

Public / Private Architecture [12.6]
Originally, the elements of the library 'Automotive' had their internal variables and methods as well as variables and methods for user setting placed on the same level. They were distinguishable on the grounds of their color only. This solution, especially by more complex elements, caused lack of clarity and difficulties.

Therefore, a new architecture has been designed. Now, user modules and standard modules are separated from each other. This means that each element of the library 'Automotive' has user modules in its top level (referred to as station level) only. All standard modules are then placed in the frame 'nw_Private' – a frame stipulated for non-public objects which is contained in almost all objects of the library 'Automotive' (see also Fig. 12.9 and the section 12.4).

12.3.2 General Principles of the Functionality

To use the objects of the modular library 'Automotive' in the right way, it is vital to follow certain rules, otherwise there is a risk of malfunction of some of the parts or the whole of the simulation model.

12.3.2.1 Object-to-Object Communication

There are two kinds of material flow objects used in the library 'Automotive', namely basic objects and multifunction objects. [12.4]

- Basic objects always transfer elements to the successor. Only after a successful transfer their exit user-method (with no default code) is called by the rear of the transferred element (so that the exit user-method is called once only).
- Multifunction objects, which run on pull principle, have a Dispo-method embedded. This method pulls elements from the predecessor of the particular multifunction object and immediately transfers them to the successor. In the event of occupation of all the successors, multifunction element stays empty until it is triggered by the first successor to be ready for receiving another element. The method m_SuccFree, which is embedded in the frame nw_Private, is responsible for the right functionality in that it is called by the ready successor. In the foreground of the Fig. 12.9 there is a frame of the object JunctionPull. The frame nw_Private (embedded as an object into the frame) has been selected. In the background of the Fig. 12.9 this frame nw_Private with the method m_SuccFree selected is displayed.

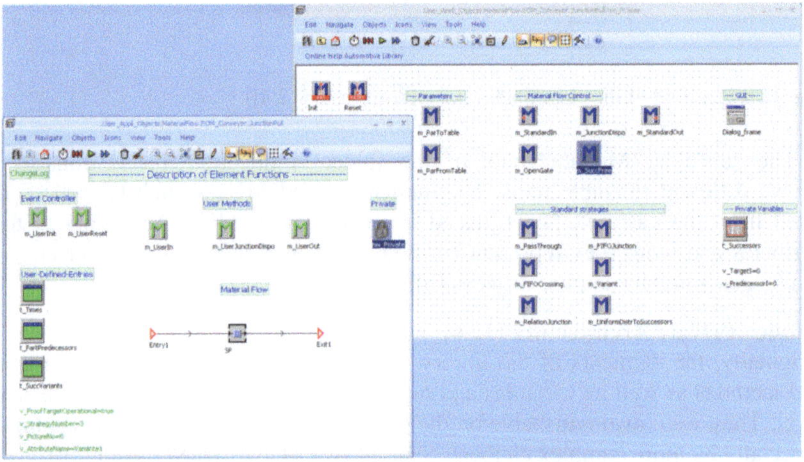

Fig. 12.9 Frame of the object JunctionPull and its nw_Private frame

12.3.2.2 Creation of User-Defined Objects

Basically it is possible to create user-defined objects, too. Yet it is necessary to follow certain rules to make sure that the user-defined object will seamlessly communicate with objects of the library 'Automotive'. These rules are: [12.4]

- for parameterization variables and methods have to be used,
- use of any user-defined object must not require any changes of other objects (built-in objects or objects of the library 'Automotive'),

- user-defined objects should be created using the objects from the folder .ApplicationObjects.BasicObjects.

12.3.2.3 Parameterization and Collecting Statistical Data

To enable a well-arranged parameterization which will match the instructions for generating process simulations in automotive industry[1] there are frames for central parameterization and for collecting statistical data in the library 'Automotive'. These frames are shown in Fig. 12.10 in which an empty frame from the experiment level is depicted. Each root (on the experiment level) should contain these frames.

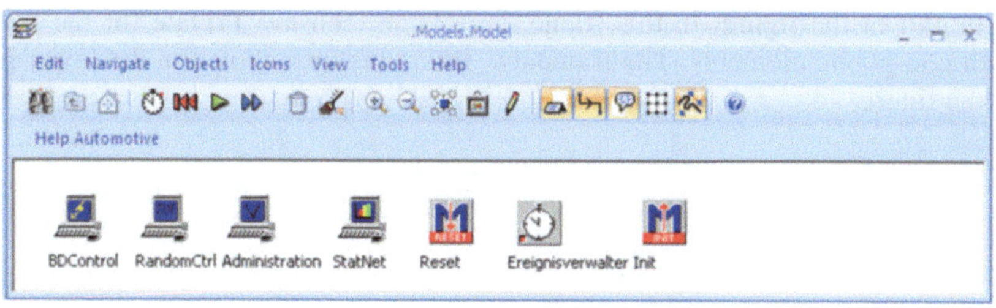

Fig. 12.10 Frames for central model management

For central parameterization the frames BDControl, RandomCtrl and Administration are available. The frames BDControl and Administration include write and read methods (each of them relevant for one of the object classes). Read methods are designed for registering the parameters (of particular instances of objects used in a model) into tables which are also included in the frames. Write methods, on the other hand, set parameters of the instances used in a model on the grounds of values registered in those tables. The frame StatNet involves other sub-frames with tables, methods and other standard objects for statistical evaluation. Finally, the frame RandomCtrl is designed for an automatic management of random number streams. It guarantees that each object gets a unique seed value. Important is that central parameterization severs the inheritance relation between the instances in the model and their classes, so that subsequent change of parameters of those instances by parameterization of the classes is not possible. [12.4]

12.4 Structure of Objects of the Modular Library 'Automotive'

The structure of most of the objects from the folder .User_Appli_Objects is uniform. The majority of those 'Automotive' objects have their own dialog window with one or more tabs for setting basic parameters. On the dialog window there is usually an 'Open element' button which opens the frame (i.e. the station level of

[1] Ausführungsanweisung Ablaufsimulation in der Automobil- und Automobilzulieferindustrie.

the object) where all base objects (public elements) used for modeling the particular 'Automotive' object are used – user methods, user variables and tables (apart from material flow objects). Usually, there is another frame named nw_Private in the public frame of the 'Automotive' object (see subsection 12.3.2.1). In the nw_Private frame there are standard methods, variables and tables which the user is not expected to modify.Tehy are triggered to ensure the right functionality and call user methods. [12.4]

The structure described above is illustrated in Fig. 12.11 where the object Facility_1St_AssemblyVar is used as an example. In the background of the upper left part of the figure there is the default icon of the object which is placed in the root frame. In the foreground there is its dialog window with the button 'Open element' marked. The button opens the frame of the object depicted in the upper right part of the figure. In this frame there is an icon nw_Private for the frame with non-public elements. The frame nw_Private is then shown in the lower part of the figure.

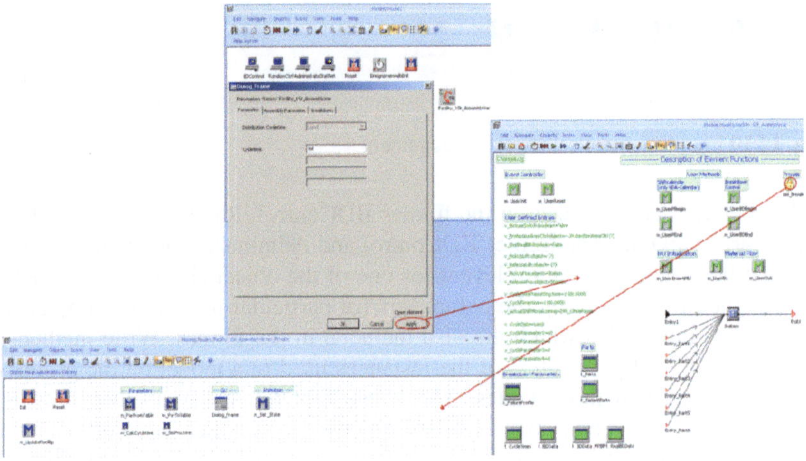

Fig. 12.11 Structure of the object Facility_1St_AssemblyVar

Basically, in the library there are two groups of objects for modeling production units. [12.4]

- The first group consists of so called generic facilities. These can be used for modeling on the protection area level. Their internal structure, i.e. structure of their station level, can be created using various (usually) basic objects interconnected between the in- and out-interface.
- Those objects which already have a ready to use structure on the station level belong to the second group. Instances of those objects are to be modified by setting the parameters displayed in their dialog window only.

In Fig. 12.12 the difference can be seen. The object Facility has been used as an example of a generic facility. Its icon in the root frame as well as its station level with no prearranged structure is shown in the left part of Fig. 12.12. The object Facility_1Station represents, on the other hand, objects with ready to use structure on their station level. Its icon and its frame can be seen in the right part of Fig. 12.12.

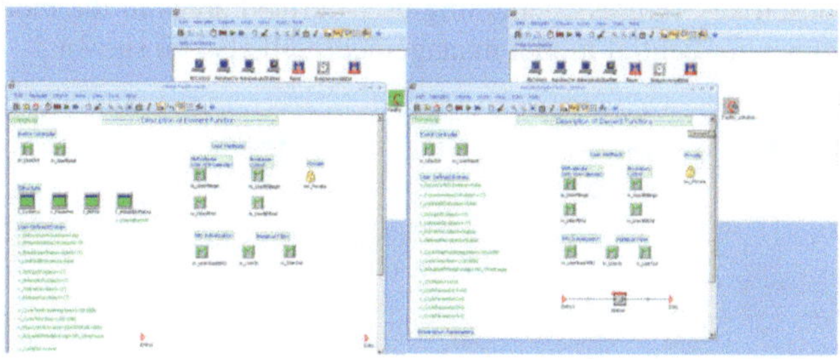

Fig. 12.12 Generic facility (left) and (pre-modeled) facility (right) – station levels and icons

12.5 Examples of Simple Models Using Point-Oriented Objects from the Modular Library 'Automotive'

In this chapter two models based on point-oriented objects of the library 'Automotive' will be presented. The focus of each model is placed on particular functionalities which can be achieved thanks to the features of the objects, their special methods, variables, tables etc.

The modular library provides the user with a range of point-oriented objects for effective simulation of various machines which can be found in body shops (laser welding machines, lifts, assembly or inspection stations etc.). An overview of these objects, which are embedded in the toolbox, is shown in Fig. 12.13.

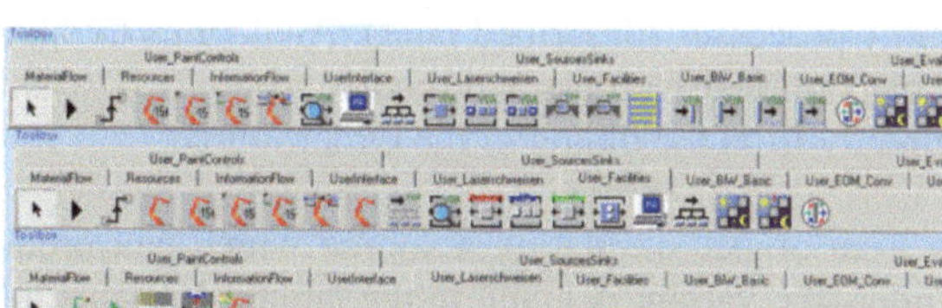

Fig. 12.13 Body shop components of the library 'Automotive'

The first model, which will be presented, illustrates modeling of a Kanban system with the use of appropriate 'Automotive' objects. The second one represents a hypothetical production process in a body shop. It comprises of other typical point-oriented objects from the library 'Automotive'.

At the beginning of each of the following sections dedicated to the models there is a list of used objects from the library 'Automotive'.

12.5.1 Model of a Kanban System

The main aim of this model (depicted in Fig. 12.14) is to show various settings of 'Automotive' objects which are used to simulate a kanban production system. Collection of chosen key figures concerning user-defined areas in the model with the use of the frames for central model management will also be shown.

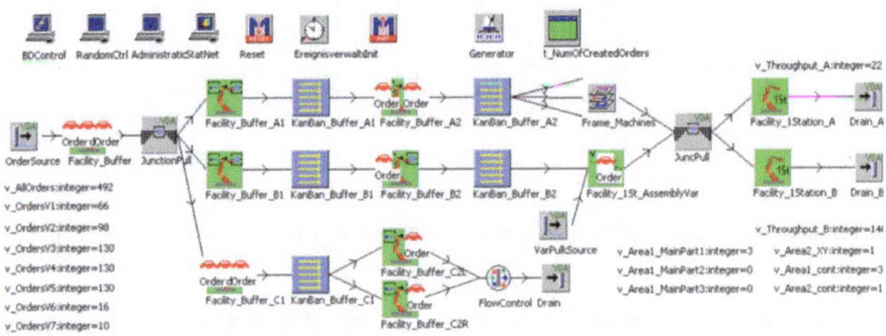

Fig. 12.14 Model of a KanBan system

Apart from built-in objects, the following objects from the library 'Automotive' have been used in the model:

OrderSource – an object simulating a kanban source. It produces MUs Order ('Automotive' object derived from the built-in object Transporter; basically an MU with a range of additional attributes) when the object KanBan_Buffer orders them. Simultaneously with the creation of a new Order it assigns the name of the ordered product into the MU-attributes Variante1 and Premid. Premid is an internal variable of type table used by the ordering mechanism and during variant-dependent assembly processes (described later). Analogous to other 'Automotive' objects, the user can optionally access and modify methods 'UserInit', 'UserReset' and 'UserOut' directly from the dialog of the OrderSource.

VarPulkSource – an object simulating a source which can produce several variants of MUs Order (the variant is given in the table t_Variants in which the user can set the size of batches and their ratio). Besides, the number of MUs to be created (-1 stands for an unlimited source) and creation of batches, compliance of a production program or random distribution of variants can be set up. In Fig 12.15 the dialog window of the source feeding the sequent assembly station Facility_1St_AssemblyVar and its table t_Variants (on the right) with two variants X and Y to be produced are shown.

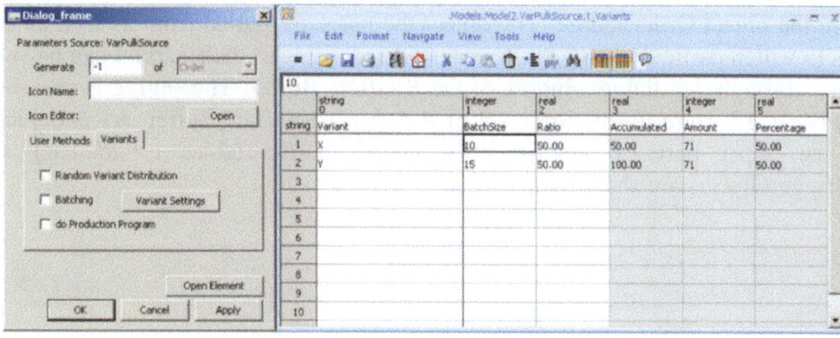

Fig. 12.15 Settings of VarPulkSource (left) and its table t_Variants (right)

Facility_Buffer – an object simulating several stations in series, where the user can set their cycle time, capacity of the buffer (i.e. number of stations in series) and a specific behavior mode – see Fig. 12.16 Furthermore, the failure profile of Facility_Buffer can be set through the dialog window (tab Breakdown).

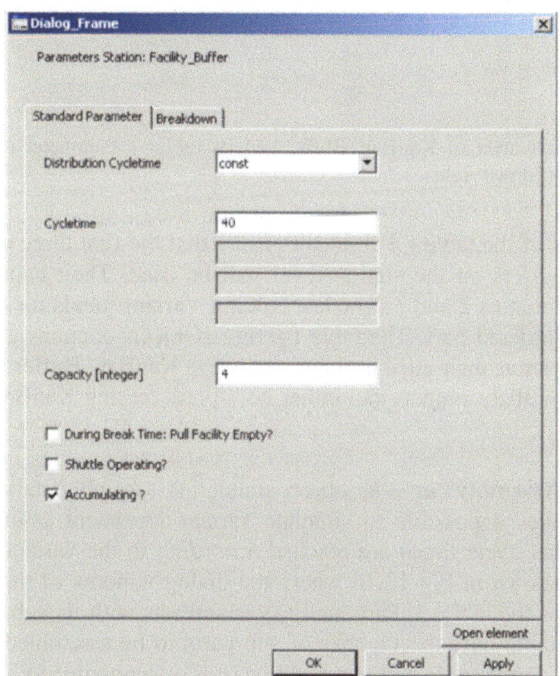

Fig. 12.16 Dialog window of Facility_Buffer, tab Standard Parameter

KanBan_Buffer - each object KanBan_Buffer, which can represent up to ten parallel buffers, orders specific parts from either another (upstream) KanBan_Buffer or (in case of the first KanBan_Buffer in each parallel line of the model) directly from the object OrderSource. The object the parts are pulled from as well as buffers to which particular parts are transferred according to their variant can be set in tables through the dialog window of the KanBan_Buffer. The setting is presented in Fig. 12.17. Here the station level of the object KanBan_Buffer_A1 is depicted in the upper part of the figure. Contents of the tables t_Parameter and t_TriggerObjects are shown, too.

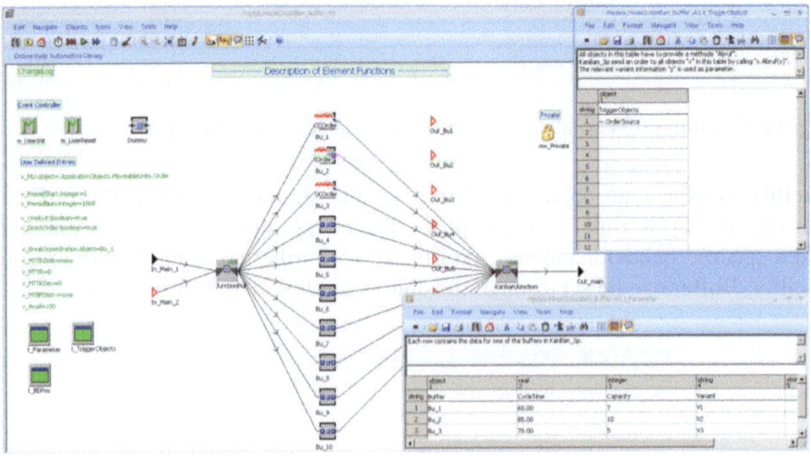

Fig. 12.17 Station level of KanBan_Buffer and its tables t_Parameter (above right) and t_TriggerObjects (down right)

The settings of the table t_Parameter ensure that the first three of the ten available parallel buffers on the station level will be used. Their capacity and cycle time is set in columns 2 and 3. The last column Variant stands for the name of the variant of the ordered part. The table t_TriggerObjects contains all objects (with the method Order as their attribute) the respective KanBan_Buffer orders the parts from. Basically, these objects can either be a predecessing KanBan_Buffer or an OrderSource.

Facility_1St_AssemblyVar – an object simulating assembly stations. Above all, this object makes it possible to simulate variant-dependent assembly processes where individual cycle times are reached according to the variants of assembled parts. This is shown in Fig 12.18 where the dialog window of this object is pictured as well as the tables t_Parts and t_VariantParts with its subtable Code_V1. In the former table only the number of sub parts to be assembled with the main part can be set. In the latter table, in addition, it is also possible to set up extra time (even as a negative value) required for assembling every single variant of sub part with the main part. Apart from this, cycle time and failure profile of Facility_1St_AssemblyVar object can be set up. Everything can be accomplished using the dialog window of this object.

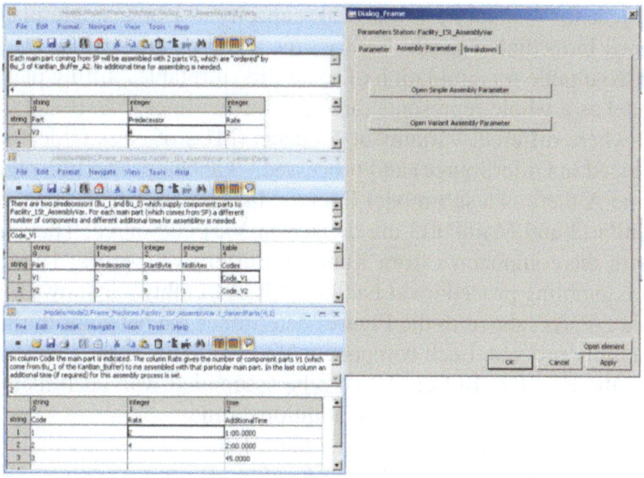

Fig. 12.18 Settings of assembly parameters according to variants to be assembled

Facility_1Station – a place-oriented object with the capacity = 1. It is used to simulate single workstations (similarly as the built-in object SingleProc). Cycle time and failures of this object can be set through its dialog window which also has the 'Open element' button for opening its station level. It can be part of a protection area or a group of stations which work in a shuttle mode.

JuncionPull and **JuncPull** – both objects can be used to simulate the function of rail-switches, transfer slide carriages or any other equipments which consume time while dividing or joining the material flow (the appropriate times are to be set in tables which are accessible from the dialog windows of these objects; by zero times the same function as the one of the built-in object FlowControl is achieved). The main difference between JunctionPull and JuncPull is that JuncPull enables to determine input and exit strategies separately. Furthermore, JuncPull can work in a kanban mode by selecting the checkbox 'Only pull if target empty and operational' on its dialog window – see Fig. 12.19.

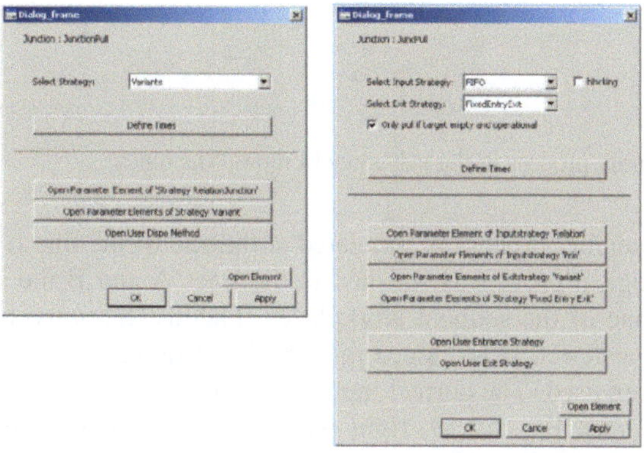

Fig. 12.19 Dialog windows of the objects JunctionPull (left) and JuncPull (right)

SP – an object for simulating a line with capacity = 1 or a single place between two elements. In its dialog window entry time and lock time can be set so that this object is also suitable for simulation of basic lifts, turntables, lifting platforms etc.

The model as a whole represents three production lines – A, B and C (from top to bottom) where different variants of products (MUs of class Order) are ordered (to be produced in OrderSource) and processed – kanban production regime.

In the line A there is a sub-model Frame_Machines in which main parts MainPart1, MainPart2 and MainPart3 are created in VarPulkSource. They are then assembled with the components from KanBan_Buffer_A2 (see Fig 12.20). Parameters of the assembling procedure at Facility_1St_AssemblyVar shows Fig. 12.18.

The table t_Parts (above in the Figure) determines the behavior during the non-variant assembly process. Each component V3 coming from the third parallel buffer Bu_3 of the KanBan_Buffer_A2 is to be assembled with each main part from VarPulkSource. This does not demand additional time.

In the table t_VariantParts (in the middle of the Figure) assembling of main parts with components V1 and V2 from KanBan_Buffer_A2 is set. Here, on the other hand, extra time is required. Subtables Code_V1 and Code_V2 in the last column of t_Variants in each line (i.e. for each component V1 and V2) contain information about the assembly process. According to the settings of the subtable Code_V1 for the component V1 (below in the Figure), two pieces of V1 are to be assembled with MainPart1 (requiring 1 minute additional time), four pieces of V1 with MainPart2 (consuming 2 minutes additional time), while the MainPart3 remains without any component, though requiring 45 seconds extra to be processed at the Facility_1St_AssemblyVar. This is how variant dependent processing can be simulated. A similar setting is applied in the subtable Code_V2 for the component V2 (from buffer Bu_2 of KanBan_Buffer_A2).

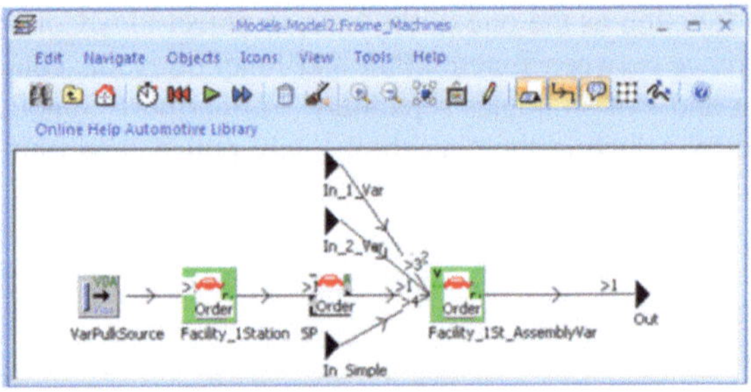

Fig. 12.20 Assembly process in the sub-model Frame_Machines

A similar assembly process is simulated at the end of the line B in the root. After the technological process is finished in the lines A and B the assembled parts are shifted to one of the sequent workplaces (Facility_Stations) according to the line they are transferred from. Since this shifting requires certain time, the object JuncPull has been used. The correct input and exit strategy has been set through its dialog window (see Fig. 12.19, right part). After the final treatment at Facility_1Station_A and Facility_1Station_B the material flow ends in drains.

The line C shows a sector of production where parts are sent from two parallel kanban buffer lines (of KanBan_Buffer_C1) to two parallel Facility_Buffers and then these parts continue to one common drain. Apart from the structure of the station level shown in the upper part of Fig. 12.17, in case of KanBan_Buffer_C1 the buffers Bu_1 and Bu_2 are directly connected with interface objects Out_Bu1 and Out_Bu2, i.e. the KanBanJunction on the station level is bypassed. From Out_Bu1 and Out_Bu2 the material flow continues separately into two sequent Facility_Buffers. This has been achieved by deselecting the check box 'OneExit' in the dialog window of KanBan_Buffer_C1.

The model also gives an overview of the number of created Orders (there are in total seven different MUs of the class Order which the OrderSource feeds the KanBan_Buffers in lines A, B and C with). In the 'UserOut' method of the OrderSource incrementation of all Order variants is carried out – the figures are recorded in respective variables. Each of these variables has an observer which triggers the method m_NumOfOrders whenever the value of the variable changes. The method m_NumOfOrdres has been created as a User-defined attribute (of type method) of the table t_NumOfOrders. This method records values of those variables in the table. The table itself is then referenced by the chart which measures the number of created Orders (see Fig. 12.21, on the right).

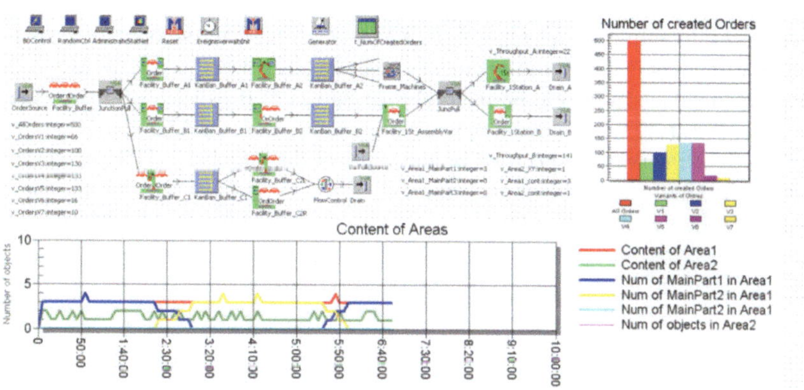

Fig. 12.21 Visualisation of statistical data

Furthermore, the functionality of the modular library 'Automotive' makes it possible (among other things) to measure the number of MUs or the throughput time in user-defined areas of the model. For central management of all relevant procedures the frame StatNet is designed – also see section 12.3.2.3 and Fig. 12.10.

To evaluate the (gross) throughput time in user-defined areas, the frame TPT_Gross located in the frame StatNet is useful – see Fig. 12.22. Its content is depicted in the lower right corner of Fig. 12.22. As there are two separate areas in the model where the throughput time is to be measured (Area1 reaching from VarPulkSource in Frame_Machines to Drain_A, Area2 from VarPulkSource in the root to Drain_B), two copies of the original frame TPT_Gross has been made

(TPT_Gross_Area1 and TPT_Gross_Area2) – they are marked with the red rectangle in Fig. 12.22. The methods m_RegisterIn and m_RegisterOut (called when entering or exiting the area respectively) are responsible for the right functionality. As one of the parameters of these methods the variant of the MU (i.e. the value of the MU`s attribute Premid, precisely Premid[1,1]) is passed.

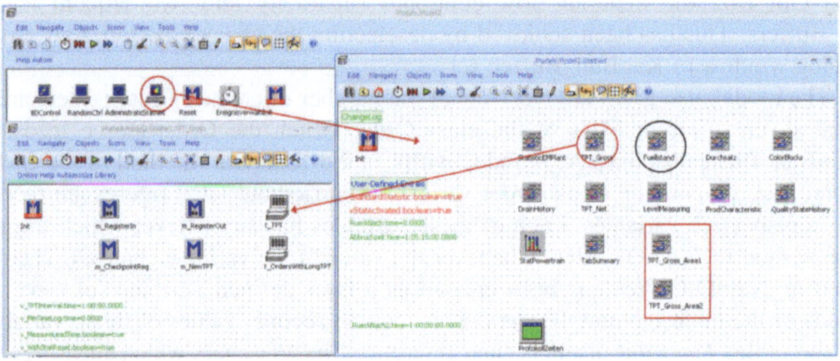

Fig. 12.22 Central statistical evaluation with the use of the frame StatNet

As a result, minimal, average, maximal and accrued throughput time and the number of entries is stored in the table t_TPT (Fig. 12.23 below). In its last column there are subtables in which the interval distribution of the throughput time is contained (in Fig. 12.23 the subtalbe 'table71' related to the MU MainPart1 is shown). The length of intervals is specified by the value of the variable t_TPTInterval in the frame TPT_Gross (here the copied frame TPT_Gross_Area1). Finally, the variable v_MinTimeLog enables the user to set a time span during which no data will be collected (e.g. during start-up time).

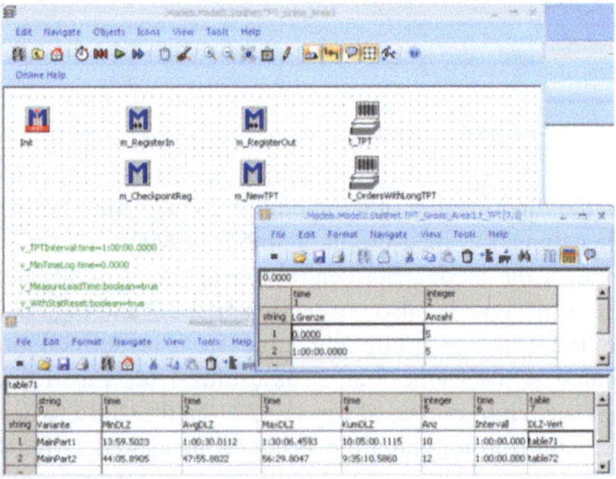

Fig. 12.23 Throughput time-related statistics in the user-defined area

For the measurement of the number of MUs in user-defined areas the frame Fuellstand (located within the frame StatNet – Fig. 12.22, marked with the black circle) is designed. There are three possibilities how (or where) the number of MUs can be observed (in single objects or frames, in user-defined areas or in objects of type Line_Buffer). Here the monitoring of the number of MUs in the same areas as in case of the throughput rate (above) is shown.

The structure of the frame Fuellstand and important tables it contains is shown in Fig. 12.24. The value 'true' of the variable TypAuswertung indicates a variant-dependent monitoring. In the table Merge (upper right corner of the Figure) each variant of MU, which is being transported through any of the area, is classified as a type (at the most four different types can be distinguished). Orders X and Y, which are generated by the VarPulkSource in line B, are classified as one common type (Type4). In the table Fuellung_Aktuell (below in the Figure) names of the observed areas and current number of MUs (as a total number in column 1 and separate values for each type in columns 2, 3, 4 and 5) in the respective area are shown. The entry mechanism is accomplished by the method Verw_Fuellst which is to be called by each MU entering or exiting the respective area. The name of the area, variant of the MU and increment are parameters of this method (theoretically the increment can be any integer number – e.g. for incrementing or decrementing all entities in a carrier integer > 1 will be needed).

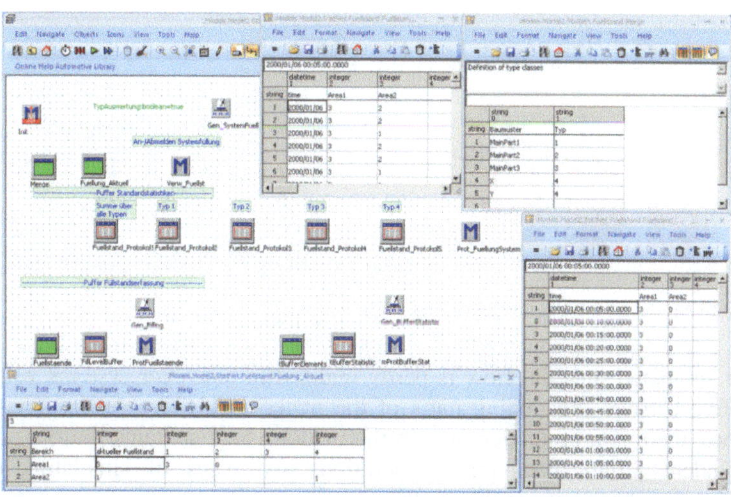

Fig. 12.24 Structure of the frame Fuellstand and tables Merge, Fuellsand_Protokoll, Fuelstand_Protokol2 and Fuellung_Aktuell

In the time intervals, which are specified by the generator Gen_SystemFuell, records from the table Fuellung_Aktuell are copied into the first empty row after the last record in tables Fuellstand_Protokoll-5. This gives an overview of time development of the number of MUs in user-defined areas. In the table Fuellstand_Protokoll1data for all types are stored altogether. In tables Fuellstand_Protokol2-5 data for each defined type are stored separately.

Additionally to this functionality of the 'Automotive' frame StatNet described above, data from tables Fuellstand_Protokoll1-5 are regularly set as values of variables in the root of the model (this is accomplished by the object Generator in the root). The time intervals of this generator are identical as those of the Generator Gen_SystemFuell in the frame Fuellstand, so that the variables immediately show the last entered data in those tables. Values of the variables in the root are referred by the chart Content of Areas (see Fig. 12.21).

12.5.2 Model of Body Shop Production Line

This model (depicted in Fig. 12.25) presents a hypothetical production line in the automotive industry. It will be shown (among further objects from the 'Automotive' library) how the functionality of further frames for central model management can be used to easily set parameters of all objects in the model and how to gain comprehensive statistical data.

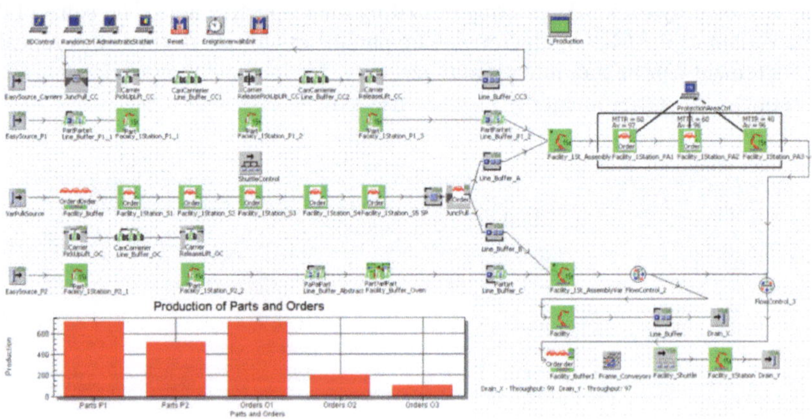

Fig. 12.25 Model of a body shop production line

The simulated material flow is partly realized in conveyor systems where the respective carriers (containing parts of the material flow) serve machines. Furthermore, there is a protection area in the model which contains stations with interdependent failure behavior. In addition, two ways of simulating shuttle processes have been used in the model.

Apart from built-in objects and some of the objects from the library 'Automotive' which are described in the section 12.5.1, the following further objects from the library 'Automotive' have been used in the model:

PickUpLift, ReleasePickUpLift, ReleaseLift – objects simulating lifts with a carrier which picks an MU of class Order or Part from a facility, feeds a facility with an MU and picks it up after the MU has been processed in this facility, or feeds a facility with an MU (respectively). These objects can be part of a closed conveyor system where a certain number of carriers circulate (the variable v_DeleteCreate on the station level of these objects is set at false) or the conveyor system is not simulated as a whole (the variable v_DeleteCreate is set at true then, which means that new carrier has to be created/ deleted each time an MUs it to be picked up from/ released in an operated facility). Furthermore, time parameters such as transport times in both directions (upward and downward), bolt time etc. can be set through the dialog windows of these objects – see Fig. 12.26. They will be automatically recalculated and entered into the table t_Times which is on the station level.

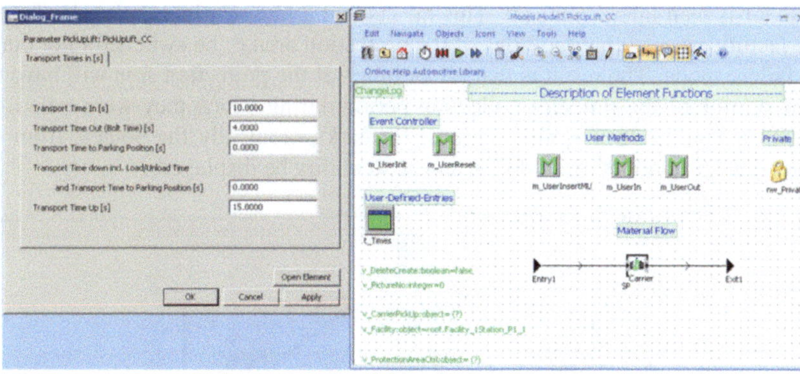

Fig. 12.26 PickUpLift – dialog window (left) and station level (right)

PickUpLift_X_To_1, ReleaseLift_X_to_1 – objects with a similar functionality as those described above. Here it is possible to pick more MUs from a facility or place more MUs in a facility at once – see Fig. 12.27.

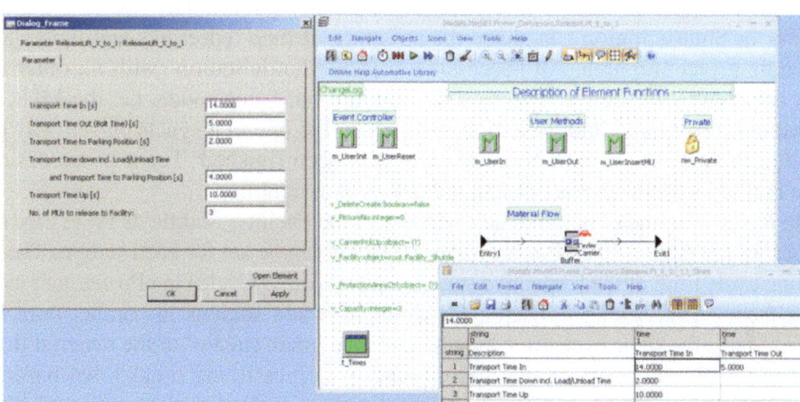

Fig. 12.27 ReleaseLift_X_To_1 – dialog window (left), station level and table t_Times

To determine the right facility the particular lift should operate, drag-and-drop mechanism can be applied with the instances of the appropriate lift objects. In this way the value of the variable v_Facility on the station level of the appropriate lift will refer to the operated facility. Simultaneously, the variables v_PickUpLift (or v_ReleaseLift) will refer to the particular lift and v_PickUpPos (or v_ReleasePos) to the station on the station level of the foperated acility.

ProtectionAreaCtrl – an object for simulating protection areas which encompass facilities (such as Facility_1Station, Facility_1St_Assembly etc.), whose failure behavior is then interdependent. This means that a failure which occurs/ ceases in one facility causes other facilities of the protection area to be switched into/ out of the failure mode. The result is that all facilities of the protection area will have the same portion of the statistics collection period during which they were failed. As shown in Fig. 12.28 it is also possible to mark graphically the objects from the protection area and let their MTTR and Availability be displayed.

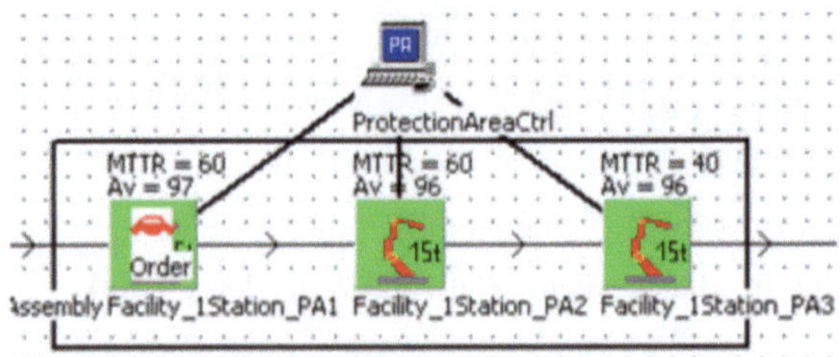

Fig. 12.28 Protection area consisting of a sequence of three stations

Facility_Shuttle – an object from the group of generic facilities. This means that its structure (on the station level) has to be created by the user. Objects Station, Buffer or ShuttleStation can shape the structure. It then represents and works as a separate protection area. All successive objects ShuttleStation (with no other object between them) define a section with shuttle operating mode, i.e. the MUs are transported from one ShuttleStation to a successive one in a synchronized way only after the process at each ShuttleStation has been finished. These processes can also include variant-dependent and independent assembling.

In the tab Parameter of the dialog window of the Facility_Shuttle its cycle time can be set. In the table t_CyclePos a cycle time factor can be set for each station. The resulting cycle time of the respective station equals then v_CycleTimePresetting / cycle time factor. This setting is useful, for example, when stations in parallel branches have to have the same cycle time as stations in the 'main stream' of the material flow.

The tab Breakdowns of the dialog window serves to set breakdown parameters of the whole Facility_Shuttle. These are then applied to the station which is referred by the variable v_BreakDownStation. Failure profiles of other stations in the Facility_Shuttle will show 100% availability, though stations which should reflect the same failure behavior as this 'leading' station referred by the variable v_BreakDownStation can be entered in the table t_BDPos. The effect is a simultaneous failure states initiation and cessation of all stations entered in the table t_BDPos according to switching instants of the 'leading' station.

Finally in the table t_Pause the behavior during pauses (in a shift plan) is set. The letter 'P' means that the station entered in the respective row will be paused. The letter 'E' means that the entry of hat station will be locked during the pause only (the exit stays unlocked). An MU which is situated in the station at the moment of the pause could leave the station. This can be useful when simulating an oven, for instance. The described station level of the object Facility_Shuttle and the tables mentioned above are depicted in Fig. 12.29.

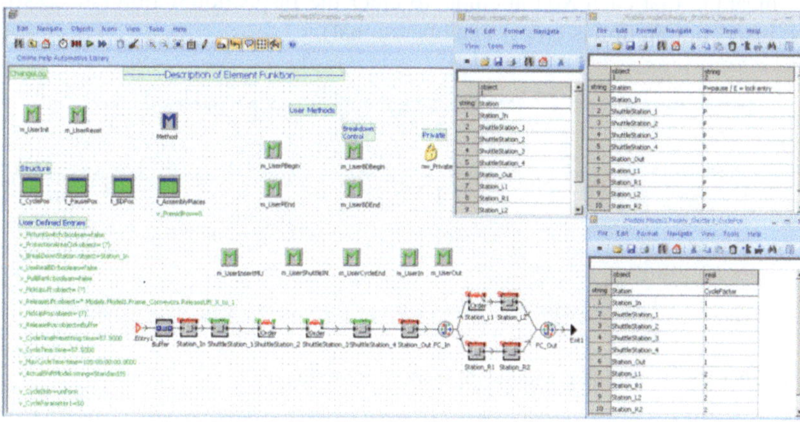

Fig. 12.29 Station level of Facility_Shuttle and tables t_BDPos (above), t_PausePos (above on the right) and t_CyclePos (below on the right)

Facility_1St_Assembly – an object for simulating simple assembly processes (independent on variants of the assembled parts/ components). Up to six components this object can be supplied with using the interfaces Entry_Part1-6 in its station level. The interface Enry 1 is reserved for the connection with the predecessor which supplies main parts. Through its dialog window cycle time and break down settings can be accomplished.

Facility – an object from the group of generic facilities. Again, its arbitrary (up to a certain extent) structure on the station level can be created by the user while the whole Facility then works as a protection area. Objects such as Station, ShuttleStation, InspetionStation (vide infra) can be incorporated in the structure – they have parameters, methods or attributes preset so that they can automatically communicate with the management of the frame of the Facility (or Facility_Shuttle), which can be regarded as a container for the whole user-defined structure.

From the dialog window of the Facility the same parameters can be set as in case of Facility_Shuttle. The internal variables, methods and tables are basically congruous, though methods for shuttle process are missing in case of Facility.

InspectionStation – an object which simulates a quality control station. Its dialog window has three tabs (see Fig. 12.30). On the tab Parameter inspection duration and on the tab Breakdown MTTR, MTBF and availability of the InspectionStation can be set. On the third tab Inspection Parameter first the Not OK Probability (between 0 and 1) can be set (i.e. with this probability, the MU's free Attribute Status will get the value NOK). Then one of three possible strategies for sending the MU from a preceding station through its second successor to the InspectionStation can be chosen with decreasing priority, respectively: 'Check nth. MU' meaning that an n-th MU is sent to the InspectionStation, 'Inspection Part' meaning that the given percentage of MUs is to be sent through the InspectinoStation or 'Next possible MU' meaning that each MU attempts to enter the InspectionStation (successfully if the InspectionStation is empty and operational).

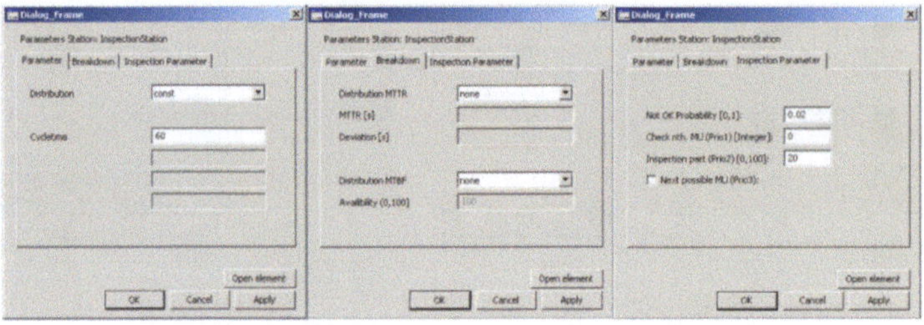

Fig. 12.30 Tabs of the dialog window of InspectionStation

Station – an object also suitable for incorporating in a structure of a generic facility. It is identical to the built-in object SingleProc, newertheless it has a range of user-defined attributes. Its entrance and exit methods are preset – e.g. the exit method tests whether the second successor of Station is InspectinoStation or not.

ShuttleStation – this object has necessarily to be incorporated in the structure of each Facility_Shuttle. It has several user-defined attributes and reference to entry and exit methods (which are in the frame nw_Private of the respective Facility_Shuttle) are preset. Especially, the position of the particular ShuttleStation in a series of interconnected ShuttleStations has to be detected – the method Init from the frame nw_Private of the Facility_Shuttle searches for the first ShuttleStation, then it loops all ShuttleStations and finally identifies the last one.

The model as a whole (see Fig. 12.25) represents a body shop section where two different components and three variants of main part are processed. The components P1 and P2 (MUs of class Part) enter the model from the sources EasySource_P1 and EasySource_P2, respectively. Both sources have their 'UserOut' methods amended so that the number of created MUs is incremented in the table t_Production (placed in the root and referred by the chart Production of Parts and Orders) whenever a new MU leaves the source. The method m_UserOut with the same code is called by MUs of class Order by exiting the source VarPulkSource which produces three variants of Orders (see Fig. 12.31). As it can be seen in this Figure, no batches are created (in contrast to the VarPulkSource in the model of a Kanban system in Fig. 12.15).

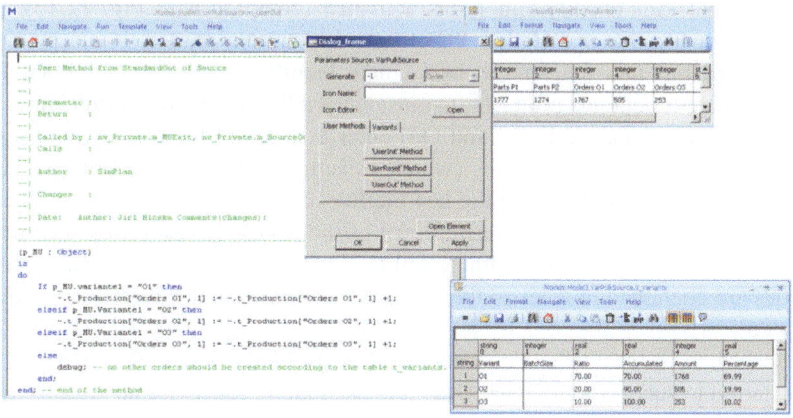

Fig. 12.31 Dialog window of the VarPulkSource (in the middle), its variant settings (below), the table t_Variants (above) and method m_UserOut (in the background)

EasySource_P1 feeds Line_Buffer_P1_1 with components P1 which are then available for Facility_1Station_P1_1. After finishing the procedure at this station P1 is picked up by a Carrier at PickUpLift_CC and transferred to Facility_1Station_P1_2. The 'Automotive' MU Carrier has been derived from a built-in MU Container (varying in its dimensions and a range of free attributes from the original). At Facility_1Station_P1_2 the component is released by ReleasePickUpLift_CC to be processed and then picked up from this facility by the same Carrier again. Finally, at Facility_1Station_P1_3 the ReleaseLift_CC releases the component while the empty Carrier goes away. Carriers are circulating in a closed system (apart from lifts there are accumulating buffers – 'Automotive' objects Line_Buffer between them). At the start of each simulation run 9 MUs of class Carrier are produces by EasySource_Carriers. They enter the conveyor system through JuncPull_CC (with a blocking FIFO input strategy) – this object is used because certain time is consumed at this junction. Since the conveyor system is closed and fed with Carriers by a source, variables v_DeleteCreate (see above) in each lift in this system have to be set at false.

Similarly, components P2 from EasySource_P2 depart at Facility_1Station_P2_1 from which they are picked by Carriers of PickUpLift_OC. Further the Carriers accumulate in Line_Buffer_OC from which they continue to ReleaseLift_OC where each Carrier releases the contained component for further process at Facility_1Station_P2_2. Since an open conveyor system is simulated here the variable v_DeleteCreate is set at true in both lifts. This means that at

PickUpLift_OC a new Carrier is created whenever a component is to be picked from Facility_1Station_P2_1 and in ReleaseLift_OC this carrier is deleted after releasing the component.

Further in this line there is an object Facility_Buffer (see section 12.5.1) named Faility_Buffer_Oven which simulates an oven. Here the behavior mode 'During Break Time: Pull Facility Empty' is activated (compare with Fig. 12.16). This leads to leaving the exit of this object unlocked during pauses. It can be useful when simulating an oven, smelter or a similar facility where it is crucial that in all circumstances contained products stay inside for a limited period of time only.

Main parts (Orders O1, O2 and O3) enter a series of facilities synchronized using the object ShuttleControl (which is actually the built-in object Cycle). The references to the first and the last facility of the balanced line have to point to the objects Station on the station level of the respective facilities. The facilities are supplied by the preceding Facility_Buffer and the balanced line the facilities are contained in leads to the object SP which separates it from the sequent JuncPull.

According to the method m_UserExitStrategy of the JuncPull, Orders O1 are transferred to successor no 1 and Orders O2 and O3 to successor no 2. Entry times depending on predecessor (here there is only one predecessor – the balanced line) and exit times depending on successor are set in table t_Times (available from the dialog window of the JuncPull – see Fig. 12.19).

Orders O1 are assembled with components P1 in Facility_1St_Assembly and then they enter a protection area to which three facilities in sequence belong. Orders O2 and O3 are routed to Facility_1St_AssemblyVar. O2 demands one component P2 while consuming 5 additional seconds for the assembly, O3 assembles with three components P2 requiring 15 seconds in excess of the cycle time of Facility_1St_AssemblyVar. This is accomplished by entering the appropriate parameters in the table t_VariantParts and its subtables (similarly as shown in Fig. 12.18). In both assembly stations the exiting MU is Main MU (i.e. the respective Order) and the assembly mode is set at 'delete MUs'.

The flow of Orders O2 and O3 is divided in the FlowControl_2 so that O2 is routed to the following FlowControl_3 and O3 is routed to the object Facility.

The structure of the Facility is shown in Fig. 12.32. The meaning of the entries in tables t_PausePos, t_BDPos and t_CyclePos, which are also depicted in the figure, has already been explained with other information about Faclity_Shuttle (see above, also see Fig. 12.29 where corresponding tables in the station level of the Facility_Shuttle are also depicted). After this Facility exiting Orders are stored in the (accumulating) Line_Buffer. Their flow terminates at Drain_X.

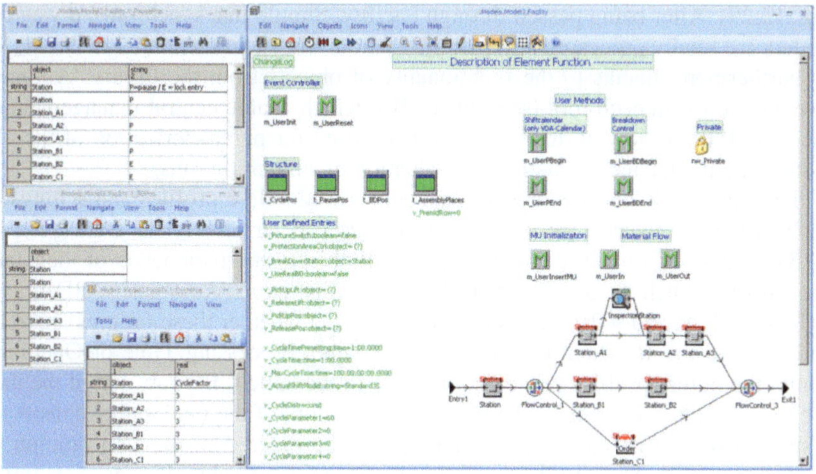

Fig. 12.32 Structure of Facility and its tables t_PausePos, t_BDPos and t_CyclePos

The flow of Orders O2 joins with the one of Orders O1 at FlowControl_3 with FIFO entry strategy. They then continue to Facility_Buffer1. At its last station StationOut on the station level three Orders are picked up by PickUpLift_X_To_1 which is placed in the instance of the frame Frame_Conveyors (see Fig. 12.33, below on the left, the station level of PickUpLift_X_To_1 with selected variable v_Facility pointing to the operated object Facility_Buffer1 is above on the left). In this frame a closed conveyor system is modeled. On the right of the Figure the station level of Facility_Buffer1 can be seen. In this the selected variable v_PickUpLiftObject refers to the station SP on the station level of PickUpLift_X_To_1 and the variable v_PickUpPos defines StationOut as the one Orders should be picked up from.

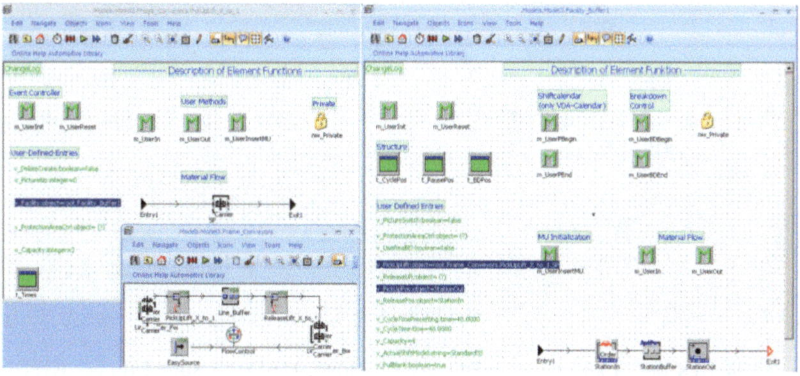

Fig. 12.33 Frame_Conveyors, station level of the PickUpLift_X_To_1 and Facility_Buffer

At ReleaseLift_X_To_1 the Orders are released at the built-in object Buffer which is placed in front of the sequent Stations and ShuttleStations in. These altogether form the structure of the generic Facility_Shuttle (see its structure in

Fig. 12.29). After being passed through Facility_Shuttle in a synchronized way the Orders are finally processed at Facility_1Station. They exit the model in Drain_X.

Furthermore, thanks to the functionality of objects from the library 'Automotive' contained in each root (see Fig. 12.10), which enable central management of the model, the following features have been used for parameterization of the objects in the model and for statistical evaluation of simulation runs.

The frame BDControl stores all data concerning failure profiles and settings of each particular object in the model. For this purpose the table t_BD_Data (see Fig. 12.34) is designed for. It contains a list of all objects and parameters of their failure profiles (such as MTTR, MTBF, availability etc.). Variables v_BDActive, v_FailureMode and v_BDTab_sorted in determine whether these failures are to be activated or not, to which time (Simulation Time, Processing Time or Operating Time) the failure profiles relate and whether in the table t_BDTab_sorted the data are to be sorted according to the frames the structure of the model may consist of, respectively. In that case these frames are to be entered in the first column of t_BDTab_sorted. Then In the second column there are subtables for each frame where the same data as in t_BD_Data will be stored, but only for objects which are contained in the respective frame in the model. Reading the parameters and entering them into the appropriate tables is accomplished by manually running the method m_Read_BD_Data. Should later any data in the respective table be changed, the method m_Write_BD_Data has to be manually run to write the up-to-date data as updated parameters to respective objects. Running the methods m_CalcAvaililibity or alternatively m_CalcDistribution ensures the consistency of MTTR, MTBF and availability of each failure profile.

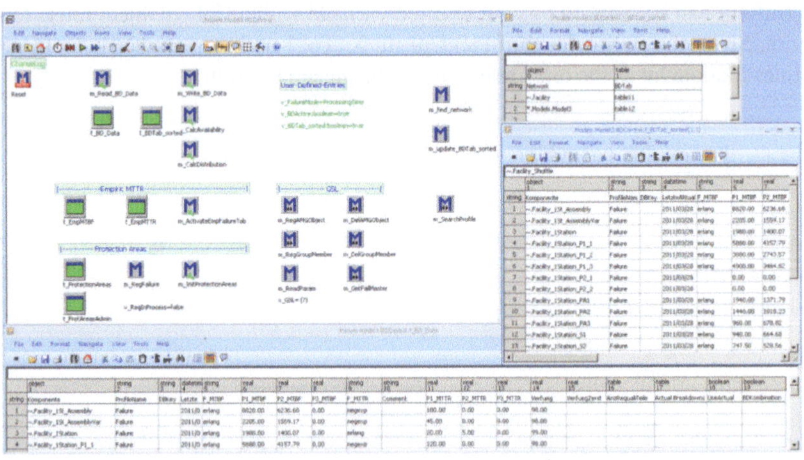

Fig. 12.34 BDControl and tables t_BD_Data (below), t_BDTab_sorted (upper right) and subtable t_BDTab_sorted [1,1] (middle right)

This enormously simplifies the management of individual failure profiles which, especially by large models with complicated structure, would be difficult. Moreover, BDControl also enables to use empirical distribution of MTTR and

MTBF or a combination of empirical and user-defined data. It also contains methods for generic standard solution. These can be used when generating a model automatically on the base of CAD-systems. In this way the failure parameters can be set by automatically created objects.

The frame RandomCtrl controls random number streams so that each object in the model has its own unique stream. The number of available streams can be changed in that the variable GenOffset (which is placed in the frame Param_BS of the Administration frame) is resized. An overview of all objects with their streams assigned can be checked in the table t_Streams of RandomCtrl.

The frame Administration is instrumental in managing parameters of point- and length-oriented objects, overhead and Power&Free conveyors and multifunction objects (turntables or transfers). In each subframe designed for one of those objects (with exception of subframes for overhead and Power&Free conveyors which have rather different functionality), a method for reading the parameters from the objects and entering them into appropriate tables and a method for applying parameters (if changed) from those tables to the respective objects is implemented.

In the model which is being described here, functionality of the subframe Param_BS has been exploited to centrally manage the settings of cycle time, breakdown parameters and shift plans for place oriented objects (see Fig. 12.35 where its icon has been marked and its internal structure is pointed at with an arrow). Here it is possible to structure the data according to the frames embodied in the model. For this the variable v_data_sorted is set at true and each frame is entered in the table t_Networks_sorted (marked with a rectangle in the structure of Param_BS depicted in Fig. 12.35 below). As a result the table t_CycleTime_sorted contains entries for each frame from t_Networks. In its subtables (one of them is shown in the bottom right corner of Fig. 12.35) not only cycle times (alternatively parameters of their random distributions) can be set but they also contain settings for capacities of buffers and shift plans. Preset shift plans are stored in the table PTMs (also included in Param_BS). Alternatively additional shift plans can be created and adapted by PTMs using the frame PTM_UGS.

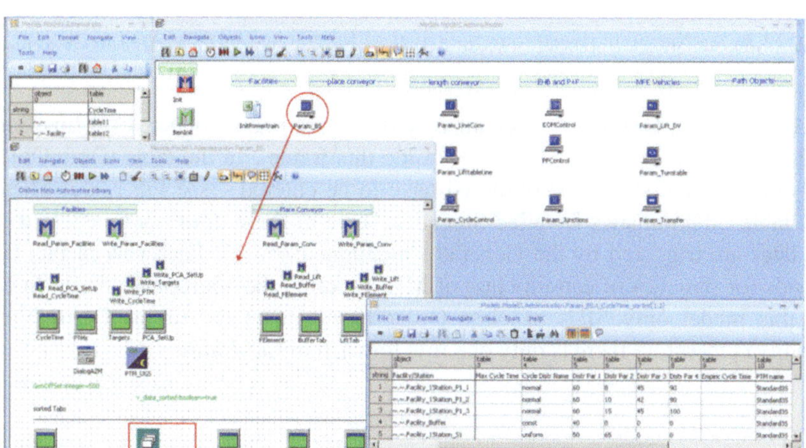

Fig. 12.35 Param_BS and tables t_CycleTime_sorted (upper left) and its subtable (below right)

Finally, it will be shown how various statistical data can be collected during and after a simulation run using the frame StatNet. The frames TPT_Gross and Fuellstand contained in StatNet were already used in the model of a Kanban system (see the subsection 12.5.1. and the Fig. 12.22 in which StatNet is depicted).

In the frame StatisticsEMPlant statistics 'Failed', 'Working', 'Blocking', 'Waiting' and 'Paused' of objects contained in t_CycleTime of Param_BS (see above) are monitored and stored in the table AntTab (see Fig. 12.36). Generally, any other object can be monitored, too in that it is additionally dragged and dropped into AntTab. In case of this model it can be found that Facility_1Station_PA1, Facility_1Station_PA2 and Facility_1Station_PA3 have identical statistics. The reason is that these elements all create one protection area. The data ale collected in intervals which are set by the Generator in StatisticsEMPlant.

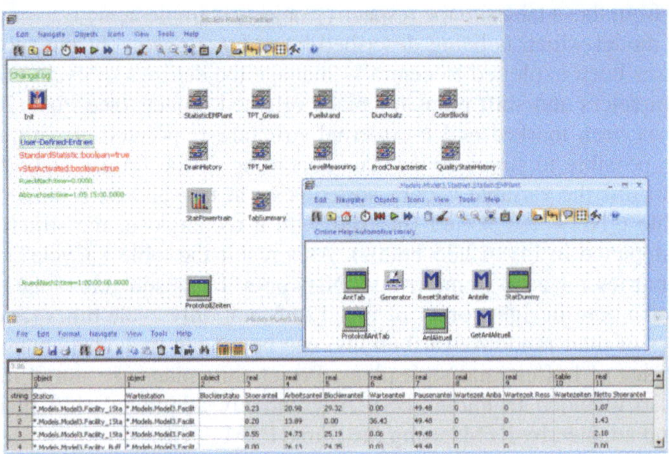

Fig. 12.36 StatNet (in the background), StatisticsEMPlant and its table AntTab (below)

In the frame DrainHistory detailed statistics are being collected (in time intervals which are set by the Generator inside this frame). In the table tDrainObjects all drains which should be observed have to be entered – see Fig.12.37. Then in the table tMUStatistics Detailed Statistics Table of these drains is transferred (the updates are triggered by the Generator mentioned above). Each row in this table stands for statistics of certain class of MUs entering the appropriate drain. In case of this model only MUs of class Order terminate at the drains Drain_X and Drain_Y, therefore there is only one row for each of the drains.

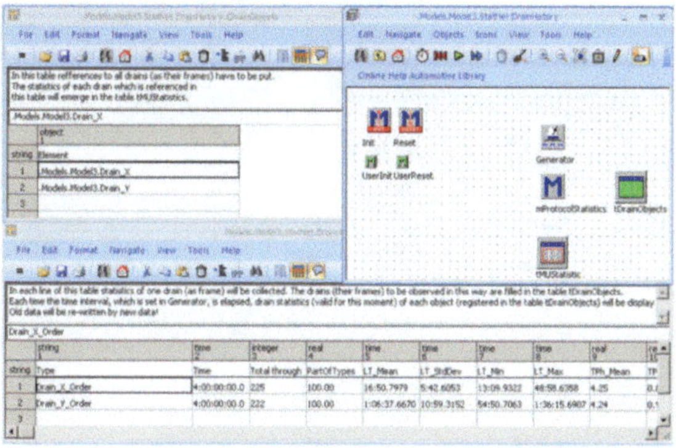

Fig. 12.37 DrainHistory (upper right) and tables tDrainObjects (upper left) and tMUStatistics (below)

In the frame TabSummary summary of statistics such as simulated availability, utilization, number of jobs per day or per hour, theoretical output etc. is entered at the end of the simulation run (see Fig. 12.38). These data refer to those objects which have been entered into the first column of the table (only those which have the object Station on their station level can be observed in this way).

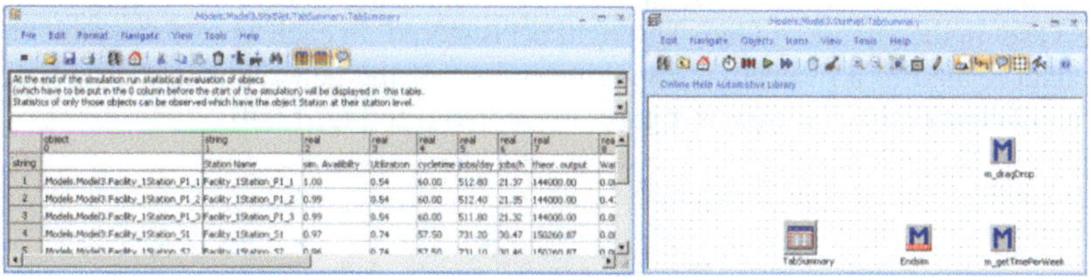

Fig. 12.38 TabSummary (left) and content of its table TabSummary (right)

In the frame LevelMeasuring occupation of groups of buffers can be displayed. For this, individual buffers have to be entered into the table BufferGroups while its first column is reserved for an arbitrary notation of the group and its second column for the batch size (larger than 1 only in case of MUs with capacity > 1 which can be useful when simulating stacking elevators or a similar equipment). In this model buffers in the open and the closed conveyor system have been grouped separately, further groups contain buffers in production lines for components P1 and P2 and for main parts (Orders) separately, buffers which separate the assembly stations from those lines and finally the last group of buffers contains the single Line_Buffer in front of Drain_X. The results (theoretical maximum of places in the buffer groups, maximal, minimal and average level etc.) are stored then in the table Buffer (see Fig. 12.39).

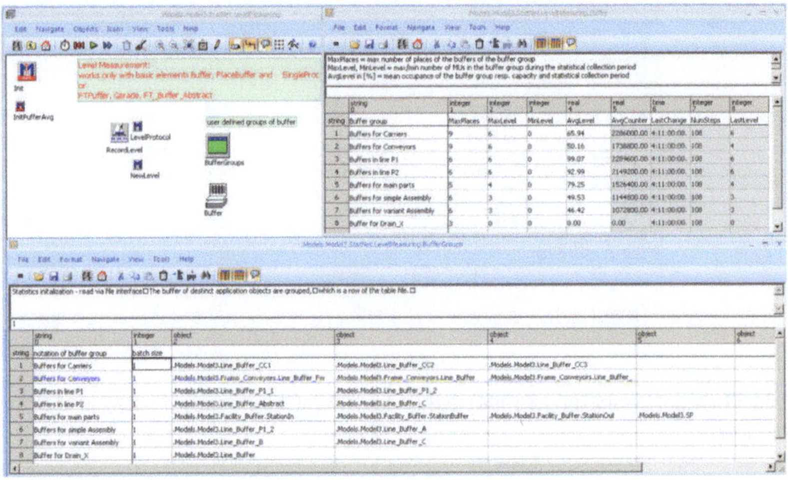

Fig. 12.39 LevelMeasuring and tables BufferGroups (below) and Buffer (upper right)

In the frame "Durchsatz" throughput of chosen objects can be observed. Automatically objects from the table CycleTime of the frame Param_BS (see Fig. 12.35) are monitored, nevertheless other objects can be added in the following tables. In the table TP_PerTimeInterval throughput over the last time interval (the length of intervals is set in the generator Gen_ThroughPut) is stored. In each row of this table there are data related to a respective triggering instant of the generator. In the table Durchsaetze cumulative throughput figures are stored (see Fig. 12.40).

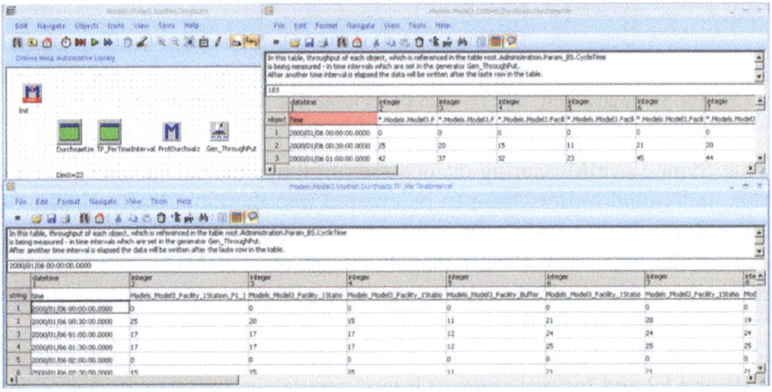

Fig. 12.40 Durchsatz and tables TP_PerInterval (below) and Durchsaetze (upper right)

Similarly, in the frame ProdCharacteristics throughput of objects, which are to be put in the table evalObjects, can be evaluated (see Fig. 12.41). Contrary to the data collection functionality of the frame Durchsatz (described above) here the throughput figures are sampled not only in hourly intervals (which is determined by the generator ProdChGen) – for these figures the table ProdCh_h is dedicated – but it is also possible to collect data at any times which are entered in the table changeShiftTime. This is meant for collecting data in intervals pertaining to certain shift plan (which makes sense when the observed objects simulate the same shift-work). Then in the table ProdCh_Shift throughput figures will be entered in time instants determined in the table changeShiftTime (usually quitting times or beginnings of shifts). Finally, in the table ProdCh_day the cumulative day throughput figures will be stored.

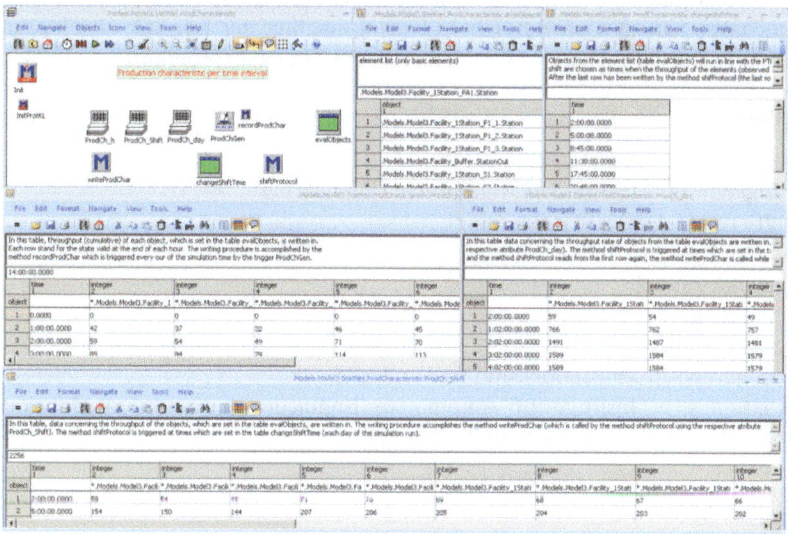

Fig. 12.41 ProdCharacteristics and tables evalObjects (middle above), changeShiftTime (above left), ProdCh_h (right), ProdCh_day (left) and ProdCh_Shift (below)

12.6 Conclusion

In this chapter point-oriented objects from the modular library ‘Automotive’ and its further elements with special functionality have been shown. Two simulation models exemplify the wide range of possible exploitation of this library, nevertheless they just confine to small part of the whole scale of this library. It can also be used for simulation of conveyor systems, logistic processes, for generic creation of models etc. Moreover, the library ‘Automotive’ is continuously being extended and improved by the Process Simulation Working Committee of the German Association of the Automotive Industry so that it meets requirement of an increasing number of users throughout the automotive industry.

Acknowledgments. I would like to express my gratitude to Mr. Carsten Pöge for his support and guidance during this project. I would also like to extend my appreciation to Mr. Steffen Bangsow for having given me the opportunity to write this contribution.

The VDA Automotive toolkit is common property of the companies which are owners of this modular library and are organized within the VDA workgroup process simulation.

Authors Biography, Contact

In 2010 the author finished the master study program Automotive and Material Handling Engineering at the Faculty of Mechanical Engineering of the Brno University of Technology with a thesis on Transformation of a simulation model from SW SimPro into SW Plant Simulation. Presently the author studies in a doctoral program Machines and Equipment at the same Faculty (the subject of study is Optimization of a material flow in the mass production using simulation methods). Within the scope of his studies he cooperates with the company Škoda Auto, a.s. Currently the author takes part in a project TiLO (Tracing intelligenter Logistik Objekte) at the University Duisburg-Essen as part of an ERASMUS study program.

Contact

Jiří Hloska
Mailing adress: Institute of Automotive Engineering
Faculty of Mechanical Engineering
Brno University of Technology
Technická 2896/2
616 69 Brno
Czech Republic
e-mail adress: yhlosk00@stud.fme.vutbr.cz
Affiliation: Institute of Automotive Engineering, Faculty of Mechanical Engineering, Brno University of Technology

References

[12.1] German Association of the Automotive Industry. German Association of the Automotive Industry in the innovations-report (c2000-2011), `http://www.innovations-report.com/html/profiles/profile-540.html` (accessed February 17, 2011)

[12.2] Clausing, M., Heinrich, S.: Mensch, Maschine, Material: die Standardisierung der Ablaufsimulation in der Automobilindustrie (2008), `http://www.virtuelle-fabrik.de/de/termine-medien/artikel/func-startdown/18/` (accessed February 17, 2011)

[12.3] German Association of the Automotive Industry, Hauptseite – VDAWiki (2011), `http://wiki.vda-ablaufsimulation.de/index.php/Hauptseite` (accessed February 17, 2011)

[12.4] Burges, U., Hilmer, F., Richter, K., et al.: AVDA-Automotive-BSK_Doku_Plant90_V012 (2010)

[12.5] Working Committee Process Simulation Ausführungsanweisung Ablaufsimulation in der Automobil- und Autozulieferindustrie. VDA UAG Ablaufsimulation (2008), `http://forum.vda-ablaufsimulation.de/attachment.php?id=218&` (accessed February 17, 2011)

[12.6] Mayer, G., Pöge, C.: Auf dem Weg zum Standard – Von der Idee zur Umsetzung des VDA Automotive Bausteinkastens (2010), `http://www.asim-fachtagung-spl.de/asim2010/papers/Proof%20103-3.pdf` (accessed March 25, 2011)

13 Using Simulation to Assess the Opportunities of Dynamic Waste Collection

Martijn Mes

In this chapter, we illustrate the use of discrete event simulation to evaluate how dynamic planning methodologies can be best applied for the collection of waste from underground containers. We present a case study that took place at the waste collection company Twente Milieu, located in The Netherlands. Even though the underground containers are already equipped with motion sensors, the planning of container emptying's is still based on static cyclic schedules. It is expected that the use of a dynamic planning methodology, that employs sensor information, will result in a more efficient collection process with respect to customer satisfaction, profits, and CO_2 emissions. In this research we use simulation to (i) evaluate the current planning methodology, (ii) evaluate various dynamic planning possibilities, (iii) quantify the benefits of switching to a dynamic collection process, and (iv) quantify the benefits of investing in fill-level sensors. After simulating all scenarios, we conclude that major improvements can be achieved, both with respect to logistical costs as well as customer satisfaction.

13.1 Introduction

The collection of waste is a highly visible and important municipal service that contributes to environmental pollution and traffic congestion, and involves large expenditures. Twente Milieu, a waste collection company located in The Netherlands, wishes to increase its corporate social responsibility and therefore searches for innovative and more efficient collection strategies. Twente Milieu is an important player in the field of waste collection and the maintenance of public areas. Its main activity is the collection of household refuse and in this area the company wants to improve the truck planning and container emptying as to save on fuel consumption, reduce CO_2 emission, and increase customer satisfaction.

Martijn Mes
University of Twente
School of Management and Governance
Dep. Operational Methods for Production and Logistics
P.O. Box 217
7500 AE Enschede
The Netherlands
e-mail: m.r.k.mes@utwente.nl

Twente Milieu operates different types of containers. The most important types are mini containers and block containers. Mini containers are located at every house and have to be emptied on pre-specified days, because residents have to put the containers along the side of the road. This is not the case with block containers, which are meant for a larger number of households and which are mostly located at apartment buildings or within the city centre. Since 2009, Twente Milieu also makes use of underground containers. At first, these underground containers mainly replace the block containers installed at apartment buildings and commercial buildings (e.g., at restaurants), but their use is now extended to all sorts of living areas. The underground containers offer several advantages: (i) they have a relative big storage capacity of 5m^3 which is roughly five times as big as the traditional block container, (ii) they are only accessible with an id-card which prevents illegal waste deposits, (iii) due to solid locking it decreases odor nuisance, and (iv) only a small part of the container is visible which makes the container suitable for use in public areas and contributes to an attractive environment.

Currently, Twente Milieu is unsatisfied with the average fill rate of the underground containers upon emptying. It is expected that, on average, the underground containers are less than 50% full upon emptying. As a result, one would expect that it is possible to reduce the emptying frequency, which results in less mileage of the trucks and less CO_2 emissions. The current planning methodology for emptying the containers is based on static and cyclic schedules. These schedules describe, for each container, at what days it should be emptied and how often, e.g., every Tuesday, or Wednesday once in the two weeks. Since deposit volumes fluctuate heavily, a static planning methodology requires a relative large amount of slack capacity. As a result, the average fill level upon emptying will be relatively low.

For the mini containers, a static planning approach is required because citizens have to place their containers at the street. However, for the underground containers, this approach is no longer necessary. Moreover, the containers are equipped with sensors that inform the company each time the container lid is opened. Twente Milieu expects that the introduction of a dynamic planning methodology, that employs this sensor information to estimate the fill levels, results in less frequent emptying and higher customer satisfaction. The additional advantage of using a dynamic planning methodology is the possibility to adapt the schedules to weather conditions or public holidays, to incorporate for example odor nuisance in warm periods, and to cope with changing patterns in deposit behavior. Finally, it is expected that additional efficiencies can be achieved by investing in fill level sensors, which provide more accurate estimates.

In this research we look at the different possibilities for a dynamic planning methodology with the aim to increase logistical efficiency and customer service. More specifically, we aim to find a method for container selection and routing that satisfies Twente Milieu's standard to save resources and to contribute to a cleaner environment. The goal of this research is the following:

> To asses in what way and up to what degree a dynamic planning methodology can be used by Twente Milieu to increase efficiency in the emptying process of underground containers in terms of logistical costs, customer satisfaction, and CO_2 emissions.

To reach this goal, we formulate the following research questions:

1. How should a dynamic waste collection strategy be designed?
2. What would be the benefits of changing to a dynamic waste collection strategy?
3. What would be the added value of investing in fill-level sensors?

To answer these questions, we use a simulation. According to Law (2007), simulation is a suitable tool to evaluate complex real-world systems which cannot be analyzed analytically and where experimenting with the real system is impossible or too expensive. Besides the fact we face such a system, we use simulation because it allows us to (i) analyze a wide range of interventions, (ii) perform sensitively analysis, and (iii) benchmark to current way of working with the proposed new planning methodology.

The remainder of this chapter is structures as follows. First, we present related work in Section 13.2 In Section 13.3 we present the underground container project in more detail. The various possibilities for a dynamic planning methodology are presented in Section 13.5. In Section 13.6, we present our simulation model and we present our numerical results in Section 13.7. We end with conclusions and recommendations in Section 13.8.

13.2 Related Work

The problem we face is related to the so-called Vehicle Routing Problem (VRP) and, more specifically, the Inventory Routing Problem (IRP). First, we provide a short review on related research in these areas after which we present related research in the area of waste management. We end this section with a statement of our contribution.

In general, the problem we consider here belongs to the broad class of Vehicle Routing Problems (VRP). The VRP involves the design of an optimal set of routes for a fleet of vehicles in order to serve a given set of customers. The VRP arises naturally as a central problem in the fields of transportation, distribution and logistics (Dantzig and Ramser, 1959). It has been studied extensively during the last few decades, with solution methodologies ranging from exact mathematical programming techniques to heuristics. For an overview of approaches developed for the VRP we refer to Toth and Vigo (2001).

A specific instance of the VRP which is related to our problem is the periodic vehicle routing problem (PVRP) where customers may require service on multiple days during a given planning horizon. The challenge is first to determine service frequencies (e.g, a customer will be serviced twice per week) or service patterns (e.g., a customer will be serviced every Monday and Thursday), and then to solve the VRP each day using the assigned customers of that day. Early formulations of the PVRP were motivated by municipal waste collection and are developed by Beltrami and Bodin (1974) and by Russell and Igo (1979). Usually, the PVRP is solved heuristically and often using a two-stage approach consisting of a construction and improvement step. Chao et al. (1995) review some of these early heuristics, and propose a new heuristic to overcome issues of poor local optima. More recently, Cordeau et al. (1997) propose a tabu search algorithm for the PVRP. In all of these works, the service frequencies are pre-determined. Variants in which the service frequency is a decision variable can be found in Newman et al. (2005), Mourgaya and Vanderbeck (2007), and Francis, Smilowitz and Tzur (2006). For a literature review on the PVRP and its extensions we refer to Francis, Smilowitz and Tzur (2008).

A distinguishing feature of our problem compared to the PVRP, is that the service frequency is not something we have to determine at the beginning of a given planning horizon. Instead, each day we have to select the customers to visit using actual sensor information. In a way, the static planning methodology as currently used by Twente Milieu, can be seen as a solution to the PVRP. The problem class that combines vehicle routing with inventory management is the so-called Inventory Routing Problem (IRP). In an IRP, the following trade-off decisions are considered:

- At which point in time should a customer be delivered to fill up its stock? (selection)
- How much ought to be delivered in that situation? (demand determination)
- What is the best order and therefore route to deliver the set of selected customers? (routing)

The IRP differs from the VRP because it is based on the usage of customers rather than just the number of customer orders. As a result, solution methodologies for the IRP are suitable for planning the emptying's of sensor-equipped waste containers. The containers, ideally, should be full upon emptying, but at the same time they should not overflow. Our problem can be seen as a reverse IRP, or an IRP where the product to be replenished is empty space (air); we collect waste by filling the containers with empty space. The most important decision here is when to serve a customer.

Solving an IRP is difficult and even gets more complicated with the number of customers (Campbell et al., 1998). A crucial decision in IRPs is the choice which customers to include in the routes of the current period. With this short-term decision, we have to take into account the long-term effects of this decision since a short-term approach might postpone as many customers as possible to the next period (Campbell et al., 1998). Therefore, Campbell et al. (1998) propose two solution methodologies, (i) an integer program with a relative long horizon where subsets of delivery routes and aggregation of time periods are used to keep the program computationally tractable, and (ii) an infinite horizon Markov decision process (MDP). Jaillet et al. (1997) take a rolling horizon approach to tackle the differences between short-term and long-term solutions. They do this by determining a schedule for two weeks, but only implementing the first week. A common heuristic approach for the IRP is to distinguish between customers that have to be served in the current period (which we indicate as MustGo's) and those that might be served (which we indicate by MayGo's). To determine which customers should be served first, Golden et al. (1984) use the ratio of tank inventory to tank size. When this ratio is smaller than some threshold, customers are excluded from service for that day. Campbell et al. (1998) use a ratio of urgency to extra time required for the selection of customers. In this chapter, we use a similar approach with MustGo's and MayGo's. For a further literature review on inventory routing, we refer to Andersson et al. (2010).

A growing amount studies are dedicated specific to waste collection strategies. As McLeod and Cherrett (2008) state, efficient waste collection strategies are not only vital from economic perspective, but also from an environmental perspective with reductions in emission and traffic congestions. The common approach to model the waste collection process is to use the VRP; see, for example, Chang and Wei (2002), Kim et al. (2006), and Nuortio et al. (2006). Nuortio et al. (2006) propose a stochastic variant, because the amount of waste in the bins is highly variable. For solving the problem, they use a node routing approach. This approach makes it possible to consider each bin separately. Kim et al. (2006) describe a VRP that uses time windows. These time windows include stops for lunch breaks and disposal operations. For solving the problem, they use a clustering based algorithm. McLeod and Cherrett (2008) describe the routing and scheduling problem as an capacitated VRP, which has constraints on vehicle capacity and working hours and they propose different ways to solve this waste collection problem, such as tabu search, a genetic algorithm, and fuzzy logic methods. Karadimas et al. (2007) also point out the importance of an efficient collection process, because 60-80% of the total costs are spent during the waste collection process. To solve the problem, they use an ant colony system. Here, artificial ants (trucks) are searching the area for the optimal route for a given set of container locations. This is done by initially random cycling through the area and leaving a "pheromone trail" in the intensity of the found solution value of travelled kilometers. A route with a high pheromone density is more likely to be followed by the other artificial ants so that better routes are found. Chalkias & Lasaridi (2009) use a geographic information system (GIS) in their optimization of municipal solid waste collection. For the formulation of a model, they collected data about roads and bin locations. They state that the success of decision making depends largely on the quality and quantity of the available data, in which the geo-database can be very helpful. One remarkable conclusion is that fuel consumption relates more to the time of operation and the number of stops than to distance travelled. The reason for this is that most of the time is spent for loading and emptying.

In our problem, the travel distances are relatively small and drivers appear to have enough driving experience within the region such that the routing aspect has a lower priority. Instead, our focus is mainly on the selection of containers to be emptied in the current period. In this area, the most closely related research as that of Johansson (2005). This work focuses on the dynamic collection of waste from 3300 containers (aboveground) in the Swedish city Malmoe. Similar as in our research, they use discrete event simulation and analytical modeling in order to access the performance of the waste collection procedures proposed. They conclude that dynamic routing decreases the operating costs and hauling distances, increases the length of the collection cycle per container, and causes a reduction in labor costs. The containers considered by Johansson (2005) have two infrared optical sensors that provide real-time access on the fill status of each container, which can be used to access a MayGo level and a MustGo level. If the inventory in a container reached its MustGo level, it should be emptied within a fixed period of time. Containers with a waste level below the MayGo level were not allowed to be included in the emptying routes. Different policies where considered varying from static to dynamic. They conclude that with relative large systems (>100 containers), the 'most' dynamic variant (dynamic scheduling, dynamic routing, and always using MayGo's) performs best. It is further concluded that the highest savings of this dynamic policy are achieved in unstable environments with high demand fluctuation.

As seen in the short summary of existing literature on waste collection, most articles are about routing problems; finding the optimal route along a set of containers. For Twente Milieu, the main emphasis is put on the selection of containers to be emptied since driving distances are relatively small and drivers are familiar with the area they drive in. This means that existing literature in the area of waste collection is less applicable to our problem. Also in the area of inventory routing, relative much attention is given to the routing aspect. Especially in dense areas, where the travel distances are relatively small, the selection of customers might even be more important than the routing decisions. The main focus of this paper is on customer selection; especially in the area of waste collection, this is a new research area. The theoretical contribution of this work is to show how models for the IRP can be used to improve the waste collection process and to quantify the benefits of such an approach.

13.3 Case Description

To be able to make a thorough suggestion about how a dynamic way of planning should look like and how it should be implemented, it is important to have a good understanding of the current way of working. This section describes the different aspects Twente Milieu deals with in relation with the process of emptying the underground containers. We start with a description of the company (Section 13.3.1) and the underground container project (Section 13.3.2). We then present the planning methodology as currently used by the company (Section 13.3.3) and end with our main findings from the data analysis (Section 13.3.4).

13.3.1 Company Description

Twente Milieu is a government-oriented enterprise owned by six municipalities located in the Netherlands. The main activity of Twente Milieu is waste collection and processing, but Twente Milieu also operates in the cleaning of streets and sewers, mowing of verges, road ice control, and the control of plague animals. Twente Milieu belongs to one of the largest waste collectors in the Netherlands when it comes to the number of household connected to their network. In 2009, Twente Milieu was serving a total population of ca. 399,000 inhabitants; the vast majority of them (77%) living in the three bigger cities Enschede, Hengelo, and Almelo. In 2009, ca. 215,000,000 kg of refusal was collected; this amount is expected to increase in the near future.

The mission of Twente Milieu is to offer high societal value at low costs and the preservation of natural resources. To do so, Twente Milieu tries to reduce waste wherever possible, encourage citizens to segregate waste, and to increase recycling opportunities in various manners. Twente Milieu has the vision to become and stay on of the pioneers in effective, fair, and societal responsible waste collection. This also forms the drive for the underground container project and the desire to become one of the first Dutch waste collectors that are actually working with a dynamic routing methodology in order to reduce costs, increase customer satisfaction, and to reduce the CO_2 footprint.

13.3.2 The Underground Container Project

The underground container project is one of the most prestigious and ambitious projects of Twente Milieu. As mentioned in the introduction, the underground containers have a number of advantages over mini containers and block containers. However, there are still several advantages that are not fully exploited yet, namely that fact that (i) we no longer need to make appointments with citizens on the timing of emptying's and (ii) the sensors provide information that enable Twente Milieu to retrieve deeper insights into the speed of the container fill process.

In March 2011, Twente Milieu operates 677 underground containers. This number increases continuously and it is expected that within a few years this number increases to 1500. Most of the containers are equipped with a motion sensor which counts the number of lid openings. Once a day, the number of lid openings is communicated, using GPRS, to a central container registration system. Furthermore, the containers are equipped with a digital lock that can only be opened by a participant-owned RFID-card. This enables the future introduction of Diftar, which stands for differentiated tariff for waste deposits. Most of the container locations have one container per location. However, there are also locations, mainly at large apartment buildings that have two or more containers at one spot.

Twente Milieu has five trucks available for emptying the underground containers. There are a number of drivers capable of driving these trucks; this requires some experience with driving a large truck through the small streets of city centers, and it requires experience with the crane that hoists the container out of the ground. Given the expected growth to 1500 containers, additional trucks will be acquired within the next few years.

13.3.3 Current Planning Methodology

Currently, the scheduling of container emptying's and the routing of trucks is done on a static bases with some deviations in the routes incorporated. The static planning methodology describes for each container on what days it should be emptied. Most containers are emptied on a weekly basis, some of them once in the two weeks, and some twice per week. Changes in the emptying schedule are rarely made, except on Fridays to avoid overflow in the weekends. Every Friday morning, a list with lid openings per container is printed to judge whether there are any additional containers that need to be emptied before the weekend. Another source of deviation from the static schedule is due to the drivers' freedom to pick more (or less) containers based on his experience. Since the resulting collection process heavily depends on personal perception and experience, switching drivers or hiring new drivers during holiday periods becomes problematic. In addition, it is difficult to cope with changes in the network, such as the addition of new containers.

Fig. 13.1 Truck emptying an underground container.

The truck driver starts his working day at 7.30 am when he receives a list with containers to empty that day. The exact order in which he empties these containers is determined by the driver himself without planning or navigation support. This is possible since drivers are familiar with the static set of customers that have to be emptied on the different days. All trucks depart from a central depot. When the driver arrives at a container location, he empties it with the use of a remotely controlled crane. At the same time as the emptying of the containers, the driver checks whether the surrounding area needs cleaning. Any possible failures or other irregularities to the container are reported to the service department; the driver does not fix these problems himself. Emptying one underground container takes around four minutes. When the waste from the container is disposed into the truck, a press is activated to reduce the volume of waste with a factor five. In the current way of working, a truck can empty, on average, close to thirty-five containers before its capacity is reached. When the truck is full or when the driver has finished his complete route, the driver goes to the waste processing centre, called Twence, to dump the waste. The truck is weighed at arrival and departure. The difference between these two is the total weight of waste collected from the containers. After a tour through one city, first a trip to the waste processing centre has to be made, before continuing to another city. This is because the different municipalities have to pay for the discarding of the waste. At the end of the day, the trucks have to return empty to the depot. On average, the trucks will visit the waste processing centre twice per day. A normal workday has eight hours from half past seven until four o'clock, with a lunch break of half an hour.

13.3.4 Data Analysis

For our simulation study, we need input data with respect to fleet characteristics, driving- and handling times, distances, and information on deposit frequencies and volumes. Most of the data is readily available. However, the information on deposit frequencies and deposit volumes requires a further analysis, which we describe in this section.

To get information on the deposit frequencies and volumes for each of the containers, we make use of the container registration systems (different systems are in use since Twente Milieu uses different types of underground containers). These registration systems record for each connected container, the number of times the lid of the container is opened and closed again. During our analysis, we found several errors and inconsistencies in the registration systems. In addition, these registration systems provide only information on lid openings whereas we also need information on the volume of waste disposals. Therefore, we had to collect more information via a number of other channels:

1. Deposits at the waste processing centre
2. One week of weighing the emptying's
3. One week of visual checks for (almost) full containers
4. Interviews and brainstorm sessions with employees

First, we retrieved the data on lid openings in 2009 for all containers from the registration systems. This gives us an idea about the waste disposal frequency for all these containers. Second, we combined this data (for a part of the network) with records from the waste processing centre. This enables us to relate lid openings with the average weight of a disposal. Third, we performed an experiment for one week with a collection truck that is able to weight the containers upon emptying. This provides another source to relate the lid openings with the average weight of waste disposals. From this analysis, it became clear that average weight per deposit differs a lot for the different containers.

Figure 13.2 shows the wide spread of the number of lid openings in relation to the weight measured. The arrow in Figure 13.2 shows the weight of waste for containers that have approximately the same number of lid openings of 75 times. We see large differences ranging from 115kg to 475kg. It should be noted that some of the deviations can be explained from the fact that some containers require a manual reset of the counter for lid openings, and sometimes drivers forget to do this. Further, it is more likely that the density is different per unit of weight, which will decrease the previously mentioned observation. Still, the huge differences would result in unreliable estimates of the container volumes solely based on the number of lid openings.

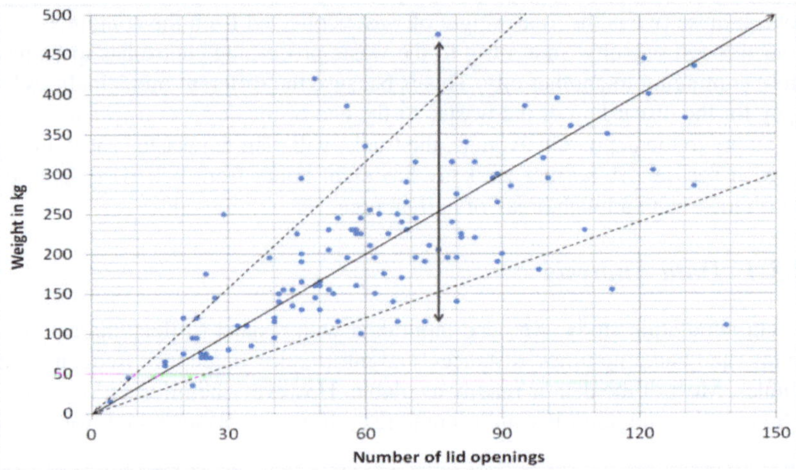

Fig. 13.2 of waste as a function of the number of lid openings.

At this point, we are still not able to relate lid openings to volume. Assuming more or less equal deposit volumes, the number of lid openings provides an indication of the fill level. Currently, the company assumes that every time the lid of the container is opened and closed again, the output ratio is raised by 1%. This means that a container is considered to be 'full' after 100 deposits. To verify this, we perform a one week field experiment in which we visually check the actual volume stored in containers that are full or almost full. This provides us with insight on the ratio between lid openings/weight/volume. From our analysis, it appears that the 1% assumption is not correct and the deposit volumes differ a lot per container. However, based on our one week field experiment, we expect that the deviations in deposit volumes are relatively small compared to the variations in weight, which means that the density varies.

To summarize our findings, we obtained the following insights:

- On average, one cubic meter of waste weighs 110 kilos, the average volume of a deposit is 48 liter, and the average fill level upon emptying is 50%. The effective capacity of 5m^3 containers is only 4000 liter because the top part of the container is filled in the shape of a cone. On average, containers are emptied 56.63 times on a yearly basis. This conclusion will be helpful in the verification phase of the system performance of the simulation model.
- There are huge differences in deposit volumes between the different container locations. For example, containers near stores have higher deposit volumes than containers for households. Also, there are huge differences in deposit frequencies between containers that are on the same location. This indicates that the alignment of the containers influences the speed of the filling process (closest to the entrance is full earliest).
- The number of deposits seems to fluctuate heavily from weekday to weekday and from week to week. Seasonal fluctuations became visible as well. Regarding

seasonal patterns, we see a weekly pattern (relative many deposits on Monday and few deposits on Sunday), a less visible monthly pattern (slightly more deposits at the beginning of the month), and huge random variation from day to day. These deviations already provide a good argument to switch to a dynamic routing approach, since a static planning approach would empty too less in peak periods and too much in other periods.

Although we do not include all the data resulting from our data analysis, we do provide some of the results in our section on input settings (Section 13.6.2).

13.4 Problem Description

We face an infinite horizon planning problem in which we have to empty N different containers at different points in time. Container i receives waste deposits with an average volume of a_i per day. Upon a given decision moment, an estimate of the actual waste volume in container i is given by v_i. The capacity of container i is Q_i. The expected number of days left before container i becomes full is given by d_i, which is given by $d_i = (Q_i - v_i)/a_i$. When the container is full, new deposits to this container are placed outside the container (overflow) which will be cleaned after emptying this container. To empty these containers, a fleet of M homogeneous trucks is available, each having a capacity of K. We introduce r as the number of routes to use, and L as a maximum on the number of containers to empty per day.

We use the common distinction between MustGo's and MayGo's. MustGo containers have to be emptied and the MayGo containers may be emptied if they can efficiently be incorporated in the routes. We introduce the following sets: (i) C consisting of all containers, (ii) C^m containing all MustGo's, and (iii) C^n containing all MayGo's. We define MustGo's as those containers i for which $d_i \leq D^m$, with D^m being a threshold on the number of working days. Here we explicitly state working days since we have to take into account the weekends since no emptying's will be done on Saturday and Friday. As an example, if $D^m=1$, we select all containers that are expected to be full before the next working day. On a Thursday morning, C^m contains all containers that are expected to be full before Friday morning. But on a Friday morning, C^m contains all containers that are expected to be full before Monday morning. The MayGo's are defined similarly, having $d_i \leq (D^m + D^n)$.

For the purpose of this simulation study, we make the following assumptions:

- Each truck can only be assigned to one job at a time and a job can be assigned to at most one truck. Each truck has finite capacity.
- Truck drivers have a maximum working time per day. Lunch breaks are ignored (we reduce the time of a workday with the time required for breaks) as well as additional trips required for fuelling. All trucks depart from the depot and return to the same depot at the end of the day.
- Containers are always entirely emptied. There are no time windows for empting the containers. When a container is full, deposits are placed next to the container, which we denote as overflow.

- All times are considered to be deterministic. This involves time for traveling, loading, and unloading at the waste processing centre.
- Costs for trucks and drivers are not taken into account. As a result, the algorithm might decide to use multiple vehicles and drivers for only a few hours per day.
- A natural approach to model the waste deposits would be to use a Poisson arrival process. However, the huge variance in deposit frequencies cannot accurately be described by a Poisson distribution. To model the arrival process, we use a Gamma distribution for the number of deposits per day, and then uniformly distribute the arrivals over the day. A chi-square test with α=0.05 does not result in a reject of our hypothesis that the number of deposits per day follows a Gamma distribution (see Section 13.6.2). The size of the deposits (deposit volumes) also follows a Gamma distribution (see Section 13.6.2).

The expectations of using a dynamic routing methodology are rather high. First, it should increase customer satisfaction and avoid waste overflow. Second, it should reduce the operational costs of emptying the containers. The initial objective was to empty the containers as close to their due dates as possible achieving an increase in service level (the percentage of containers emptied on time). However, emptying a container that is far from full might still be efficient when a truck just passes this container. Therefore, the main objective is to reduce the mileage of trucks in the long run, the total working time required to empty all containers, and to increase customer satisfaction with respect to waste overflow.

Variability in the waste disposal pattern has to be minded in the new approach, since it is already expected that the true demand of waste collection might vary strongly because of external events, like weekly, monthly, and seasonal patterns, special occasions, and holidays.

Given the problem description and the assumptions made in this section, we now present the planning approaches themselves.

13.5 Planning Methodologies

Independent on the type of planning methodology we use, we always create a planning at the beginning of a day for the whole day. The advantage of this that it is relatively easy to execute and truck drivers know the work they have to do on that day (just as in the current situation). However, we still need to be able to perform re-planning since plans might become infeasible during the day. The latter might be the result of travel delays or of collecting more garbage than expected, which requires scheduling a trip to the waste processing center earlier. In next two subsections we describe the static planning methodology as currently used by the company (Section 13.5.1) as well as our proposed dynamic planning methodology (Section 13.5.2).

13.5.1 Static Planning Methodology

Currently, a static planning methodology is used. This policy is described in Section 13.3.3. For the purpose of our simulation study, we need to model this policy. We have to model it in such a way that static plans are created automatically, based on experimental settings such as the number of containers, deposit frequencies, and volumes.

In essence, static plans are based on a desired time between emptying's (delivery frequencies). This time between emptying's is computed by dividing a target fill level upon emptying by the expected daily deposit volume. Then, this time is rounded to half a week, one week, or two weeks. So, containers are emptied twice per week, once per week, or once in the two weeks. Next, all emptying's are assigned to specific days in such a way that the workload is spread equally over the days insofar possible. Finally, manual adjustments are being made to cope with negative aspects of rounding the time between emptying's. In addition, at the beginning of each day, a planning employee might add additional emptying's to avoid overflow. Especially on Fridays, this manual adjustment takes place to avoid overflow in the weekend.

Obviously, the above description, which involves human intervention, is difficult to model in a simulation environment with changing settings for deposit behavior and number of containers. Therefore, we choose to model it slightly different. First, we determine the time between emptying's as done before. However, we do not round this time. Instead, after each emptying, we set the time for the next emptying equal to the current time plus this desired time between emptying's. Second, to resemble a balanced workload, we determine a fixed number of containers to empty per day, given by L. Each day, we sort the containers based on their planned emptying time, and then select the first L of these containers to empty. The result is that the required time between emptying's is not used exactly, but this procedure avoids explicitly taking into account the weekends. The main difference with the real static plan is (i) that the time between emptying's is not rounded and (ii) no manual adjustments take place. We expect that this model still provides a good match with the current way of working since the manual adjustments in reality are just necessary to cope with the rounding of time between emptying's.

13.5.2 Dynamic Planning Methodology

In the dynamic planning option, we daily select containers based on their estimated fill levels. For solving this problem, we might use an exact approach such as a Mixed Integer Linear Program. However, our problem has some characteristics which make a successful application of such an exact approach very unlikely. First, our problem involves multiple vehicles (up to 7 trucks), multiple depots (2 parking areas and 1 waste processing centre), and a large number of customers (expected to grow to 1500 containers within a few years). Further, our problem requires a long-term planning horizon, since a short-term approach will postpone deliveries to the next period (Campbell et al., 1998). Finally, we face dynamic environment with stochastic travel times and stochastic waste disposals which may require re-planning during the day. In addition, we also have to deal with weekly and monthly patterns, and special days (e.g., holidays). Given the scale and complexity of our problem, exact approaches are not suitable and we decided to use a heuristic approach. An illustration of this heuristic can be found in Figure 13.3.

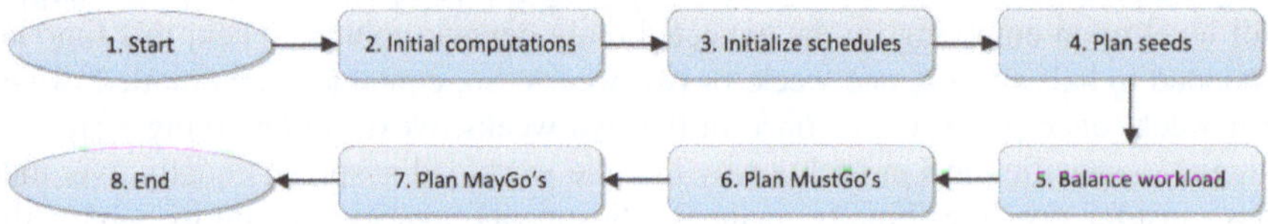

Fig. 13.3 - Steps in our heuristic.

We now explain the different steps of our heuristic.

1. The planning heuristic is started by two events: (i) initial planning in the morning and (ii) re-planning during the day. Re-planning can be triggered by several events, e.g., after each collection, after visiting the waste processing centre, periodically, or when the deviation from the planning exceeds some threshold (deviation with respect to volume or time).
2. The initial computation involves, (i) estimation of the days left d_i for all containers, (ii) determination of MustGo's C^m, (iii) determination of the number of trucks to use W ($W \leq M$), and (iv) determination of a lower bound on the number of routes to use, $r = \lceil \sum_{i \in C^m} v_i / (W * K) \rceil$.
3. For all trucks we decided to use (see Step 2), we initialize their schedules. For trucks currently having an empty schedule (typically the case with planning at the beginning of the day), their initial schedule would be from the parking area, to the waste processing centre, and ending up at the parking area again. For trucks with a non-empty schedule (we are doing re-planning during the day), we empty their schedule in a non-pre-emptive way. To keep the initial schedules feasible, a collection job will be followed by a visit to the waste processing centre, and a visit to the waste processing centre will be followed by a return trip to the parking area. We use these initial schedules to insert new collection jobs in the next steps.
4. We extend the initialized schedules (see Step 3) with seed customers. We decided to use one seed customer per route and divide the routes over the trucks. Seed customers are chosen based on the largest minimum distance from the depot and the other seed customers. We use the seed customers to (i) spread the trucks across the area, (ii) realize insertion of collection jobs from containers close as well as far from

the depot, and (iii) balance the workload per route to anticipate the insertion of MayGo's.

5. As an optional step, we assign MustGo's to the trucks in a balanced way. This means that we loop over all trucks and assign jobs to them one by one. Obviously, this will not be the most efficient way with respect to the MustGo insertions. However, it becomes particularly useful when we are going to extend the routes with relatively many MayGo's (see Step 8). MustGo's are added to the routes according to the cheapest insertion heuristic (see Campbell and Savelsbergh, 2004), where the insertion costs depend on the additional time required for the insertion. Note that additional visits to the waste processing centre are scheduled automatically, when necessary, the time required for these additional visits is also included in the insertion costs. The first time we do not find a feasible insertion for some truck we stop this procedure and continue with Step 6.
6. For all remaining MustGo's, we try to assign them using the same cheapest insertion heuristic as used in Step 5, but now by considering all insertion positions for all trucks and routes.
7. When all MustGo's are scheduled, there may be some space left in the trucks to empty other containers. By adding MayGo's, we make use of this free capacity to improve the routing efficiency. Also the MayGo's are scheduled using the cheapest insertion heuristic. However, this time we use another cost criterion which we explain later on. A high value for D^n has the benefit that we can choose between a large number of MayGo's. However, emptying them all will not always be the most efficient option. Therefore, we use the limit L on the number of emptying's per day.
8. We execute the planning and perform replanning when needed (see Step 1).

To schedule the MayGo's, a common choice in Inventory Routing Problems is to use a ratio of the additional travel time required to empty this container and its volume, see for example Golden et al. (1984). The problem with using such a ratio is that the more remotely located containers are unlikely to be considered as a MayGo. The criteria we use here is the relative improvement of the before mentioned ratio compared to a smoothed historical average of this ratio. A large positive value represents a one-time opportunity we should take. We do this based on a ratio of the additional time it takes to empty the container (both travel and handling time) to the waste volume in the container. A small ratio indicates a high amount of waste compared to the additional time; this means the smaller the ratio, the better. This procedure selects containers for which the emptying costs today are expected to be lower than at some later point in time.

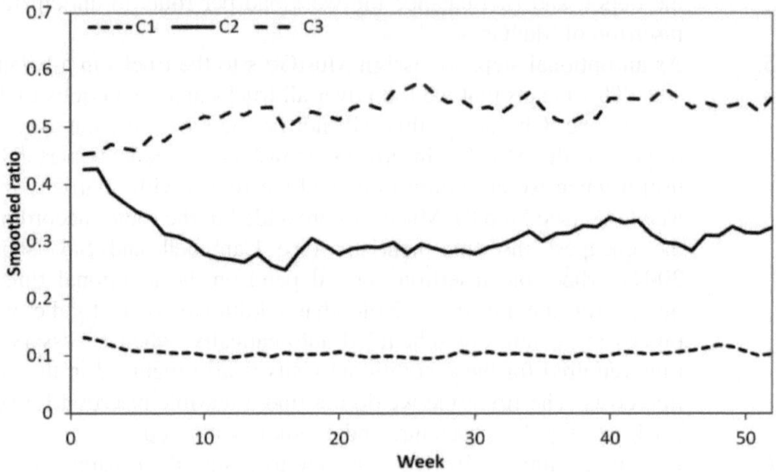

Fig. 13.4 Smoothed ratios for three containers.

Figure 13.4 shows the smoothed ratios for a number of underground containers. Here, C3 is an isolated container. C1 is a container close to the waste processing centre, and C2 somewhere in between. The ratio of a container at a favorable location is much lower compared to one at an isolated location. This makes sense since containers at a location with more containers in the neighborhood require less additional driving than containers at remote locations. This results automatically in smaller ratios. Figure 13.4 also support our choice to select MayGo's based on their improved ratio. Otherwise, containers at a remote location would never be selected, while the costs for emptying these containers might be relatively low today. Finally, Figure 13.4 also indicates we need at least several weeks as warm-up period for our simulation (see Section 13.6).

13.6 Simulation Model and Experimental Design

In this section the simulation model will be described that will be used to evaluate different routing and container selection methods as presented in the previous section. Subsequently, we present the structure of the simulation model (Section 13.6.1), the experimental settings (Section 13.6.2), experimental factors (Section 13.6.3), performance indicators (Section 13.6.4), and the replication-deletion approach (Section 13.6.5). We end with some notes on the verification and validation of our model in Section 13.6.6.

13.6.1 Structure

A schematic view on the structure of our simulation model can be found in Figure 13.5. The simulation is driven by the object "Citizens" which generates waste disposals. The planning and scheduling of emptying's is done with the object

"Waste collection company". The events upon which both objects operate are controlled by the "Event controller". The actions of citizens (waste disposals) and of the waste collection company (trucks emptying the containers) are displayed on an animated network. The object "Waste collection company" contains the methods that actually execute all steps necessary to develop an emptying schedule. This object needs the input of the experimental settings, keeps track of the performance of the different planning methodologies, and provides this as output data. The input of the simulation will be discussed in Section 13.6.2. The output, in the form of performance indicators, will be discussed in Section 13.6.4.

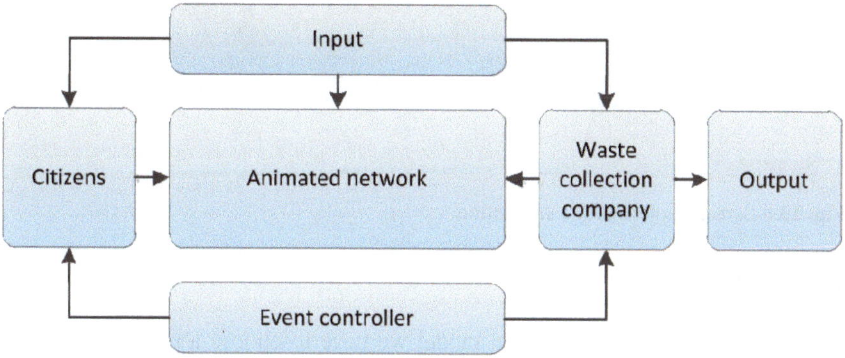

Fig. 13.5 Structure of the simulation model.

To make the simulation model more accessible for usage, we added visualization in the form of an animated network. This does not contribute to the actual output of the model, but it increases the understanding of the operation of the model.

The animated network consists of a map of the area Twente Milieu operates in. The underground containers are all marked on that map. Displaying a part of a 3D globe on a 2D map requires some transformations. For this, we use the Universal Transverse Mercator coordinate system (UTM) to transform the GPS coordinates of all containers into XY coordinates. In our case, this projection is somewhat easier, because all container locations are in the same zone (32U). In addition, also the planned routes are displayed on this map, although this is done based on straight lines. We use separate colors for the different routes. Also, MustGo's are displayed in red whereas the others are displayed in black. A screen capture of our simulation model can be found in Figure 13.6.

We implemented our discrete-event simulation model in the software package Tecnomatix Plant Simulation. Tecnomatix Plant Simulation is a computer application developed by Siemens PLM Software for modeling, simulating, analyzing, visualizing and optimizing production systems and processes, the flow of materials, and logistic operations (Plant Simulation, 2011).

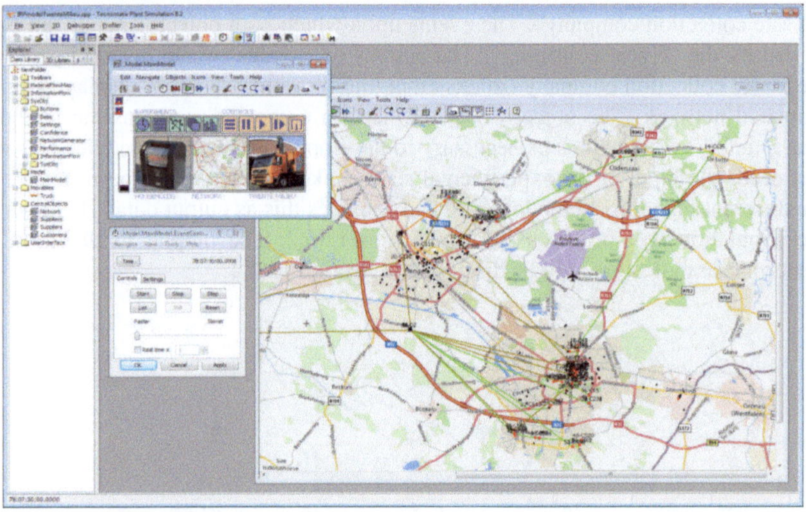

Fig. 13.6 Screen capture of the simulation.

13.6.2 Settings

For the settings of our simulation model we have to choose a reference point in time since new containers are installed on a weekly basis. For this use the situation as is was at the end of March 2009. At that moment, Twente Milieu operated in total 520 underground containers. From these containers, only 378 are equipped with sensors. In the near future, all containers will be equipped with sensors. But for now, we limit ourselves to the 378 containers for which historical sensor data is available. In our simulation experiments, we also consider a situation with 700 containers, as we discuss later on.

For every container, we need the following input: (i) the two parameters for the Gamma distribution for generating the number of deposits per day, (ii) the two parameters for the Gamma distribution of the volume of each deposit, (iii) the capacity, and (iv) the handling time. Obviously, these settings will differ per container. Instead of showing the settings of all containers, we here show the results for a typical container:

- Deposit frequency: Gamma (1.62, 5.88)
- Deposit volume: Gamma (248.78, 0.17)
- Capacity: 4000 liter
- Handling time: 4 minutes

As default scenario, we use two trucks (M=2) which we use every workday (W=2) independent on the amount of emptying's for that day. The capacity of these trucks is 18.000 liter of compressed waste. Given the compression factor of 5, this comes to a capacity of 90.000 liter of uncompressed waste. The handling time at the waste processing centre is 15 minutes. Workdays are Monday till Friday from 7:30am to 3pm where we subtracted the lunch breaks from the end of the workday (see Section 13.3).

The travel times between each of the container locations are derived from the Google Maps API, using the GPS coordinates of the 378 containers as input. The main assumption here is that the truck speed is equal to the speed of passenger cars. In the urban areas Twente Milieu operates in, this assumption is reasonable. To give an idea about the network, the average travel time between two container locations is 14 minutes with a maximum of 43 minutes. The largest distance between two containers is 51 kilometers.

In the static planning approach, the planned time to empty a container depends on the last time this container was emptied. As a result, we need to do something at the start of the simulation. At the start of the simulation, we randomly fill the containers (see Section 13.6.5) and calculate the days left d_i for each container. As long as there are containers that have not been emptied before, we give priority to these containers, starting with those having the lowest value of d_i. For the static planning approach, we further use a target fill level of 75%.

For the dynamic planning methodology, we have to determine the threshold D^n and D^m for respectively the MustGo's and MayGo's. Based on some preliminary experiments, we choose $D^n=1$ and $D^m=5$. As mentioned in the next subsection, we also consider a dynamic policy that only empties the MustGo's. For this policy we use $D^n=2$.

As mentioned in the beginning of Section 13.5, all planning options might require re-planning during the day and we mentioned several possibilities for this. In our simulation we choose the following. After each emptying that is not immediately followed by a visit to the waste processing centre, we look if the effective capacity of the next container to empty still fits in the truck. If not, we perform rescheduling only for this truck. To avoid excessive re-planning, we work with a truck slack capacity of 5000 liter in our planning methodology.

As mentioned in Section 13.3.4, deposit frequencies fluctuate heavily. We saw large random fluctuations per day as well seasonal patterns. To simulate the seasonal patterns, we multiply the mean deposit volume for each day with some factor. This factor follows a sinus curve with a given amplitude FA and a period of 4 weeks. We assume that the company is not aware of this sinus curve. Hence, within one period, there will be 2 weeks in which the company overestimates the deposit volumes and 2 weeks in which it underestimates these volumes. To simulate the random fluctuations, we further multiply the mean deposit volumes with a factor uniformly drawn from $[1\text{-}FR,1+FR]$ with $FR\leq 1$. To mimic the current situation, we use $FA=0.05$ and $FR=0.7$.

As default value for the maximum number of jobs per day (L), we use 22% of the number of containers. For our reference point, this gives $L=0.22*378=83$.

The final setting is related to the time we use between updating the smoothed ratio (see Section 13.5.2). For this we use a week. So, at the end of each week, we compute the average emptying ratio's (required additional travel time to empty this container divided by the volume of waste in this container) for each container and smooth these values, using $\alpha=0.05$, with the smoothed historical average.

13.6.3 Experimental Factors

To see how a planning methodology performs, we will test its behavior under varying circumstances. We chose the following factors for our simulation experiments:

- Number of containers (*N*): 378 and 700. At our reference point, 378 where in use. We extend this number to 700 by randomly selecting new container locations from the current locations.
- Planning methodologies (Policies): Static, MustGo, Dynamic. The MustGo policy is a dynamic planning methodology in which we only empty the MustGo's.
- Fill-level sensors: with and without. Without fill-level sensors, we estimate the fill levels by multiplying the number of lid openings with the expected mean deposit volume. With fill-level sensors, we have a perfect estimate of the actual fill level. We denote the use of fill-level sensors in combination with the three previously mentioned policies by StaticS, MustGoS, and DynamicS.
- Factor amplitude in sinus fluctuations (*FA*): [0, 0.5].
- Factor mean deposit volumes (*FM*): [0.5, 1.5]. Here, we multiple the mean deposit volumes every day with a factor *FM*.
- Factor expected deposit volumes (*FE*): [0.75, 1.25]. Here, we multiply the expected deposit volumes with *FE*. The expected deposit volumes are used to estimate the fill-level of the containers (and hence the days left) based on the number of lid openings. A value of 1 means that our expectation is accurate. However, the actual deposit volumes might still fluctuate due to the random fluctuations (*FR*) and the seasonal fluctuations (*FA*).
- Factor maximum number of emptying's (*FL*): here we multiple the maximum number of emptying's *L* with a factor *FL*.

13.6.4 Performance Indicators

With our simulation model, we evaluate the performance of the dynamic planning methodology and benchmark it against the current way of working. As key performance indicator in this analysis, we use a weighted combination of transportation costs, handling costs, and penalty costs for emptying too late. As weights, we use ct for the travel costs per time unit, c^h for the handling costs per time unit, and cp for the penalty costs per volume overflow. The key performance indicator *CL* gives the total costs per liter:

$$CL = \frac{c^t(travel\ time) + c^h(handling\ time) + c^p(volume\ overflow)}{volume\ collected},$$

where time and volume are measured over the whole simulation run.

With this objective function, we aim to minimize the travelling costs, while at the same time ensuring the service level by penalizing when a container is emptied too late. In agreement with the company, we set the parameters as follows: $c^t = 1$, $c^h = 0.5$, and $c^p = 0.7$. Here, the travel costs are considered to be the most influential with respect to the overall performance; one time unit of travelling costs twice as much as spending one time unit on loading\unloading. The penalty factor is also relatively large to maintain customer satisfaction.

As secondary performance indicators we consider:

- *CT* = average travel time per day (hours)
- *CH* = average handling time per day (hours)
- *CP* = average volume of overflow per day (m^3)
- *VC* = average volume of collected waste per day (m^3)

13.6.5 Replication/Deletion Approach

Given we are facing a non-terminating simulation we use the replication/deletion approach (see Law, 2007). This involves a number of replications each having a certain warm-up period. The warm-up period indicates after which time the system comes into a steady state. In our case, the warm-up period is necessary for (i) learning the smoothed ratios used for adding MayGo's and (ii) to create realistic fill levels in the underground containers. We initialize the simulation by filling the containers uniformly between 0 and 80% of the container capacity. For calculating the warm-up period, we use Welch's graphical procedure as described in Law (2007). The use of 10 replications and a window of 40 days results in a warm-up period of around 8 weeks. Given the warm-up period of 8 weeks, we use a run length of 24 weeks (excluding the warm-up). Next, we calculate the number of runs using the sequential procedure (Law, 2007). With a confidence level α=0.05 and relative error γ=0.05, the required number of runs for different experimental settings fluctuates around five replications. To avoid inaccuracies, we use 10 replications for all our experiments.

13.6.6 Model Verification and Validation

To verify our model, we tested each module separately. In addition, we kept the management of the company in the process to establish credibility. After completing the simulation model, we validated our model to see whether the reality was approached accurately enough. Validation is done using data from the actual operations and it is intended to prove that the assumptions incorporated in the model were well chosen. For the case of the Twente Milieu, the validation process is relatively difficult to execute, since the current collection process involves human decision making which we had to approximate with our static planning model (Section 13.5.1). We distinguish the following validation and verification criteria:

- From the interviews conducted with the planning department, it became clear that, on average, each working day 22% of the containers are emptied. Upon the chosen reference point in time, the total number of containers is 378, which result in 83 emptying's per day. The emptying's are done by two trucks, one of them being utilized for 50%. This corresponds with the average workload of 55 containers which we found during our data analysis using observations from half a year around the reference point. This data analysis also revealed an average of 412 emptying's per week which confirms the results from our interviews (5*83=415 emptying's).

- From the interviews conducted with the planning department, it became clear that, on average, the amount of garbage that is collected from a container is 2500 liters. Our data analysis revealed that the daily deposal volume is 148.070 liters and 415 emptying's take place weekly. This translates to an average of 7*148.070/415=2498 liter per emptying which confirms the expectations of the planning department.
- Criteria such as deposit and emptying frequencies cannot be used to validate our simulation model since we use them as input. A useful validation criterion we can use here is the time required for the collection process, which depends on the travel times, handling times, and the routing efficiency. From the interviews conducted with the truck drivers, it became clear that, under normal circumstances, emptying 55 containers can be seen as the maximum workload for one truck on one day. Under ideal circumstances (no traffic delays and many containers to empty close to each other), a maximum workload of around 70 emptying's can be achieved. To validate our simulation model, we used (i) the static planning methodology without the fixed maximum of 83 emptying's per day and (ii) the dynamic planning methodology without a maximum and using $D^m=1$ and $D^n=5$. The results of these experiments can be found in Table 13.1. Here, we use as maximum in our simulation experiments the 97.5 percentile. Clearly, the static planning approach provides a perfect match with respect to the normal maximum amount of daily emptying's. This amount of emptying's is higher in case of dynamic planning due to the insertion of MayGo's which are normally chosen such that they require limited additional travel time. With respect to the maximum number of emptying's that can be achieved under ideal circumstances, we see a perfect match with the dynamic planning approach. In reality, this maximum is only achieved with human intervention where one deviates from the original static plan, thereby including additional containers that are closely located to the current routes. This is exactly the reason that the static planning approach yields a lower maximum in our simulation.

Table 13.1 Validation results.

	Reality	**Static**	**Dynamic**
Normal	55	55.4	61.7
Maximum	70	61.8	69.7

The verification and validation steps described above convince us that our simulation model provides an accurate representation of the real system. The numerical results from this simulation model are presented in the next section.

13.7 Results

In this section we present the results of our simulation study. First, the results for the sensitivity analysis are shown (Section 13.7.1) and then the results for expected network growth (Section 13.7.2). We end with a benchmark of the current way of working, thereby providing an indication of the savings that can be achieved by Twente Milieu when switching to a dynamic planning methodology (Section 13.7.3).

13.7.1 Sensitivity Analysis

In our sensitivity analysis, we vary the following things: the mean deposit volumes, the maximum number of emptying's per day, the deviation of the expectation, and the amplitude of the sinus pattern of daily deposit frequencies. The results can be found in Figure 13.7.

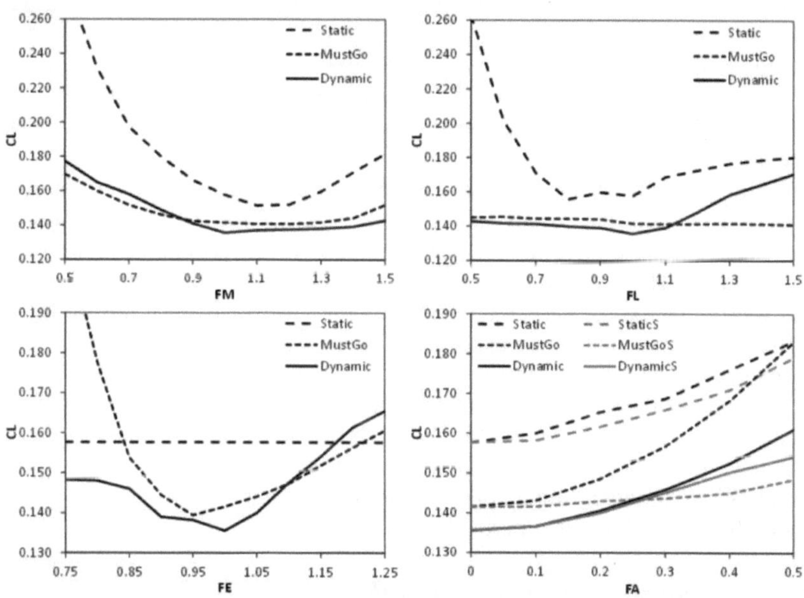

Fig. 13.7 Results sensitivity analysis.

We draw the following conclusions. First, with a varying factor *FM* for the mean deposit volumes, the different policies have their lowest costs *CL* around 1.1. Obviously, an increase in deposit volumes will result in higher costs. However, the cost per liter initially decreases. With further increase in deposit volumes, the penalty costs will raise. We also observe that with relative low deposit volumes, Dynamic will be outperformed by MustGo. The reason for this is that the policy Dynamic is bounded with the maximum of 83 emptying's (0.22*378) per day. With low deposit volumes, this bound is too low. The consequence of this is that Dynamic is simply doing too many MayGo's which results in relative high costs per liter collected.

If we look at varying factor *FL* for the maximum number of jobs, we see the following. First, the policy MustGo is not sensitive to this maximum. We also see that for a low maximum, the difference between the performance of MustGo and Dynamic becomes smaller, since the ability of adding MayGo's decreases. With increasing maximum, we see that MustGo outperforms Dynamic. Again, the explanation is that Dynamic is using too many MayGo's. Finally, we observe that the minimum of Static is attained in the area 0.8-1, which provides an indication that the choice of emptying 22% of the total container population daily, seems to be a good choice in combination with the weights of the three costs factors (travel time, handling time and overflow). The number of emptying's is a bit on the save side, which indicates that in reality the company puts even more weight on the penalty costs and hence on customer satisfaction.

If we look at varying factor *FE* for the expected mean disposal volume, we observe the following. First, Static is not sensitive to this value since it does not estimate the average fill levels (although it does require to determine the time between emptying's which we assume to be known in this study). Obviously, the dynamic policies are influenced by this. If we underestimate the deposit volumes, we will incur more penalty costs. If we overestimate the deposit volumes, we are doing more emptying's then necessary. Overestimation will be worst for dynamic since it uses too many MayGo's.

Finally, we consider the factor *FA* for the amplitude in sinus pattern of deposit frequencies. Obviously, for all policies, the costs increase with increasing amplitude. This is because there will be periods of heavy over estimation as well as under estimation. However, with increasing amplitude, the added value of using fill level sensors increases. Particularly for the policy MustGo since this policy only empties the containers that are expected to be almost full. MustGo without sensors will eventually be outperformed by Static. Remarkable here is that MustGo with sensors will eventually outperform Dynamic with sensors. The explanation for this is that, if we perfectly know the fill levels, the value of adding MayGo's decreases. Finally, the policy Dynamic heavily depends on the choice of parameter levels D^n and D^m. With increasing amplitude, these parameters will be too low in some periods and too high in other periods.

13.7.2 Analysis of Network Growth

We now compare the original network with 378 containers with an increased network consisting of 700 containers. The results can be found in Table 13.2. It appears that the costs for the three policies practically remain the same. Note that these are the costs per liter. Since more liters are collected with 700 containers, the total costs will obviously be higher. Further, if we look at the individual costs components, we see that all costs are higher with 700 containers, especially the penalty costs. The reason for this is that we still use the same amount of trucks (2 trucks).

Table 13.2 Results of growth in number of containers

N	Policy	CL	CT	CH	CP	VC
378	Static	0.1576	5.53	6.33	2.05	207.54
378	MustGo	0.1416	5.18	4.92	2.66	207.47
378	Dynamic	0.1356	4.53	6.29	0.73	207.57
700	Static	0.1587	6.23	8.42	33.22	383.37
700	MustGo	0.1384	6.27	8.42	21.90	383.23
700	Dynamic	0.1352	6.67	8.14	18.84	383.35

Next, we vary the mean deposit volumes. The results can be found in Figure 13.8. Here we clearly see that two trucks are sufficient to cope with an increase in deposit volumes whereas this is no longer the case with 700 containers. With 378 containers, increasing volumes will reduce the costs per liter since there is a situation of overcapacity. In case of 700 containers, an increase in mean deposit volume will results in an increase in penalty costs.

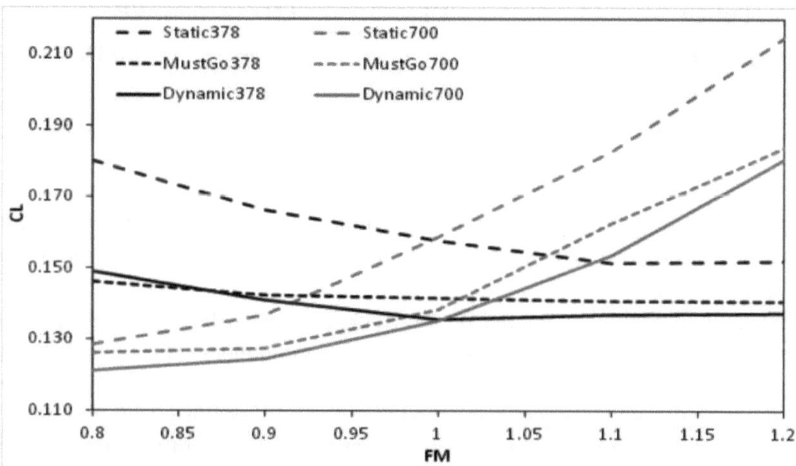

Fig. 13.8 Varying mean deposit volumes for 378 and 700 containers

13.7.3 Benchmarking

In the last experiment, we compare the performance of the dynamic planning methodology with the static planning methodology as currently used by the company. For this we use the settings with both periodic and random fluctuations. The results can be found in Table 13.3

Table 13.3 Benchmarking results

Policy	CL	CT	CH	CP
Static	0.1687	5.70	6.40	4.42
StaticS	0.1656	5.57	6.33	4.37
Dynamic	0.1468	4.73	6.24	3.29
DynamicS	0.1434	5.02	5.85	1.78

We clearly see that the travel costs as well as the penalty costs can be decreased significantly. To make it more clearly, we also present the savings of all policies compared to the static planning methodology. These results can be found in Table 13.4.

Table 13.4 Relative savings

Policy	CL	CT	CH	CP
StaticS	1.81%	2.31%	1.09%	1.08%
Dynamic	12.96%	17.07%	2.42%	25.63%
DynamicS	14.95%	11.94%	8.61%	59.74%

The total savings of switching to a dynamic planning methodology would be almost 13%. When we also switch to the more reliable fill level sensors, the savings increase to almost 15%.

The savings obviously depend on the truck capacities. In the current situation, Twente Milieu uses two trucks to empty the 378 containers. This is a situation of overcapacity. We have seen earlier (Section 13.7.1) that the dynamic planning methodologies will require less capacity and therefore still perform well with increasing network size. To study this effect, we vary the maximum number of empty-ing's per day. The relative savings compared to the static policy are displayed in Figure 13.9.

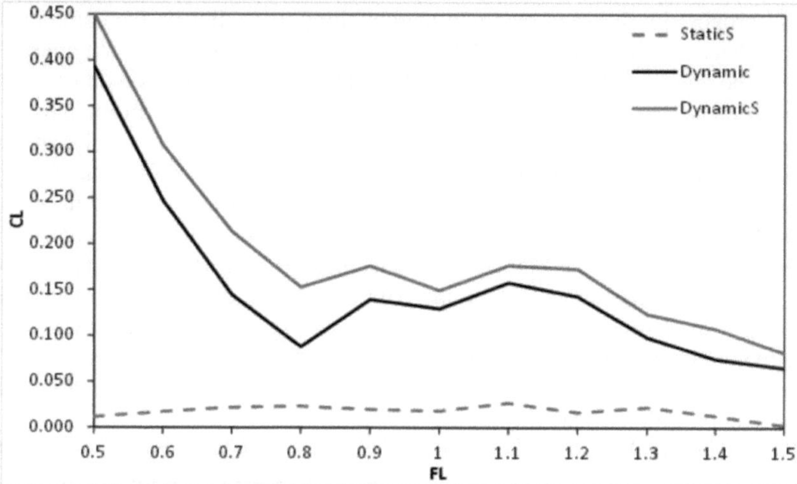

Fig. 13.9 Varying maximum workload

We clearly see that savings increase with decreasing capacity. For example, if trucks are allowed to do only 50% of their regular workload (resembling the case with 50% less tucks or 50% shorter working days), the relative savings of the dynamic planning methodology are close to 40%. Again, additional savings can be achieved by using fill level sensors, which yields savings of up to 45%.

Even though the performance of the dynamic policy seems promising, there is still room for improvement. One specific weakness of the dynamic policy is its strong sensitivity to used parameter settings, i.e., the values of D^m, D^n, and L. As a result, we need to tune these parameters first. This also means that with changing deposit patterns (such as the simulated seasonal and random fluctuations in deposit volumes) we continuously need to adapt our parameters. This also explains our earlier observations that in some cases Dynamic is outperformed by MustGo (situations in which Dynamic is doing too many MayGo's). We also observed (results not shown here) that the right choice of parameter values also heavily depends on the day of the week. As a result, we need to tune (D_t^m, D_t^n, L_t) for $t=1,..,5$, with t being the day of the week. Moreover, there are also several dependencies between these parameters, e.g., a high value for D_t^m or a low value for L_t, reduces the impact of D_t^n. In principle, we could optimize over these parameters, in this case over a 25 dimensional function which we measure using simulation. This simulation optimization approach is part of our future research.

13.8 Conclusions and Recommendations

In this chapter, we analyzed the options to use a dynamic planning methodology to increase efficiency in the emptying process of underground containers in terms of logistical costs, customer satisfaction, and CO_2 emissions.

We proposed a dynamic planning methodology that relies on the common distinction between MustGo's and MayGo's. The MustGo's are those containers that have to be emptied. The MayGo's are those that might be emptied depending on how efficient they can be incorporated in the current routes. From the set of MustGo's, seed customers are selected to (i) spread the trucks across the area, (ii) realize insertion of collection jobs from containers close as well as far from the depot, and (iii) balance the workload per route to anticipate the insertion of MayGo's. The algorithm then schedules all remaining MustGo's using the cheapest insertion heuristic. Then the MayGo's are scheduled as long as there is capacity left. We proposed a new criterion for the MayGo's depending on the relative improvement over an historical smoothed average ratio of the additional travel time required to empty a specific container and its volume. The algorithm is used in the morning to plan all emptying's for that day. During the day re-planning takes place to cope with inaccurate estimates of the fill levels.

We evaluated the performance of our dynamic policy using simulation. We conclude that, for our reference point in time, the benefits for Twente Milieu, of switching to a dynamic waste collection policy is a cost reduction of 13%, which consists of a reduction in travel costs of 17% and a reduction in penalty costs of 26%. We further conclude that with increasing deposit volumes or decreasing truck capacities, these savings increase. In other words, by switching to a dynamic collection policy, investments in additional trucks can be postponed. We showed that the savings per liter remain almost the same with an expanded network of 700 containers. This means that the absolute savings increase drastically with increasing network size. We also analyzed the added value of investing in fill-level sensors. Obviously, the higher the (unforeseen) fluctuations in deposit volumes, the higher the potential benefits of using fill-level sensors. For our reference point in time, the dynamic policy with fill-level sensors result in cost savings of 15%, which consist of a reduction in travel costs of 12% and a reduction in penalty costs of 60%.

We end with suggestions for further research. We made several simplifying assumptions which have an impact on the reliability of our simulation model (and to some extend the usability of our planning methodology). First, we assumed deterministic travelling times. Of course, in reality, the time to travel from one container to another is stochastic. Although we can use a deterministic algorithm to make decisions in a stochastic environment, it would be nice to study the impact of stochastic travel- and handling times in our simulation model as it will definitely impact the need for re-scheduling. Next, we looked at each container individually. However, at many locations there are multiple underground containers placed. It is arguable to look at all containers at the same location together. Only when all containers of the group are almost full, the group is eligible for emptying. The final direction for further research is the simulation optimization approach as mentioned at the end of Section 13.7. The methodology we have in mind for this is based on the hierarchical knowledge gradient policy as described by Mes et al. (2011).

Authors Biography, Contact

Martijn Mes is an assistant professor in the department of Operational Methods for Production and Logistics at the University of Twente, The Netherlands. He holds an MSc in Applied Mathematics (2002) and a PhD in Industrial Engineering and Management at the University of Twente (2007). Martijn's main expertise is simulation; more specifically, simulation of logistics systems. Martijn has developed numerous simulation models for industry and gives a master course on simulation at the University of Twente. Martijn's research interest involves simulation, simulation optimization, stochastic optimization, dynamic vehicle routing problems, multi-agent systems, port logistics, transportation, and health logistics.

The University of Twente is an enterprising university with a clear focus: High tech, human touch. Scientists and other professionals are working together on cutting-edge research, innovations with real-world relevance, and inspiring education.

Contact

Martijn Mes
University of Twente
School of Management and Governance
Dep. Operational Methods for Production and Logistics
P.O. Box 217
7500 AE Enschede
The Netherlands
m.r.k.mes@utwente.nl

References

Andersson, H., Hoff, A., Christiansen, M., Hasle, G., Løkketangen, A.: Industrial aspects and literature survey: Combined inventory management and routing. Computers & Operations Research 37(9), 1515–1536 (2010)

Angelelli, E., Bianchessi, N., Mansini, R., Speranza, M.G.: Short term strategies for a dynamic multi-period routing problem. Transportation Research Part C 17(2), 106–119 (2009)

Engelelli, E., Speranza, M.G., Savelsbergh, M.W.P.: Competitive analysis for dynamic multi-period uncapacitated routing problems. Networks 49(4), 308–317 (2007)

Beltrami, E.J., Bodin, L.D.: Networks and Vehicle Routing for Municipal Waste Collection. Networks 4, 9–32 (1974)

Berbeglia, G., Cordeau, J.F., Laporte, G.: Dynamic pickup and delivery problems. European Journal of Operational Research 202(1), 8–15 (2010)

Campbell, A., Clarke, L., Kleywegt, A., Savelsbergh, M.: Inventory routing. In: Crainic, T., Laporte, G. (eds.) Fleet Management and Logistics. Kluwer Academic Publishers, Boston (1998)

Campbell, A.M., Savelsbergh, M.: Efficient insertion heuristics for vehicle routing and scheduling problems. Transportation Science 38, 369–378 (2004)

Chalkias, C., Lasaridi, K.: A GIS based model for the optimisation of municipal solid waste collection: the case study of Nikea, Athens, Greece. WSEAS Transactions on Environment and Development 10(5), 640–650 (2009)

Chang, N.B., Wei, Y.: Comparative study between the heuristic algorithm and the optimization technique for vehicle routing and scheduling in a solid waste collection system. Civil Engineering and Environmental Systems 19(10), 41–65 (2002)

Chao, I.M., Golden, B., Wasil, E.: An improved heuristic for the period vehicle routing problem. Networks 26, 25–44 (1995)

Cordeau, J.F., Gendreau, M., Laporte, G.: A tabu search heuristic for periodic and multi-depot vehicle routing problems. Networks 30(2), 105–119 (1997)

Dantzig, G.B., Ramser, J.H.: The Truck Dispatching Problem. Management Science 6(1), 80–91 (1959)

Francis, P.M., Smilowitz, K.R., Tzur, M.: The Period Vehicle Routing Problem with Service Choice. Transportation Science 40(4), 439–454 (2006)

Francis, P.M., Smilowitz, K.R., Tzur, M.: The period vehicle routing problem and its extensions. In: Golden, B.L., Raghavan, S., Wasil, E.A. (eds.) The Vehicle Routing Problem: Latest Advances and New Challenges, pp. 73–102. Springer, New York (2008)

Golden, B., Assad, A., Dahl, R.: Analysis of a large scale vehicle routing problem with an inventory component. Large Scale Systems 7(2-3), 181–190 (1984)

Golden, B., Raghavan, S., Wasil, E.: The Vehicle Routing Problem: Latest Advances and New Challenges. Springer, New York (2008)

Jaillet, P., Huang, L., Bard, J.F., Dror, M.: A rolling horizon framework for the inventory routing problem. Working paper. University of Texas, Austin (1997)

Johansson, O.M.: The effect of dynamic scheduling and routing in a solid waste management system. Waste Management 26(8), 875–885 (2006)

Jozefowiez, N., Semet, F., Talbi, E.G.: Multi-objective vehicle routing problems. European Journal of Operational Research 189(2), 293–309 (2008)

Karadimas, N.V., Papatzelou, K., Loumos, V.G.: Optimal solid waste collection routes identified by the ant colony system algorithm. Waste Management & Research 25(2), 139–147 (2007)

Kim, B.I., Kim, S., Sahoo, S.: Waste collection vehicle routing problem with time windows. Computers and Operations Research 33(12), 3624–3642 (2006)

Lacomme, P., Prins, C., Sevaux, M.: A genetic algorithm for a bi-objective capacitated arc routing problem. Computers & Operations Research 33(12), 3473–3493 (2006)

Larsen, A., Madsen, O.B.G., Solomon, M.M.: Recent developments in dynamic vehicle routing systems. In: Golden, B.L., Raghavan, S., Wasil, E.A. (eds.) The Vehicle Routing Problem: Latest Advances and New Challenges, pp. 199–218. Springer, New York (2008)

Law, A.: Simulation Modeling and Analysis, 4th edn. McGraw-Hill, New York (2007)

McLeod, F., Cherrett, T.: Quantifying the transport impacts of domestic waste collection strategies. Waste Management 28(11), 2271–2278 (2008)

Mes, M.R.K., Powell, W.B., Frazier, P.I.: Hierarchical Knowledge Gradient for Sequential Sampling. Journal of Machine Learning Research 12, 2931–2974 (2011)

Mourgaya, M., Vanderbeck, F.: Column generation based heuristic for tactical planning in multi-period vehicle routing. European Journal of Operational Research 183(3), 1028–1041 (2007)

Newman, A.M., Yano, C.A., Kaminsky, P.M.: Third party logistics planning with routing and inventory costs. In: Geunes, J., Pardalos, P.M. (eds.) Supply Chain Optimization. Springer, New York (2005)

Nuortio, T., Kytöjoki, J., Niska, H., Bräysy, O.: Improved route planning and scheduling of waste collection and transport. Expert Systems with Applications 30(2), 223–232 (2006)

Pacheco, J., Martí, R.: Tabu search for a multi-objective routing problem. Journal of the Operational Research Society 57(1), 29–37 (2007)

Plant Simulation (2011), `http://www.plm.automation.siemens.com/`

Russell, R.A., Igo, W.: An Assignment Routing Problem. Networks 9, 1–17 (1979)

Tan, K.C., Chew, Y.H., Lee, L.H.: A hybrid multi-objective evolutionary algorithm for solving truck and trailer vehicle routing problems. European Journal of Operational Research 172(3), 855–885 (2006)

Toth, P., Vigo, D.: The Vehicle Routing Problem. SIAM, Philadelphia (2001)

14 Applications of Discrete-Event Simulation in the Chemical Industry

Sven Spieckermann and Mario Stobbe

Production processes in the chemical industry are to a large extend not discrete but continuous. Hence, the application of discrete-event simulation (DES) in this field is not as widespread as in discrete manufacturing. In order to apply DES methodology to chemical production processes, continuous aspects have to be covered sufficiently. This contribution briefly introduces and discusses combined discrete-continuous simulation approaches and illustrates the potential of the methodology using three cases of a leading German chemical company from supply chain optimization to the shop floor.

14.1 Introduction

The chemical industry is to a large extent not a “typical” domain for the application of discrete-event simulation. An extensive literature review on simulation in business and manufacturing by Jahangirian et al. (2010) refers to only two out of more than 200 papers with a connotation to the chemical industry. An earlier review by Smith (2003) with a sample size of 188 papers is more or less focused on discrete manufacturing industries. Discussing simulation applications in discrete product manufacturing, batch production, and continuous production, Mehra et al. (2006) make the observation that the majority of studies relate to discrete products.

All in all, the footprint of the chemical industry in the scientific simulation literature is relatively small compared to the economic impact of the according companies which contribute more than 10% to European GDP according to Eurostat (cf. Stawinska 2009, p. 19). To a large extent, this mismatch is explained by the fact that the most common simulation technique in the manufacturing context is discrete event simulation (DES; cf. Smith 2003 and Jahangirian 2010) and DES does have some limitations when it comes to the modeling of specific process characteristics as we will discuss in Section 14.2. The approaches to overcome these limitations by combined simulation techniques and to tackle the industry specific challenges as well as the state-of-the-art in terms of tools and applications in the

Sven Spieckermann
SimPlan AG, Edmund-Seng-Str. 3-5 D 63477 Maintal, Germany

Mario Stobbe
Evonik Industries AG, Edmund-Seng-Str. 3-5 D 63477 Maintal, Germany

field are discussed in Section 14.3. Subsequently, Section 14.4 presents some case studies. The article finishes with a short summary and some conclusions.

14.2 Specific Challenges in the Chemical Industry

DES is applied to approach manufacturing issues on different operational layers from supply chain down to shop floor level and in different phases of the planning process as illustrated by Fig. 14.1 and shown by the cases in this book. Considering the chemical industry, DES may basically be applied on the same operational levels (cf. Schulz and Spieckermann 2008). Depending on the operational level there are, however, more or less specific characteristics of the industry which modeling approaches need to take into account.

With respect to the analysis of supply chains, there are almost no differences between the chemical industry and other industries. On this operational layer, in almost all cases discrete processes such as scheduling orders, dispatching transports, or planning resources (warehouses, production capacity, capacity for transportation) need to be considered. Thus, DES is an appropriate means to support supply chain studies in the chemical industries as the example in subsection 14.4.1 will demonstrate.

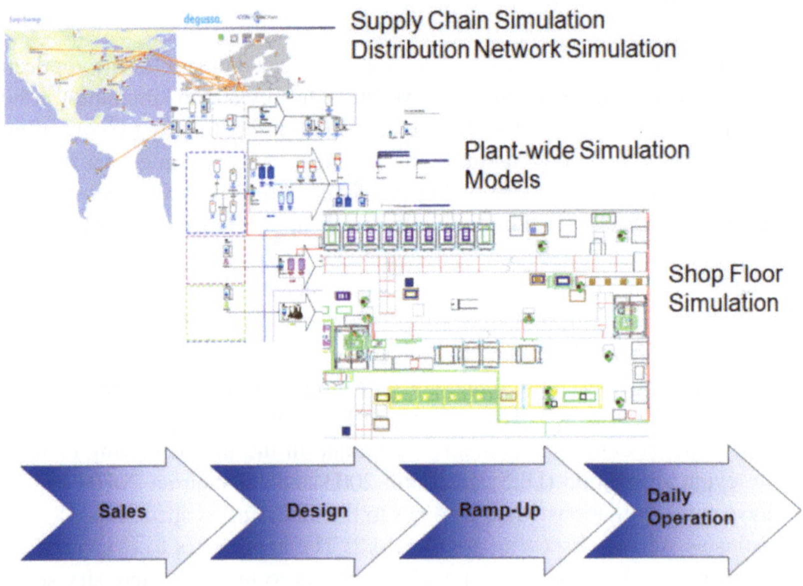

Fig. 14.1 Application areas of DES

However, as soon as the processes within one plant or within a selected part of a plant are subject to a study, some specific characteristics of processes in the chemical industry have to be taken into account. Günther and Yang (2004) summarized several of these characteristics with respect to production planning: constraints to batch sizes, shared intermediates, limited predictability of processing times and yields, product specific storage devices, finite intermediate storage, cyclical material flows, detailed quality controls, complex packaging and filling operations, use of multi-purpose resources, usage of secondary resources such as energy or steam, blending and mixing processes, sequence and usage dependent cleaning operations, changing proportions of input and output goods, production of by-products.

Watson (1997) presents a similar list additionally comprising items like floating bottlenecks, limited number of specially trained plant operators, or products with rather short shelf lives.

A production environment like this makes production planning as well as modeling on plant and shop floor level a very challenging task. Considering the modeling of chemical shop floor operation from the perspective of discrete modeling, this task is made even harder by the fact that a significant part of chemical production processes are not discrete by nature. Instead, there are continuous or batch processes in place. Following the terminology given by ISA-S88 (1995), p. 18, in a *discrete parts manufacturing process* the parts move in a specified quantity through the production system and each part typically maintains its unique identity. In a *continuous production process* materials are moving in a continuous flow through production equipment. A *batch production process* somewhat combines elements of discrete and continuous processes: within a so-called batch, given input materials are converted into one or more products by applying a defined sequence of processing actions requiring a given set of equipment. The dedicated set of input materials and input material quantities as well as the required process equipment together with process parameters (heat, pressure, process times etc.) are summarized in a recipe. A batch production typically is producing a sequence of different batches, i.e. after one batch has been processed according to its recipe, some cleaning or set-up activities take place and afterwards the next batch is produced following a different recipe and resulting in different products.

Whereas each process action is either discrete or continuous by nature, the batches may well be considered as discrete items. Also, the process actions are defined as a finite number of steps following a given sequence. Thus, an appropriate modeling approach will have to take discrete and continuous facets into account.

DES software tools do have some shortcomings with respect to continuous or batch processes modeling. The methodology is dedicated to model discrete manufacturing processes and not to handle continuous flow. Some ways to close the gap between DES and the requirements of modeling chemical production processes appropriately will be discussed in the next section.

14.3 State-of-the-Art and Solution Approaches

Traditionally, DES and continuous simulation are separate methodologies and approaches. Where DES schedules events and considers discrete elements with a fixed number of states, continuous simulation approaches model processes by means of differential equations.

Fahrmann (1970) was among the first to suggest a combination of both methodologies resulting in what is called combined simulation. As Cellier (1986) explains in detail and Bauer et al. (2008) summarize, there are different approaches for combined simulation: integration of DES in continuous simulation, integration of continuous simulation in DES, and approaches designed to combine DES and continuous simulation. The literature describes applications of each of the three approaches as the three following examples of combined simulations illustrate: Sharda and Vazquez (2009) present the analysis of a tank farm using the DES simulation software ARENA which does in addition to discrete building blocks offer some elements to model continuous processes. Mušič and Matko (1998) discuss an integration of a discrete petri-net based modeling approach into the continuous simulation tool Matlab-Simulink. The bottleneck analysis of a batch-conti-process (a process where batch processing steps and continuous processing steps are mixed) can be found in Sharda and Bury (2010). They are using the simulation software ExtendSim, a tool that was designed from the beginning to be also used for combined simulation.

However, if DES software is the starting point to model batch processes, there are (as alternative to combined simulation) two more ways to handle the continuous process elements: The first way is to consider the batch process from the batch level, i.e. each batch is modeled as one transaction moving through the simulation model (cf. Alexander 2006). While this way of modeling batches may well suite those kinds of batch processes which are following a more or less linear structure of process steps, it comes to its limits when a lot of the characteristics described in Section 14.2 apply. For example, if process actions generate by-products and these by-products need to be stored in specific tanks before they are used in a different production process, or if batch-conti-processes need to be involved in the model, it gets very hard to find a sufficient mapping between the process and product flow on the one hand and transactions on the other hand. It might get even harder to tackle batch processes coming from the DES side depending on the complexity of the equations used to describe the continuous aspects (cf. Barton and Pantelides 1994 and Wöllhaf et al. 1996).

A second approach to cover the characteristics of continuous processes in DES software is to discretize the continuous flow by breaking it down into adequate units of volume or weight. As Chen et al. (2002) explicate in their case study on a silo and filling system in a chemical plant, the adequacy of the discretized units is somewhat the critical point for this approach. If the units are too large in size or volume, the model might not be accurate enough. But since the number of events in a DES grows with number of transactions (moving units), the computational performance of the simulation experiments might suffer badly if the units are too small.

The variety of approaches in the literature indicates that there currently is no such thing as *the* one way to tackle the challenges imposed by the characteristics of chemical production processes from the technological standpoint. However, when it comes to success factors for sustainable use of simulation in a commercial context, technology is only one aspect. As Mayer and Spieckermann (2010) show for the automotive industry and Sharda and Bury (2010) confirm for the chemical industry organizational factors are at least as important for long-term success and benefit of simulation use as technology. The long-term conservation and exploitation of simulation know-how within a companies group of simulation experts is identified in both sources as being crucial.

The chemical firm Evonik Industries (formerly known as Evonik Degussa and Degussa) has made a similar experience. Since more than a decade, Evonik is continuously and successfully working with DES to improve its production processes (cf. Splanemann 2001). The examples presented in the following section all come from engineering projects supported by Evonik engineers with DES.

14.4 Examples

The examples in this section are organized following the operational levels described in section 14.2. The first case deals with the optimization of a global supply net of a product division. The second example describes a model that has been used to support the design process of a large new production site and the third example is about selected production processes within a larger production facility. In all three cases, the DES software PlantSimulation from Siemens (Bangsow 2010) has been used. This does not necessarily mean that this commercial simulation tool is the best technological choice in every case but is simply propelled by the fact that it has been the standard for DES at Evonik for many years now. In combination with the sound expertise of the companies' engineers in applying the software it makes the use of the tool very effective and efficient.

While the same basic simulation tool was used throughout the examples, the add-ons elements to tackle the challenges of the specific applications were different. The details will be explained in the following subsections.

14.4.1 Study of a Global Supply Net

As a survey of Terzi and Cavalieri (2004) shows there are several ways to apply simulation in a supply chain context. It can be used for supply chain design as well as for support of supply chain operation. In supply chain design, simulation can e.g. support decisions on ramp up or shut down of locations as well as on the capacity of production and warehouse sites. In supply chain design and in daily supply chain operation simulation may help to adjust control parameters such as safety stock or lot sizes and to assess transport.

The case considered in the following was part of a strategic initiative for one business unit of the company. Hence, the objectives were rather strategic such as analysis of the existing supply chain structure, identification of bottlenecks and weaknesses within this structure, improved understanding of cost drivers for stock and transportation costs, and improved understanding of lead times and causes for stock out.

Figure 14.2 gives some indication of structure and complexity of the considered network. It is showing only the part of the supply chain which is directly governed by the company, i.e. all nodes are either warehouses or production sites run by the analyzed business unit. On top of these more than 700 locations, suppliers and customers in 56 countries or regions had to be included into the model sufficiently. Over 220 products with more than 1,200 potential allocations of products to locations and hundreds of different supply options had to be incorporated and the simulation model processed a total of 7,000 orders per evaluated year.

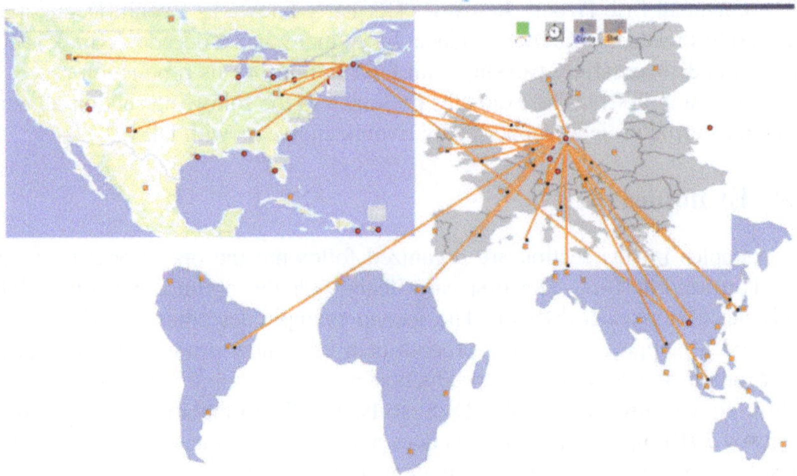

Fig. 14.2 Sample Supply Network (Simulation Model Screenshot)

The simulation model was used to evaluate the consequences of the integration of two new production locations into the supply chain and several ideas with respect to changes in product allocations coming with the new sites. The criteria to assess alternative supply chain configurations were cost (for transportation, stock, and production), service level, and utilization of production resources in the different locations.

As in many cases of supply chain simulation it turned out to be very painful (and costly) to finally generate a valid data model of all supply chain operations. However, the insights gained with the model were so fruitful that it was not only used to assess the planning of the new locations, but to integrate it into tactical planning decisions of the involved business unit.

With respect to simulation technology, the Evonik DES standard tool was used, and the building blocks were taken from a library dedicated to model supply chains for discrete manufacturers. However, since the units considered on this level are production orders and lots and transports, i.e. discrete units, this approach is absolutely adequate for a chemical supply chain as well.

14.4.2 Support of New Site Design

While the first example was considering several sites all over the world and a whole network of warehouses and plants, the second example is focused on one site. A couple of years ago, there has been a decision for a major investment in a completely new production plant in Asia.

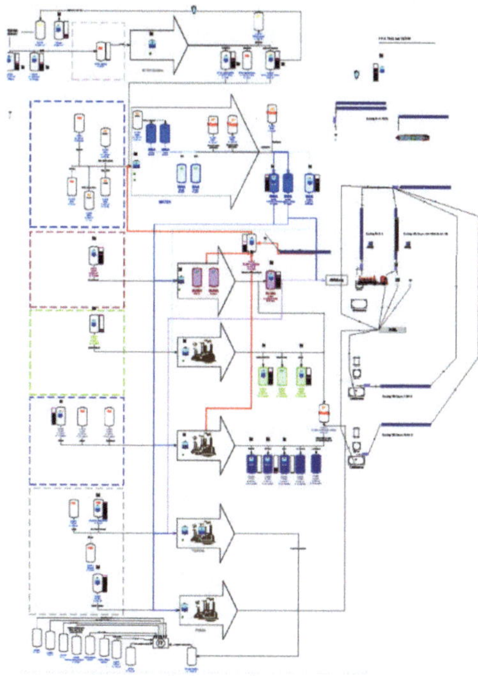

Fig. 14.3 Screenshot of a simulation model on site level

The chemical processes and products were the nearly the same as already established elsewhere in the world, however, the dimensions in terms of yield per year, the customer structure (number of orders, ordered quantities), and some of the transport options (more sea vessels, less tank cars) were different to previous experiences.

In order to limit the risks associated with such an investment with respect to production and transport logistics a DES simulation model was set up. Fig. 14.3 shows a screenshot of the simulation models and is meant to convey some idea of the included number of tanks for reagents and products and the number production lines (indicated by the large arrows). On the right hand side of the screenshot some drumming and filling stations for IBC containers, tank cars (truck and rail) and sea vessels are sketched out.

Major objective of the simulation model was to ensure that the envisioned production volume per year could be handled by the site in an efficient and effective manner, i.e. to test that the production capacity is sufficient, that the tank capacity is adequate without being dispensable, and that the capacity of the filling stations does fit the needs as well. Maintenance and quality related breakdowns were included as well as fluctuations in demand and, e.g. in sea vessel arrival times.

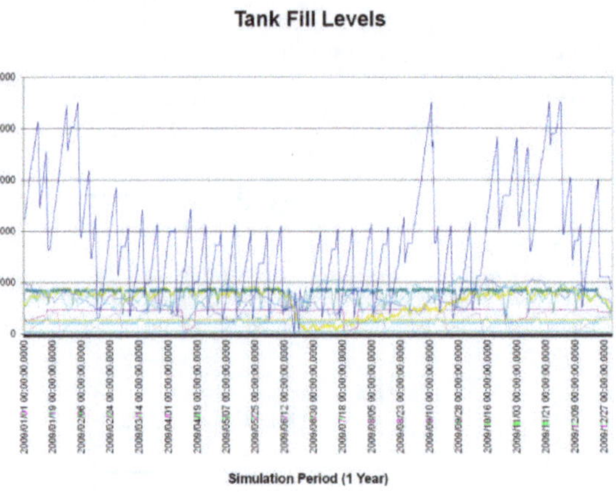

Fig. 14.4 Tank Fill Levels over one Year

The findings were discussed using for example charts like the one shown in Fig. 14.4 which shows the fill level of some of the tanks over a period of one year based on a forecasted customer orders for this year and a dedicated campaign policy for the production lines. As a result of the simulation model several adjustments were made to the original site design: some tank number and tank dimensions were adjusted, guidelines for production planning were derived with respect to upper and lower bounds of campaign sizes, and rules for maintenance were evaluated. All in all, the simulation activities went along with the plant design for about almost two years and some scenarios were re-visited after the ramp-up of the plant.

The simulation methodology was DES with integrated continuous aspects on a very basic level, i.e. the behavior of processes and tank fill levels was calculated using simple approximations based on sums and differences of linear equations which turned out to be absolutely sufficient.

14.4.3 Capacity Analysis of Selected Tanks

The third and last example presented here is compared to the two preceding cases rather clearly arranged: within a larger production site, the engineers had to make decisions with respect to a three step batch production process (process steps P1 – preparation, FP – filtering, and generation of recylate) and the size of the assigned tank capacities. The screenshot in Figure 14.5 illustrates the scope of the model.

It might be arguable to use simulation instead of, e.g. analytical approaches in cases like this. However, random breakdowns of equipment, variable campaign sizes, sequence dependent changeover times and other details had to be taken into account. Furthermore, experienced simulation modelers are able to set up a model like the one shown in this example within hours rather than in days, and to perform analysis which would not be applicable or a lot more complicated with other approaches.

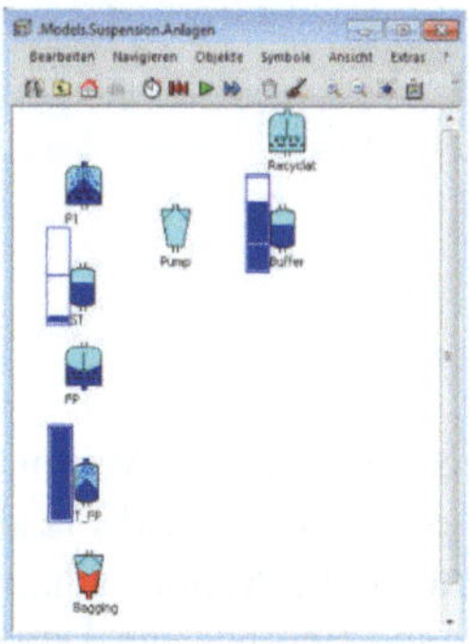

Fig. 14.5 Screenshot of a small scale simulation model

The simulation methodology (DES), the simulation tool, and the building blocks (tanks, processes etc.) used in this example are exactly the same as in the significantly more comprehensive example presented in section 14.4.2.

14.5 Summary and Conclusions

Even though the chemical industry is a somewhat difficult area for discrete event simulation as the introduction of this chapter made clear, there are quite a few technological approaches available to tackle the related challenges. The presented examples give some insight in the great bandwidth of possible applications from supply chain design to evaluations on the level of single processes.

The fact that all these applications were selected among dozens of successful simulation projects conducted within on major German chemical company since 1993 makes clear, how economically beneficial simulation technology can be if it is embedded in the right organizational framework.

Authors Biography, Contact

SVEN SPIECKERMANN, Ph.D., is Chief Executive Officer at SimPlan AG, Maintal, Germany, mainly working as a senior consultant and project manager in simulation projects for several industries. SimPlan is one of the leading simulation service providers worldwide with offices in Germany, Austria, Slovakia, and Shanghai. Since 1992, Sven Spieckermann has been participating in over 200 simulation projects. Additionally, he has been giving lectures in simulation at the Technical University of Braunschweig since 1995 and at the Technical University of Darmstadt since 2008. He has published several papers on simulation, simulation-based optimization and related topics. His e-mail contact is sven.spieckermann@simplan.de

Contact

Sven Spieckermann
SimPlan AG
Edmund-Seng-Str. 3-5
D 63477 Maintal
Germany

Mario Stobbe started his career at Degussa AG, one of the predecessors of Evonik Industries which is today one of the world's leading speciality chemical companies. Since 2008 he is head of the Supply Chain & Production Management Group within the Technology and Engineering Department. He holds a degree in chemical engineering and has a professional background in logistics simulation and operations research. Besides his experience as senior consultant and project manager, he has given lectures and published papers on supply chain related topics including logistics simulation in the chemical industry.

References

Alexander, C.W.: Disrete Event Simulation for Batch Processing. In: Perrone, L.F., Wieland, F.P., Liu, J., Lawson, B.G., Nicol, D.M., Fujimoto, R.M. (eds.) Proceedings of the 2008 Winter Simulation Conference, SCS International, San Diego, pp. 1929–1934 (2006)

Bangsow, S.: Manufacturing Simulation with Plant Simulation and SimTalk. Springer, Heidelberg (2010)

Barton, P.I., Pantelides, C.C.: Modeling of Combined Discrete/Continuous Processes. AIChE Journal 40(6), 966–979 (1994)

Bauer Jr., D.W., McMahon, M., Page, E.H.: An Approach for the Effective Utilization of GP-GPUS in Parallel Combined Simulation. In: Mason, S.J., Hill, R.R., Mönch, L., Rose, O., Jefferson, T., Fowler, J.W. (eds.) Proceedings of the 2008 Winter Simulation Conference, SCS International, San Diego, pp. 695–702 (2008)

Cellier, F.E.: Combined Continuous/Discrete Simulation Applications, Techniques, and Tools. In: Wilson, J., Henriksen, J., Roberts, S. (eds.) Proceedings of the 1986 Winter Simulation Conference, pp. 24–33. ACM, New York (1986)

Chen, E.J., Lee, Y.M., Selikson, P.L.: A Simulation Study of Logistics Activities in a Chemical Plant. Simulation Modelling Practice and Theory 10(3-4), 235–245 (2002)

Fahrmann, D.A.: Combined Discrete Event Continuous Systems Simulation. Simulation 14(2), 61–72 (1970)

Günther, H.O., Yang, G.: Integration of Simulation and Optimization for Production Scheduling in the Chemical Industry. In: Proceedings of the 2nd International Simulation Conference, Malaga, Spain, pp. 205–209 (2004)

ISA-S88. Batch Control Part 1: Models and Terminology. ANSI/ISA-88.01-1995, ISA, North Carolina, USA (1995)

Jahangirian, M., Eldabi, T., Naseer, A., Stergioulas, L.K., Young, T.: Simulation in manufacturing and business: A review. European Journal of Operational Research 203, 1–13 (2010)

Mayer, G., Spieckermann, S.: Life-Cycle of Simulation Models: Requirements and Case Studies in the Automotive Industry. Journal of Simulation 4(4), 255–259 (2010)

Mehra, S., Inman, R.A., Tuite, G.: A simulation-based comparison of batch sizes in a continuous processing industry. Production Planning & Control 17(1), 54–66 (2006)

Mušič, G., Matko, D.: Simulation Support for Recipe Driven Process Operation. Computers & Chemical Engineering 22(suppl. 1), S887–S890 (1998)

Schulz, M., Spieckermann, S.: Logistics Simulation in the Chemical Industry. In: Engell, S. (ed.) Logistic Optimization of Chemical Production Processes, pp. 21–36. Wiley, Chichester (2008)

Sharda, B., Bury, S.J.: Bottleneck Analysis of Chemical Plant Using Discrete Event Simulation. In: Johansson, B., Jain, S., Montoya-Torres, J., Hugan, J., Yücesan, E. (eds.) Proceedings of the 2010 Winter Simulation Conference, SCS International, San Diego, pp. 1547–1555 (2010)

Sharda, B., Vazquez, A.: Evaluating Capacity and Expansion Opportunities at Tank Farm: A Decision Support System Using Discrete Event Simulation. In: Rossetti, M.D., Hill, R.R., Johansson, B., Dunkin, A., Ingalls, R.G. (eds.) Proceedings of the 2008 Winter Simulation Conference, SCS International, San Diego, pp. 2218–2224 (2009)

Smith, J.S.: Survey on the use of simulation for manufacturing system design and operation. Journal of Manufacturing Systems 22(2), 157–161 (2003)

Splanemann, R.: Production Simulation – A Strategic Tool to Enable Efficient Production Processes. Chemical Engineering & Technology 24(6), 571–573 (2001)

Stawinska, A. (ed.): European Business – Facts and Figures 2009 edition, Eurostat, Office for Official Publications of the European Communities (2009)

Terzi, S., Cavalieri, S.: Simulation in the Supply Chain Context: A Survey. Computers in Industry 53(1), 3–16 (2004)

Watson, E.F.: An Application of Disrete-Event Simulation for Batch-Process Chemical - Plant Design. Interfaces 27(6), 35–50 (1997)

Wöllhaf, K., Fritz, M., Schulz, C., Engell, S.: BaSiP – Batch Process Simulation with Dynamically Reconfigured Process Dynamcis. Computers & Chemical Engineering 20(suppl. 2), S1281–S1286 (1996)

15 Production Planning and Resource Scheduling of a Brewery with Plant Simulation

Diego Fernando Zuluaga Monroy and Cristhian Camilo Ruiz Vallejo

In the brewing industry the quantities to be produced for each product are specified in weekly meetings of planning. Afterwards, manually detailed planning and resource scheduling is carried out by a production specialist, who usually takes basic applications developed on MS Excel as a planning support.

A disadvantage of this planning process is the high spending time because of the high complexity of hundreds of constraints involved in the production process and the need of having expertise to understand how the process might change along the week because of many related biological processes.

Through application of simulation with Plant Simulation for production planning and resource scheduling the stated disadvantages can be avoided. This simulation is oriented to be a planning tool that automatically generates the production schedule on the basis of current stocks, master production schedule and minimum lot sizes, which are included taking into account all manufacturing restrictions.

The user can configure plant parameters, stock levels and week production orders. The scheduling tool thus generates the production schedule in a really short time and makes possible evaluating several production scenarios and makes the best decision to optimize plant utilization, stock levels and throughput times.

In this chapter is presented a scheduling tool for breweries based on a simulation of a real plant, the development and the benefits achieved using it in the real process of planning.

15.1 Introduction

The dynamic environment of a brewery is represented by a simulation model which constitutes a Digital Factory solution. It integrates physical layout, production processes, human resources and shifts, material flow, inventory management, maintenance schedules and utilities consumption.

The digital factory is expanded with additional components like optimization algorithms that may be applied in the solution of planning and scheduling

Diego Fernando Zuluaga Monroy . Cristhian Camilo Ruiz Vallejo
COO at OptiPlant Consultores S.A.S
Diego Fernando Zuluaga Monroy
OptiPlant consultores Calle 23 14-15 Ed. Parquesoft
Armenia, Quindío. Colombia
e-mail: dzuluaga@opc.com.co

problems. They together conceive a powerful scheduling tool that generates production schedules that does not violate constraints related to the limited availability of resources in the brewery.

Resource constraint violations or conflicts can be resolved automatically by the scheduling tool, the user can interactively modify the schedule and mix automated and manual scheduling to formulate a production plan that is feasible and satisfies the company objectives.

The user can configure plant parameters, stock levels and week production orders. The scheduling tool thus generates the production schedule in a really short time and makes possible evaluating several production scenarios and makes the best decision to optimize plant utilization, stock levels and throughput times.

15.2 Case of Study

15.2.1 Structure of the Brewing Process Related to the Digital Factory

An abstraction of brewing process is shown in Figure 15.1. The process named as brew house includes transportation of raw materials, milling, mashing (process where the malt grain is made ready for brewing), lautering and wort boiling.

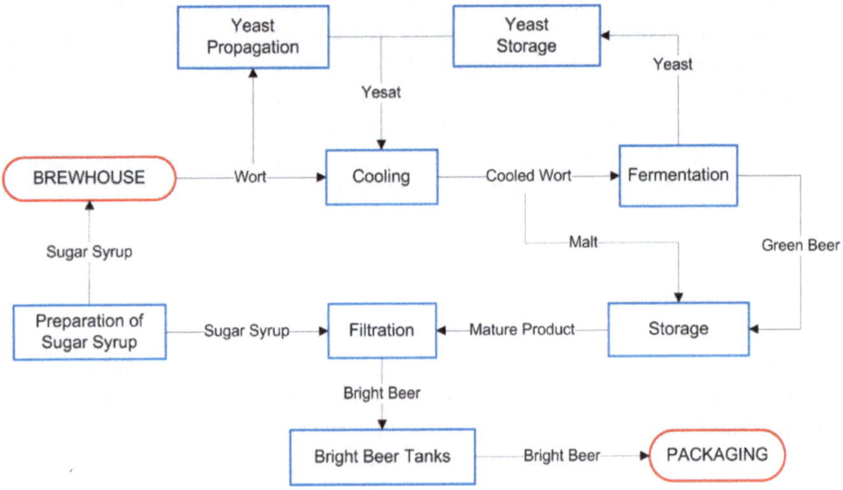

Fig. 15.1 Most relevant phases of brewing process for simulation.

Mashing is the process that converts the starch of malt, into sugars. The result of mashing then strained through the bottom of the mash tun to separate the residual grain and the wort.

In the boiling process, the wort is moved on to a kettle and mixed with hops and high maltose corn syrup (HMCS). When the wort is ready, it must be cooled down in order to avoid harming the yeast and start the fermentation accurately.

In the simulation for the digital factory, the process begins with the cooling and its inputs are wort and yeast. The source of wort makes deliveries in batches usually smaller than the fermentation vessels (FV), so several batches are needed to fill a FV.

In the fermentation stage, the yeast metabolizes the sugars of the malt into alcohol and carbon dioxide. The duration of this process is very variable due to its biological nature, thus in the simulation was included a statistical distribution that generates the time of fermentation; the FVs are represented as buffers.

Yeast propagation and yeast recovering are some of the most important steps in brewing process, because if the yeast is not recovered on time (less than 42 hours) it must be discarded and a new yeast strain has to be propagated. This new strain propagation takes more time than the standard fermentation lead time. Another constraint related to the yeast is the number of times that the yeast may be used in fermentation. Due to that it is quite important to extend the yeast life cycle and it is easier using the scheduling tool.

As the fermentation ends, the yeast is removed and the beer is moved on to the storage vessels to mature the beer during several days with temperatures below zero degrees centigrade. Given that nonalcoholic beverages made from malt do not need fermentation, they are storaged without yeast.

Beer filtration is to remove rests of the yeast and any solids like grain particles. Besides that, it makes the beer bright. More hops and cane sugar syrup is added during filtration to obtain the final flavor and the needed conditions according to its container (e.g., glass, PET, draft, etc.).

The filtered beer is stored in tanks known as bright beer tanks (BBT) prior to the packaging process.

The simulation represents the movement of beer between fermentation - maturation vessels, filtration process and the most important issue: yeast handling. The bottling lines are represented as sinks, thus it lets know the bottling sequence according to the availability of product in BBTs.

15.2.2 Production Planning and Execution

Figure 15.2 shows the flow of information in breweries until production is executed. The ERP system links the sales department with the production department and generates automatically the inputs necessary for the master production schedule (MPS) like forecast demand, production costs, inventory costs, customer orders, transportation costs, inventory levels, supply, lot size, production lead time and capacity. The result MPS may include amounts to be produced, staffing levels, quantity available to promise, and projected available balance.

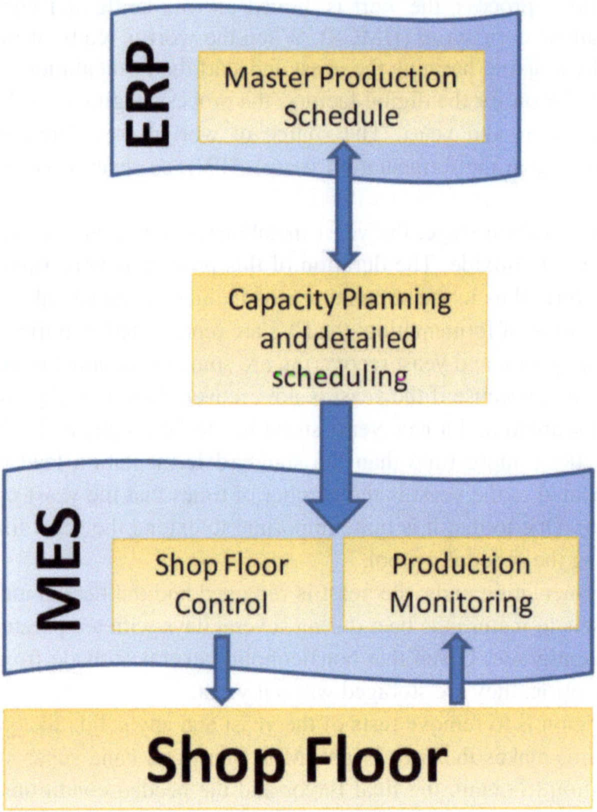

Fig. 15.2 Production planning and execution in breweries

Capacity planning and detailed scheduling is carried out by a production specialist, who usually takes basic applications developed on MS Excel as a planning support. The production specialist generates a detailed schedule that coordinates the manufacturing activities in order to meet organizational objectives, and to anticipate potential performance obstacles (e.g., delays in sub-processes), thus to minimize the disturbing effects on the factory operation.

The control and execution system generally named as the Manufacturing Execution System (MES) of the production system controls the physical system (shop floor), by translating the scheduled tasks to commands to the physical system. Furthermore, it receives reports about the execution state of the schedule. This level does not involve any complex decision-making function but a close connection to the real manufacturing facilities of the factory.

15.3 The Scheduling Tool

15.3.1 Architecture of the Scheduling Tool

To build a digital factory of a brewery, a simulation model is developed, for modeling the overall behavior of the system, including control methods and reflecting the physical system by modeling the resources. As shown in Figure 15.3, the digital factory integrates the manufacturing execution system and the shop floor.

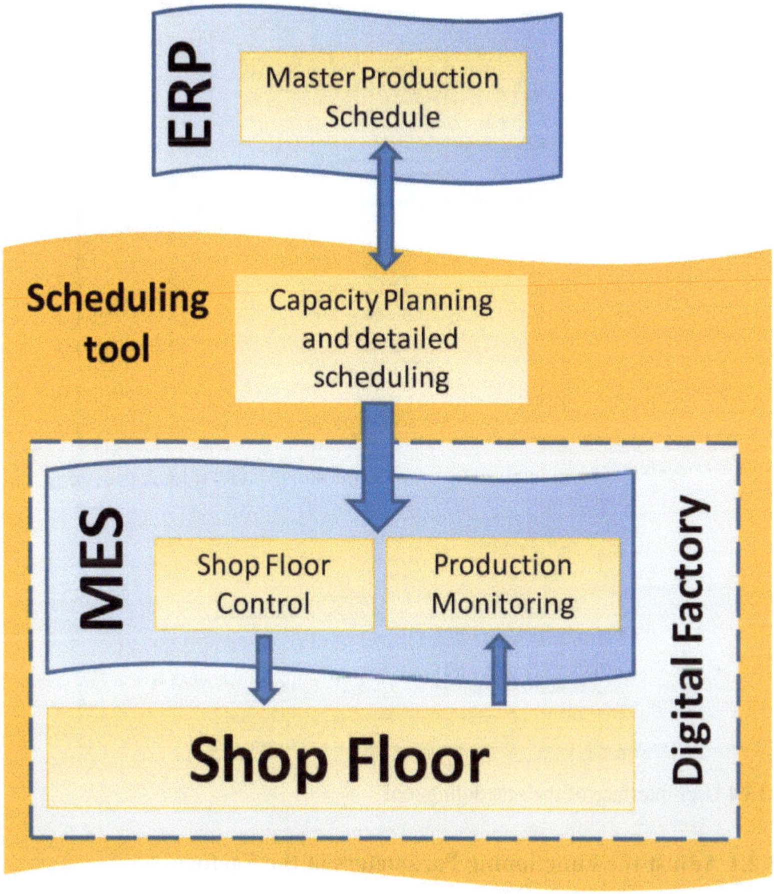

Fig. 15.3 Digital Factory as a part of a scheduling tool.

The digital factory is applied as an evaluation function (schedule evaluator) of optimization algorithms. These algorithms reside in the simulation system as an integrated sub-module that automatically performs tests and validates production plans to generate a detailed schedule of production from the Master Production Schedule.

15.3.2 User Interaction

The scheduling tool has a user interface developed with Plant Simulation as shown in Figure 15.4. The user takes the following steps to obtain an optimal production schedule:

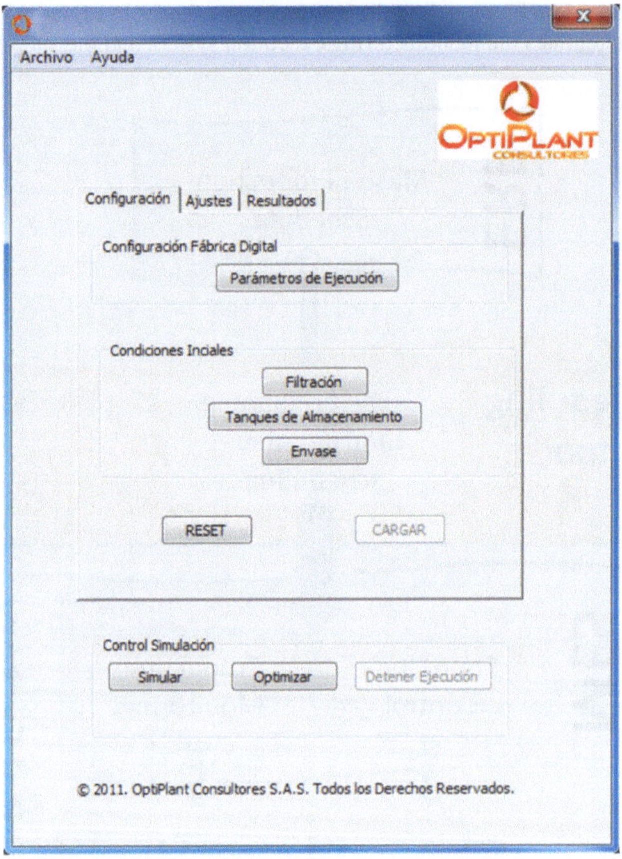

Fig. 15.4 User interface of the scheduling tool.

15.3.2.1 Adjust the Functioning Parameters of the Factory

The user sets the digital factory according to the functioning of the real factory, entering parameters (see Figure 15.5) like velocity of transportation, time for cleaning in place, efficiencies, rate of temperature, fermentation time, filtration speed, bottling speed, etc.

15.3.2.2 Load the Initial State of the Factory

The user loads the initial state of the factory from the production database. The state of the factory makes reference to the levels of WIP in every phase of the brewing process and what the processes are doing when the scheduling is taking place. The data can be adjusted in case of deviations due to outdated information.

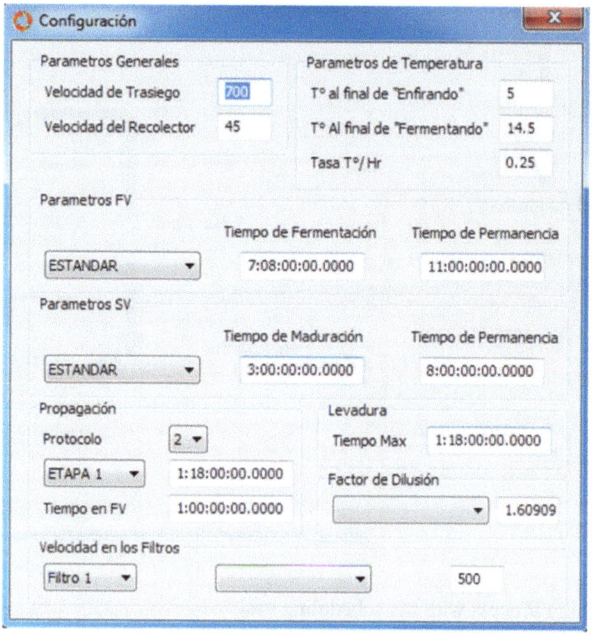

Fig. 15.5 Panel for functioning parameters of the digital factory.

15.3.2.3 Load the Master Production Schedule

The user loads the MPS of 4 weeks. It is very important to plan 4 weeks because of the processing lead time of beer. Thus, the user can track the product since brewing until packaging through simulation.

15.3.2.4 Planning Process with the Scheduling Tool

The scheduling tool has all the information about the real factory necessary to start iterating and find the best possible production schedule.

The top level algorithms of the scheduling tool execute the steps shown in Figure 15.6. In step 4, the scheduling tool modifies the production schedule for each iteration and makes decisions based on priorities related to the brewing process (e.g., avoid stops in bottling lines, extend yeast lifespan, reduce CIP efforts, etc.).

The user verifies the production schedule through Gantt diagrams (see Figure 15.7) and makes modifications if it is necessary by applying enhancement strategies through the user interface.

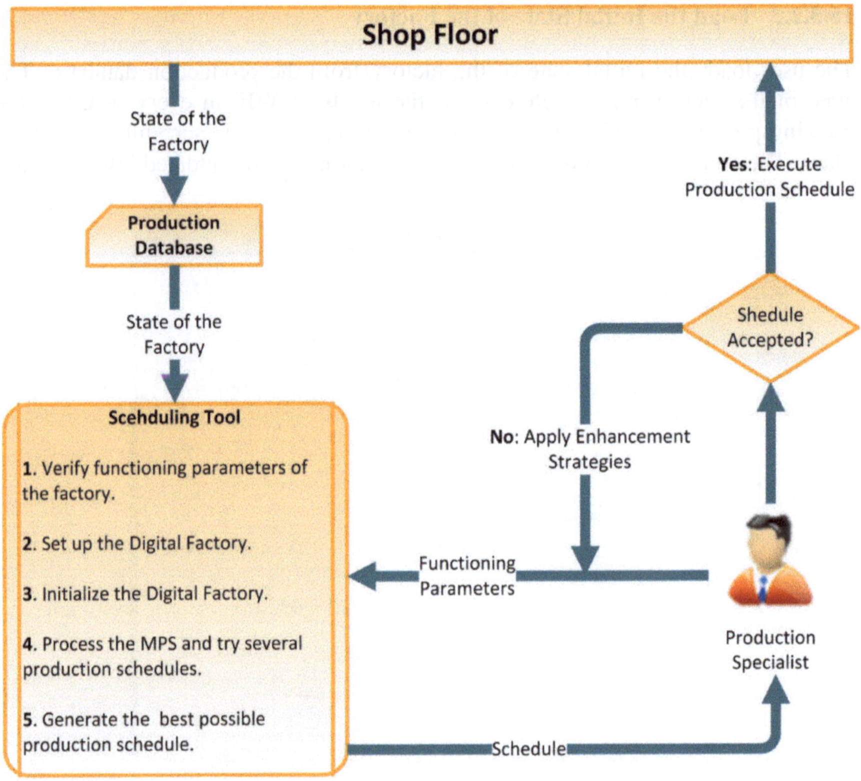

Fig. 15.6 Planning process with the scheduling tool.

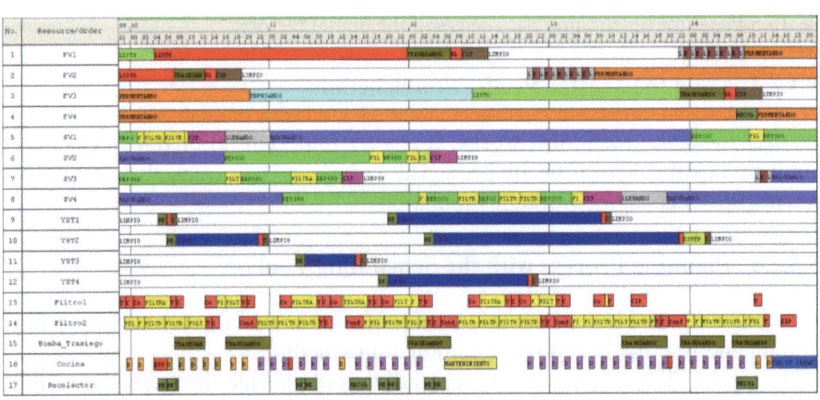

Fig. 15.7 Example of Gantt diagram for verification of the production schedule generated.

When the production specialist accepts the schedule, the scheduling tool generates the instructions in MS-Excel in a format that operators (shop floor) can understand. An example of instructions is shown in Table 15.1.

Table 15.1 Instructions for brew filtration and bottling.

Event	Start	End	BBT	Vol.	Brand	Line	Botlling	BBT free
Conformando	22/08/2011 10:20	22/08/2011 12:20						
Filtrando	22/08/2011 12:20	22/08/2011 15:31	BBT7	2200	MARCA1	L1	23/08/2011 02:04	23/08/2011 08:05
Filtrando	22/08/2011 15:31	22/08/2011 18:44	BBT5	2200	MARCA3	LPET	23/08/2011 08:06	23/08/2011 14:17
Filtrando	22/08/2011 18:44	22/08/2011 21:57	BBT3	2200	MARCA1	L1	23/08/2011 14:19	23/08/2011 20:19
Filtrando	22/08/2011 21:57	23/08/2011 01:09	BBT6	2200	MARCA2	L2	23/08/2011 19:15	24/08/2011 07:33
Agua Fría	23/08/2011 01:09	23/08/2011 02:09						
CIP A	23/08/2011 02:09	23/08/2011 03:29						

15.4 Benefits of Digital Factory as a Scheduling Tool

Traditional scheduling tools work with static data about the factory (e.g., aver-age capacities) and rarely integrate the uncertainty of multiproduct facilities what may render plans and schedules inadequate.

Using digital factory as part of a scheduling tool, companies can achieve efficient behavior of factories according to production demands, while effective responsiveness to changes in operating conditions is increased. Following are presented some more benefits:

- Avoid conflicts with resources that can only be used by only one procedure at a time
- Increase the use of resources
- Reduce the time of production planning
- Increase the detail and accuracy of planning thanks to the use of digital factory
- The interaction of different products and processes that share equipment and resources can be analyzed and optimized.
- Beer macroloss decreases.
- Equipments usage efficiency increases due to the scheduling optimization.

Authors Biography, Contact

Diego F. Zuluaga holds bachelor's degree in Electronic Engineering (University of Quindío, 2009). He is cofounder of OptiPlant Consultores. Since 2009, he works as CEO at OptiPlant Consultores.

Contact

Diego Fernando Zuluaga Monroy
OptiPlant consultores
Calle 23 14-15 Ed. Parquesoft
Armenia, Quindío.
Colombia
dzuluaga@opc.com.co

Cristhian C. Ruiz holds bachelor's degree in Electronic Engineering (University of Quindío, 2010). He is cofounder of OptiPlant Consultores.
Since 2009, he works as COO and R&D Director at OptiPlant Consultores.

OptiPlant Consultores
Founded in 2009, OptiPlant Consultores is pioneer company in Colombia developing solutions based on digital factory concept using Plant Simulation of Siemens PLM. Major experiences have taken place in designing customized scheduling tools for Colombian breweries and graphic industries.

16 Use of Optimisers for the Solution of Multi-objective Problems

Andreas Krauß, János Jósvai, and Egon Müller

The book chapter presents two case studies that consequently use the computer-aided simulation in combination with the optimization. The optimization follows the search of the best solution for a given optimization problem. Case study 1 introduces a special procedure for the determination of the number of machines in production systems. Thereby the optimization is combined with a cost simulation. It shows that, with this procedure, very good solutions can be found automatically and concerning a specific problem. Case study 2 deals with the order controlling in Car Assembly with the Aid of Optimizers. The modeling had to consider that a lot of flexible parameters were needed to ensure enough planning roam. A main goal was to determine the computational achievable "right" production sequence. The hand-made production program should be optimized by the simulation. Both case studies present possibilities and potentials of the computer-aided simulation combined with the optimization.

16.1 Strategies and Tendencies of Factory Planning and Factory Operation

Today more than ever, production enterprises are influenced by an entrepreneurial world affected by evolving dynamics. Therefore the following developments, among others, are responsible for the accelerated change enterprises are confronted with ([16.3], p. 5; [16.6], p. 29 ff.; [16.15]; [16.22], p. 13; [16.25]; [16.31]; [16.35]:

1. globalization of the economy, a globally unequal wage level, increasing cost pressure
2. rapid development and spreading of new information and communication technologies
3. an increasing demand for individual customers' needs and quality requirements
4. a more complex network of the flow of goods and capital
5. an uncertain forecast of the customer's needs with increasingly temporal fluctuations of the order quantities
6. a Shortage of resources and primary energy carriers and consequently dramatically increasing prices for materials, additives, tools etc.

Andreas Krauß . János Jósvai . Egon Müller
Professur für Fabrikplanung und Fabrikbetrieb
Technische Universität Chemnitz D-09107 Chemnitz Germany

In order to meet increasing environmental requirements in the future, the production enterprises use, among others, the following methods ([16.3], p. 5; [16.6], p. 29 ff.; [16.15]; [16.22], p. 13; [16.25]; [16.35]; [16.31]):

1. A rapid modification of the scope of production, shortening product life cycles and customized manufacturing and assembly
2. Increase of the model variety/number of variants
3. Increase of the advantage and complexity of the product
4. Increase in innovative efforts and rapid development and diffusion of new technologies
5. Introduction of new technologies within the production systems (especially new information and communication technologies) and increasing automation
6. An Increasing number of methods rationalizing, improving and adjusting the product systems

These entrepreneurial methods cause goal conflicts between different target variables such as, e. g. time, efficiency and quality. Furthermore, the following challenges arise for production systems planning:

1. An increase in planning frequency.
2. A necessity of shortening the planning times.
3. A necessity to reduce planning effort.
4. An increase in planning complexity.

In regards to the increasing complexity of the planning tasks, a necessary shortening of the planning time combined with a reduction of the planning efforts it will be difficult to develop a high quality planning solution by using the classical methods and tools of production systems planning. Therefore, the development of new methods and tools of the production systems planning is essential for the changing conditions of today's industrial environment.

16.2 Basics of Methods for Simulation and Optimization

16.2.1 Simulation and Costs

The characteristic feature of the simulation are its' dynamics, meaning the system changes over the course of time.

The previous section highlighted the target conflicts during the planning and the operation of production systems. During the planning stage of the production systems, variants and planning alternatives are developed. For the selection of a preferred variant an evaluation is needed ([16.19], p. 4). As maximizing the efficiency (in the context of the value-added process) is the topmost objective of an enterprise ([16.8], p. 1), it can act as a target system and evaluation basis in terms of arising target conflicts. For this reason, a monetary evaluation and selection of a preferred variant in the context of the production systems planning, means that it is important to take cost data into account ([16.37], p. 9).

Most commonly taking place (in the context of conventional simulation strategies) is the examination of technical-logistical parameters, for example e. g. the capacity utilization of the plant, the processing time, and the buffer allocation, the use of the capacity or the disturbance reaction. The cost level and structures are often ignored, causing target conflicts to be irresolvable ([16.37], p. 9).

The cost simulation additionally becomes a decision-making help in order to respect contrary target figures in complex decision-making processes. Furthermore, the users of simulation tools are sensitive to economic aspects and are able to economically and comprehensively interpret alternative solutions of the production systems planning at an early stage. In terms of the planning of production systems and the various interdependencies between the single elements of the system, it is possible by means of the cost simulation, to examine the impacts methods have on the whole production system. For example, this includes the examination investments on the output of production systems and the associated efficiency parameters such as (eg.) the payback period. The cost simulation supports the optimization methods of production systems, especially in terms of changing parameters like the demand alteration, product structure, product mix, targeted output, machinery, vertical range of manufacture, operational procedures and working time models and the analysis of the consequence of running the production system. This also includes the depiction of the cost per unit in order to achieve a simultaneous: optimal operating point at minimal cost, minimal running time at maximal power all at the same time. ([16.32], p. 2, 10-11, p. 12; [16.37], p. 10) The simulation-aided order costing systems can be distinguished between integrated (in-line) and downstream (off-line) systems. ([16.32], p. 3-4; [16.37], p. 45)

Integrated cost simulation modules calculate and allocate the cost data and permanently interpret them as a component of the processing simulator during the process of a simulation ([16.32], p. 3; cf. figure 16.1). In terms of an integrated cost simulation, there is the necessity to extend the majority of the components of the simulation model by cost-specific parameters. In doing so, the resource costs are proportionate to movable components representing the products. This adds up to the rucksack-principle. Regarding this principle the residing time of products on resources are calculated on the basis of the entry and exit times and then multiplied by the respective time-related cost rates of the resource and charged to the product. Additional to the rucksack principle, the different cost types are often collected by means of cost type tables in order to be able to depict the accumulated total costs at any point of time regarding the different cost types. ([16.37], p. 46-47)

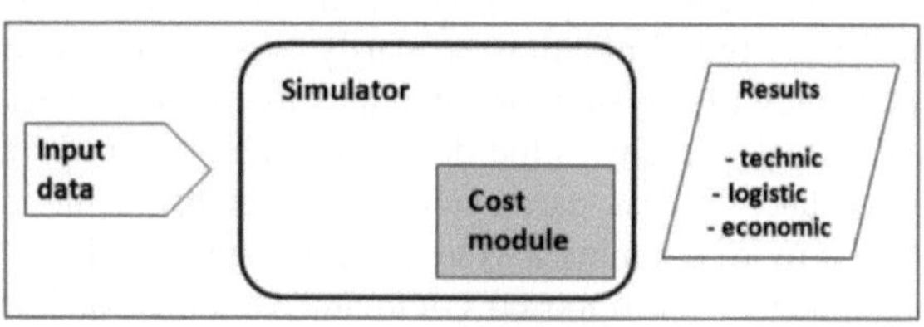

Fig. 16.1 Integrated cost simulation modules ([16.32], p. 3).

The advantage of integrated system is the situational, cost-relevant decision-making ability during the simulation process. The disadvantage is the necessity of two simulation runs for questions such as, (eg.) the calculation of dynamic machine-hour rates. Furthermore, the integrated cost simulation modules require more effort in terms of the model construction and the realism of the simulation. A serious disadvantage is indicated by WUNDERLICH which mentions that the calculation algorithm has to run through at least two levels. On one hand, the costs have to be collected on the level of the single elements of a model and on the other hand, they have to be consolidated again within the super ordinate sum function. In terms of alterations of the calculation algorithm, this leads to multi-level alteration operations. Due to the distributed calculation procedure, a high error probability arises regarding the consistency of the total cost accounting system. ([16.37], p. 48-49)

The cost calculation in downstream systems is a two-stage calculation. At first, a so-called "trace"-file is generated within a classical processing simulation. The "trace"-file includes all the results of the simulation run. Afterwards, a downstream cost module calculates based on the "trace"-file, all the cost-relevant data and analyses them. The advantage of the downstream cost simulation module is that the cost module is not totally linked to a simulation tool, but can be linked to different simulation tools. ([16.32], p. 3-4; cf. Figure 16.2).

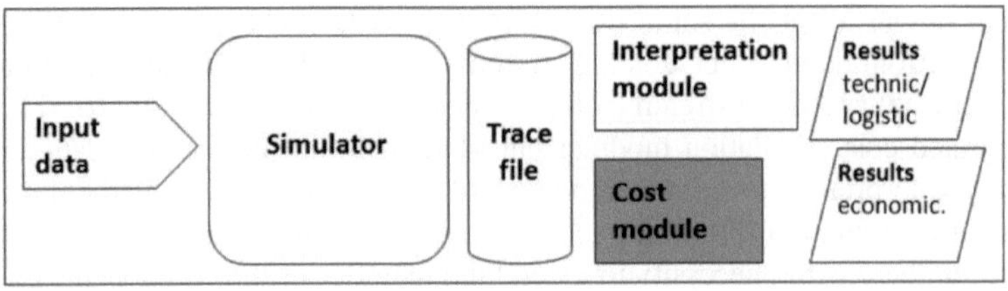

Fig. 16.2 Downstream systems for the cost calculation ([16.32], p. 3-4).

16.2.2 Simulation and Optimization

According to the simulation, findings can be found by means of highly realistic, experimental models. A wide-spread misbelief (and often a reason for the failure of simulation projects) is the assumption that the simulation itself can solve planning problems, thereby easing the planners creative planning work ([16.31], p. 6). However, it cannot, as the simulation primarily serves to explain the complex, built-in interdependency of a system or calculates its' duration periods ([16.18], p. 171). The planner's task thereby is the design of the system and the variation of structure-, resource and process parameters to preferably achieve a good target ([16.19], p. 10). The more complex a system is, the more difficult the parameterization of variables due to the opposed command variables ([16.19], p. 10). The optimization can help to find a solution to the question. The computer-aided optimization supports the search of the best solution of a given optimization problem regarding a certain quality factor by using a computer-aided optimization procedure ([16.9], p. 2). Therefore, a fully automatic solution of the problem can be found, the planner will be unburdened and ideally find a qualitatively better solution than with manual variants ([16.9], p. 2).

The optimization problem is a problem that „can be traced back to the selection of the best element of a quantity regarding one quality factor" ([16.9], p. 8). The optimization problem is characterized by the quantity as a cross-product of the domain of the amount of decision variables and by the quality factor of the real-valued target function. The optimization procedure is an algorithmically described procedure for solving optimization problems and is limited by means of process parameters in its use regarding a certain degree of freedom. The optimization procedure is based on a general, characteristic concept: the optimization principle. ([16.9], p. 8)

Optimization problems are classified into different problem categories due to their target function and their decision variable. Because of the linearity of target functions and auxiliary functions it is possibly to distinguish between linear and non-linear optimization problems. In terms of the simulation-aided optimization the classification of the optimization problem regarding the linearity often is difficult due its complexity. ([16.15], 295; [16.19], p. 21)

In the context of the digital plant, KÜHN distinguishes between optimization problems with parameter optimization, with sequence optimization and selection optimization. In terms of optimization problems with parameter optimization, a parameter-based target function can be found. Optimization problems with sequential optimization comprise elements to be brought in optimal order. The problems with selection optimization focus on the optimal selection of elements from a total quantity. ([16.18], 174)

The optimization procedures are differentiated into exact and heuristic procedures. After a certain time, exact procedures finally achieve an optimum of the optimization task or prove the task to be insoluble. When trying to find the solution, heuristic procedures also ignore the potential solutions of the problem in favor of the time, and can grant the achievement of the global optimum. ([16.4], p. 14; [16.15], p. 296-297)

KRUG & ROSE optimization procedures vary in deterministic, random, threshold, evolutionary and genetic procedures as well as in permutation procedures ([16.19], p. 22). In terms of deterministic procedures, objective function targets are calculated for a starting point and its neighboring points. The point with the best objective function target is the starting point of the search of another neighbor point with a better objective function target. Thus a determined and quick search of good solutions can be achieved. The disadvantage of this procedure is the low probability to find a global optimum[1]. If the search by means of a deterministic procedure starts near a local optimum[2], the search will be going towards the local optimum without achieving a global optimum. The random procedure or the stochastic procedure produces random starting points within the total solution space. Subsequently, objective function targets are produced for those starting points in

[1] A global optimum represents the best objective function target within the solution space.

[2] A local optimum represents the best objective function target within a section of the solution space.

order to search for better solutions in the surrounding points with especially good objective function targets. The advantage opposite to the deterministic procedure is a higher probability to randomly find a starting point near the global optimum.

The disadvantage is a longer calculating time due to a great and necessary number of calculations of the objective function targets. Also, threshold procedures are characterized by a random search within the solution space, but when searching for randomly chosen points of the solution space in the surrounding area, a certain deterioration of the target value is acceptable in order to prevent a fast convergence towards a local optimum. By gradually minimizing the predetermined threshold the process will be converted into a local search procedure. An example for the threshold procedure is the simulated annealing. Depending on the parameterization the threshold procedure can be in need of high calculating times. Evolutionary procedure are based on observations of the natural evolution of living organisms and by means of a random generator, they start producing a parental quantity. In the following evolutionary stage, by means of different mutation and/or recombination procedures, only a certain number of children are crossed from the parental quantity. Afterwards there will be an evaluation and a selection of the best individuals for the next parent generation. A subset of the evolutionary procedures is the genetic procedures. In terms of a genetic optimization procedure, the individual evaluation occurs according to a fitness value representing the extent of the adaptability to the environment, thus individuals with a high fitness value reproduce themselves with a higher probability. Genetic optimization procedures can quickly lead to good solutions especially in terms of production planning problems. The disadvantages of the evolutionary procedures are the high number of calculations due to a multitude of solution points and often an unclear optimization speed. Permutation procedures are used as heuristic procedures during the simulation of production processes of the semiconductor industry for automatic parameter variations. ([16.15]; p. 298; [16.18], p. 176-181; [16.19], p. 22-26)

The decision within the optimization is made according to HARDER 's four steps:

- Description of the system on which the problem is based.
  This step includes the development of an acceptably precise model describing the dependencies between the input parameters (variable) and the output parameters (target variables) of the system.
- Determination of the solution requirements.
  This step includes the determination of the minimal requirements (auxiliary conditions) a solution must meet.
- Determination of a criterion for the quality of solutions.
  In this step the determination of a quantitative quality criterion (target function) is made to compare the different solution with each other. According to HARDER costs, profits or efficiency are considered the topmost criteria for most of the technical and economic systems.
- Choosing the best solution
  In this step, using an appropriate strategy, the best solution out of all acceptable solutions is selected.

Although a multitude of optimization procedures and algorithms has been developed and described, generating a simulation model and parameterizing the optimization algorithms is still considered to be an individual configuration. A generally accepted and standardized procedure for the simulation-aided optimization is so far non-existent. ([16.19], p. 6)

16.3 Case Studies

16.3.1 Case Study 1: Dimensioning of Plants with the Aid of Optimizers (by Andreas Krauß)

16.3.1.1 Concept of the Status-Controlled Dynamic Dimensioning of Production Systems in the Case of Varying Capacity Requirements

The following section deals with the dimensioning of resources. The dimensioning of resources is part of the production systems planning.

According to SCHENK&WIRTH the dimensioning is defined as the quantitative determination of the required resources, the staff and the surface as well as the costs for the future production system. The balance sheet approach is the basic calculating approach for the dimensioning. The balance sheet approach contrasts the load capacity (which has to be installed) with the expected load capacity. Thereby the load capacity to be installed is larger than or equals the expected load capacity. In contrast to the static dimension, the dynamic dimensioning considers how the load changes over time. ([16.25], p. 248)

The calculation of the required quantity of resources z^*_{BM} generally affects the context of the static dimensioning and results from the quotient of the required performance (capacity, load) P_{BM} and the available installed performance (capacity, load capacity) of the resource P_{BMv}. ([16.25], p. 248)

$$z^*_{BM} = \frac{P_{BM}}{P_{BMv}} \quad ([16.25], \text{p. } 248) \tag{1}$$

z^*_{BM} is generally rounded up to an integer z_{BM} (even though in case of a possible overload of resources a partial rounding down is possible as well). The quality of the dimensioning is described with the help of the temporary workload of the resources n_{BM}:

$$n_{BM} = \frac{Z^*_{BM}}{Z_{BMv}} \quad ([16.25], \text{p. } 248) \tag{2}$$

Different reference parameters can be used for P_{BM} and P_{BMv}, for example, time (time unit per reference period), mass (mass per reference period) or quantity

(quantity per reference period). In the context of the production systems planning the commonly used reference parameter is time ([16.25], p. 248)

In terms of the statistic dimensioning, time-dependent changes are not considered and an equal distribution of the required and available capacity is assumed. However, such basic conditions are not given concerning practical problems. Furthermore, the complex temporal dependencies within the production process are not considered in terms of the static dimensioning. While considering the complex production processes and the temporal influences and their dynamic interdependencies, there is the huge advantage of the dynamic dimensioning. Regarding the dynamic dimensioning, the dimensioning results are derived from the calculated load of the means of production during the term of the production system.

The concept of rule-based dynamic dimensioning, as described (based on a defined production program) aims at defined production methods, production process and the chosen resources, which gathers knowledge on the required amount of resources and on calculating the resulting costs. The main focus is on the examination of whether the production system (in terms of fluctuating capacity demand) needs to be provided with a higher amount of resources and a higher quantitative flexibility or not. Furthermore, appropriate points for activating and de-activating the certain resources need to be determined. In order to be able to depict dynamic connections during the whole planning period, the planning period is divided into intervals. At the beginning of the first interval, there is a decision point to be found where an amount of resources is determined. At the beginning of the following intervals, there are more decision points to be found where the decision on activating or de-activating the resources is made. At the end of the last interval of the planning period, this variant will be evaluated in order to determine the benefit of the variants and to compare different variants.

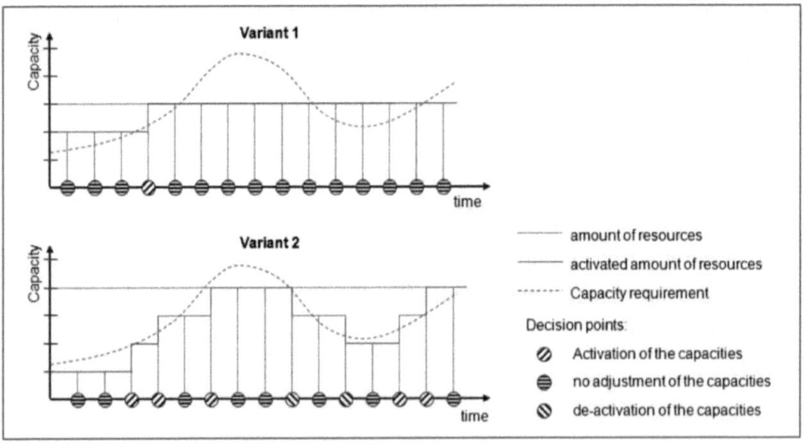

Fig. 16.3 Realization of a status-controlled dimensioning method in two special concepts.

The large amount of variants deriving from the multiplication of all possible decision alternatives of all decision points is problematic. Having ten different machine types, 100 intervals and the three decision alternatives of the resource activation, deactivation and no alteration, there are already possible $2{,}2 \times 10^{472}$ variants.

The idea of the rule-based dynamic dimensioning method is then to make a decision on the decision points depending on the condition of the production system. The decision rules form the basis for these needs, depending on certain conditions of the production system. The following method is based on one which was developed by KOBYLKA [16.16] and deals with the processing time-oriented resource-shift and the backlog-oriented gradual resource-shift.

The method of the processing time-oriented resource shift is decided at the decision points on the basis of the urgent process time derived from the cumulative process times of the orders, if their latest possible starting time in relation to the observance of the given processing time lies before or within the following interval. This method is based on the following states:

$Z_1 : t_d > t_k$, Z ... state

$Z_2 : t_d = t_k$, t_d ... urgent process time

$Z_3 : t_d < t_k$ t_k ... available process time of the resource type

and the following rules:

if Z_1, then increase in capacity (+1),

if Z_2, then no alteration of the capacity (0),

if Z_3, then decrease of capacity (-1).

The advantage of this method is the high degree of adhering to the processing time. The disadvantage of this method is the possibility, that despite an order backlog, resources can be de-activated as high backlogs of non-urgent orders have no influence on shifting resources.

In terms of the method for backlog-oriented gradual resource shifts, a strict gradual shift of resources depending on a pre-defined specific shift-backlog will take place at the decision points. By means of the shift-backlog the backlog intervals of the respective resource types are the result. These are then compared to the order backlog of the latest capacity level. The method is based on the following states:

$Z_1 : t_{ab} > t_{bi}$, Z ... state

$Z_2 : t_{ab} = t_{bi}$, t_{ab} ... order backlog

$Z_3 : t_{ab} < t_{bi}$ t_{bi} ... backlog interval of the resource type

and the rules R_n

R_1: if Z_1, then increase in capacity (+1),

R_2: if Z_2, then no alteration of the capacity (0),

R_3: if Z_3, then decrease of capacity (-1).

By determining the shift backlog it can be decided if the resources are offensively or defensively activated or de-activated. However as the backlog amount itself (and not the composition of the backlog) refers to the urgency of the orders as basis for the resource shift, not adhering to the processing time of certain orders or when avoiding them, can lead to a generally small backlog level in combination with an overcapacity.

Respectively, these two methods both use one parameter in order to make a decision on the adjustment of the production system. For this reason, the disadvantages and the necessity to describe the state of the production systems with several state variables, are the result in order to unite the advantages of the described method into one method ([16.16], p. 117).

The method of the processing time-oriented resource shift uses the state variable t_d (urgent process time), the method of the backlog-oriented gradual resource shift uses the state variable t_{ab} (order backlog). Both state variables can be consolidated in a matrix (cf. table 16.1)

	Possible states of the state variable t_{ab}			
Possible states of the state variable t_d		$Z_{1,1}: t_{ab} > t_{bi}$	$Z_{1,2}: t_{ab} = t_{bi}$	$Z_{1,3}: t_{ab} < t_{bi}$
	$Z_{2,1}: t_d > t_k$	$Z_{1,1;2,1}$	$Z_{1,2;2,1}$	$Z_{1,3;2,1}$
	$Z_{2,2}: t_d = t_k$	$Z_{1,1;2,2}$	$Z_{1,2;2,2}$	$Z_{1,3;2,2}$
	$Z_{2,3}: t_d < t_k$	$Z_{1,1;2,3}$	$Z_{1,2;2,3}$	$Z_{1,3;2,3}$

Z_n ... state

t_d ... urgent process time

t_k ... available process time of the resource type

t_{ab} ... order backlog

t_{bi} ... backlog interval of the resource type

Nine possible states are the result. For every state a decision needs to be made regarding the following possible actions:

HM_1: activation of the capacities[3]

HM_2: no adjustment of the capacities

HM_3: de-activation of the capacities[4]

[3] E. g .putting machines into service.
[4] E. g. shutting down machines.

As there are three possible actions for nine possible states, theoretically 19683 variants can be formed.
Further parameters are

t_{sz} ... length of the interval,

t_{sb} ... backlog of the shift,

n_{rapb} ... amount of resources at the beginning of the period and

n_{raa} ... activated amount of resources at the beginning of the period.

The length of the interval specifies the time between two decision points. The amount of resources at the beginning of the period expresses the amount of the resources installed within the production systems. The activated amount of resources at the beginning of the period defines how many resources are to be existent at the beginning of the planning period in the activated state.

By using optimization methods for the parameterization and the selection of appropriate decision rules, the objective of finding an acceptable solution within a justifiable time is sought. Genetic methods which form a subset of the evolutionary method are used. In terms of genetic optimization methods, the evaluation of an individual is affected by means of a fitness value which depicts environmental adaptability, whereby individuals with a higher fitness value tend to reproduce themselves with a higher probability. By parameterizing and selecting appropriate decision rules, different variants are selected representing the individuals. For the evaluation of the variants, a fitness value must be derived to represent the quality of the variant or the individual. As maximizing the efficiency can be considered the main profit objective in the context of the value-added process ([16.8], p. 1), it can serve as a target system and as an evaluating basis in terms of objective conflicts.

The efficiency can be understood as a relationship of evaluated output und input. The evaluation of the output and input is based on cost items of the internal accounting. For depicting the cost items, cost types of different cost type main groups, having the same reference parameter, are combined.

The following product-related cost items are used for the evaluation method:

Material and Procurement Costs

The material and procurement costs include the following cost types:

- Raw materials: material component, procured pre-products as an essential part of the end product (cost type main group: material costs),
- Auxiliary materials: unessential parts of the end product (cost type main group: material costs),
- Packing materials (cost type main group: material costs) and
- Mailing, cargo (cost type main group: costs for procured services)

The material and procurement cost rate [€/item] is set and used for every product (reference parameter) in the production system.

Storage and Capital Costs

The storage and capital costs consist of the cost types

- Imputed interest for the temporarily stored products (cost type main group: capital costs) and
- Imputed interest and calculatory depreciation for the storage area and the storage equipment (cost type main group: capital costs)

The storage and capital costs rate [€/year] is used for the calculation multiplied by the retention period of every product (reference parameter) in the production system.

The different conditions of the resources resulting from the activation and deactivation of the resources are depicted in figure 16.4.

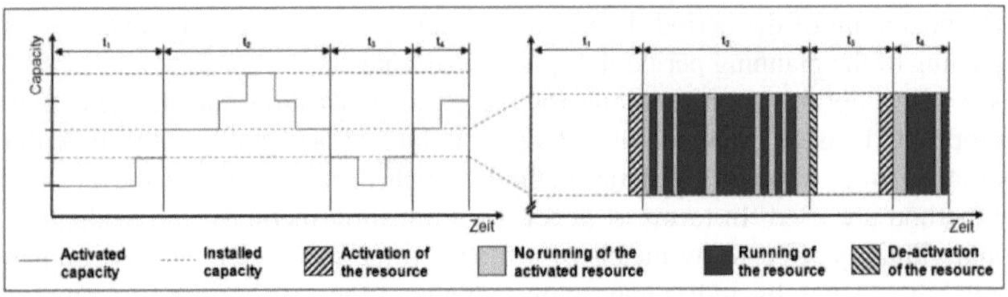

Fig. 16.4 Conditions of the resources in the context of the quantitative flexibility of the production systems.

The different cost items of the resources are derived from the different states of the resources. For example, a frequent firing (start-up) and cooling (shut down) of an oven leads to high energy costs. In this case it should be taken into account if a permanent activation is advantageous. Concerning the different states of the resources, the following cost types are:

Costs for Activating and De-activating

The costs for the activation consist of the following cost types:

- Lubricants: commodities for the production not included in the product (cost type main group: material costs),
- Energy costs (cost type main group: material costs),
- Maintenance (cost type main group: costs for procured services) and
- Direct and indirect labor costs, wage, statutory and voluntary social charges (cost type main group: personnel costs and social cost)

The cost rate for the activation [€/h], the activating time (reference parameter) of the resource type, and the amount of the activations (reference parameter) is used for the calculation of the activation.

Additionally, analogical to the activation of the resources, special costs can result from the de-activation of resources. The cost rate for the de-activation [€/h], the de-activating time (reference parameter) of the resource type, and the amount of the de-activations (reference parameter) is used for the calculation of the de-activation.

Fixed Costs of the Resources

The fixed costs of the resources consists of the following cost types:

- Imputed interest, calculatory depreciations (cost type main group: capital costs) and
- Maintenance (cost type main group: costs for procured services)[5]

The calculation of the fixed costs of the resources is based on the fixed costs rate of the resources [€/hour] multiplied by the reference period or the planning period (reference parameter).

Operational Readiness Costs

The costs for the operational readiness occurs during the period between activating and de-activating resources, if no processing of products takes place, and consists of, among others the following cost types:

- Energy costs (cost type main group: material costs) and
- Maintenance (cost type main group: costs for procured services)[6]

The calculation of the operational readiness costs of the resources is based on the cost rate of the operational readiness of the resources multiplied by the time of the operational readiness of the resources (reference parameter).

Variable Costs of the Resources

The variable costs of the resources occurs during the processing of the products and consist of the cost types

- Lubricants: commodities of the production not included in the product (cost type main group: material costs)[7] ,
- Energy costs (cost type main group: material costs) and
- Maintenance (cost type main group: costs for procured services)[8]

The calculation of the variable costs of the resources is based on the variable cost rate of the resources multiplied by the process time of the resources (reference parameter).

The focus of this article is the dimensioning of machines and plants. The dimensioning of the personnel could generally be pursued by means of the developed concepts; however this is not examined in this report. This is why the personnel and social costs (labor costs) are not explicitly considered. The same applies for the dimensioning of the areas and the resulting costs.

---

5 The maintenance costs are assigned to the fixed costs of the resources as the maintenance or the servicing happens independently from the use of the resources.

6 The maintenance costs are assigned to the operational readiness costs of the resources, if the maintenance or the servicing happens independently from the period of the operational readiness of the resource.

7 E. g. tool wear, auxiliary and operating materials.

8 The maintenance costs are assigned to the variable costs of the resources, if the maintenance or the servicing happens dependently from the period of the use of the resource.

Default Costs

Default costs occur if the given delivery dates are not kept. The calculation of the default costs is based on the product-related default cost rate multiplied by the time of the delayed delivery of the products.

For the evaluation of the efficiency and as a fitness value the total proceeds are used resulting from the sales revenue of the total of the produced products less all costs.

16.3.1.2 Systems for the State-Controlled Dynamic Dimensioning of Productions Systems in Terms of Varying Capacity Requirements

For the realization of the concept of the state-controlled dynamic dimensioning of production systems in terms of varying capacity requirements, a system was designed and realized by means of software. The core of this system is the database including all the project and master data. The database underlies a data model describing the production system with its essential criteria. Following definition given by SCHMIGALLA [16.26], a data model has to include the element set, the processes, and the structures of the production system. Furthermore, the boundary structures (meaning the input and output of the production system) must be described. Besides the description of the production system and its' boundary structures, the database also includes the actuating and command variables in order to enable the evaluation of the variants based on the evaluation method, by means of an evaluation component. The realization of the database happens in MS Excel®.

Based on the database, an automatic generation of the production system model will take place. Thereby, the expenses for the development of the model can be reduced, and the method can be accessed as well by people unfamiliar with the simulation and a flexible usage can be guaranteed. The production system model includes all the model components which the production system describes in an appropriate degree of abstraction. The primary support field of the work is the planning stage of the dimensioning. Based on the defined production methods and process and the chosen resources, the developed concept aims at determining knowledge for the necessary amount and the resulting costs. The focus of this work does not lie in the planning of the structure of the production system. Thus, the physical structure of the production system or the layout is not depicted. The depiction of transport processes can therefore only be affected by means of defined transition periods. The simulation system used is the software Plant Simulation® by the company Siemens Product Lifecycle Management Software GmbH.

The management of the simulation experiment is the task of an optimizer automatically performing (based on the optimization method) the selection of the variants, meaning the parameterizing and the selection of appropriate decision rules. The optimizer used is the optimizing tool GAWizard which is integrated into the Software Plant Simulation®. The optimization thereby takes place by means of genetic algorithms. The dimensioning requires a dimensioning component to be realized, based on the state-controlled dimensioning method, corresponding to the stored decision rules the interpretation of capacities on the decision points.

The results of the dimensioning are stored in the database and added to the variant evaluation. Figure 16.5 outlines the system for the dynamic dimensioning of production systems in terms of varying capacity requirements.

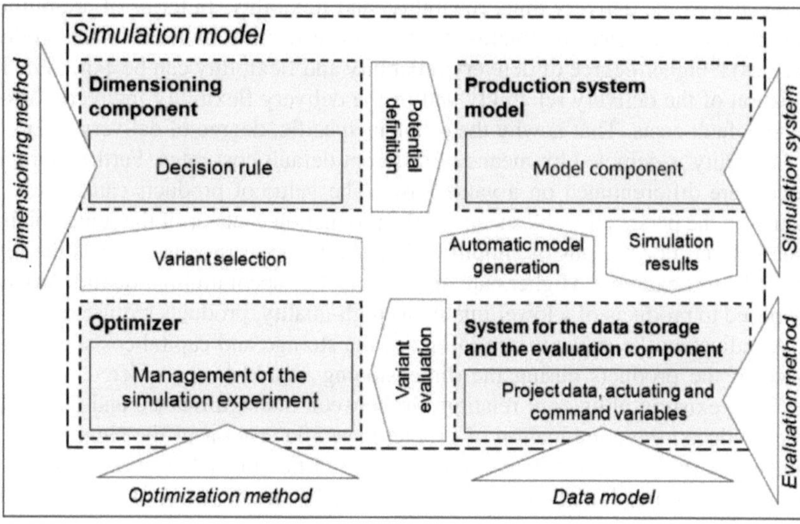

Fig. 16.5 Systems for the dynamic dimensioning of production systems in terms of varying capacity requirements in the context of the production management.

16.3.1.3 Use of the Concept of State-Controlled Dynamic Dimensioning of Production Systems in Terms of Varying Capacity Requirements

The test database is based on the problem developed by FISHER&THOMPSON ([16.5], p. 225-251) and often used in literature for comparative and test purposes. The FISHER&THOMPSON database is characterized by its quadratic structure. This includes 10 orders, each with 10 process steps occurring on 10 means of production in a different sequence, which have to be executed. Therefore, every order is processed exactly one time on one resource type. The text example on the basis of the FISHER&THOMPSON database is a basic model of a production system producing 10 different product types each with 10 different process types and 10 different resource types. Based on a given production program, the production systems should be designed in a capacitive way. The production program defines the production output, the order lot size and the reference period. Regarding the simplification, the same minimal varying order lot size has been determined for all products.

For the examination of the use of the concept regarding different problems, different products, resources and system load curves are devised. The focus is not on the elaboration of a practice-identical test example, it is more about a general depiction of differences between planning objectives and their impact on the planning results.

A rough orientation for deriving different products are given by different strategy types of the strategic production management [16.38], such as (eg.) the premium strategy, the differentiation strategy, the cost leadership strategy or the least costly products' strategy. Adjusting the capacities helps influence non-technical quality criteria such as the delivery time, reliability, and flexibility. In terms of premium or differentiation strategies (in contrast to the cost leadership or least costly products strategies) a higher degree of delivery reliability and flexibility can be expected. The evaluation of the delivery reliability and of the delivery flexibility indirectly affects via the default costs. That is why the problem-specific, degree of delivery reliability and flexibility is depicted by means of different default cost rates. Furthermore, the products are differentiated on a value basis. The value of products can be set, depending on the perspective, by means of the production costs or of the price obtained on the market. For reasons of simplification, the value of products equals the revenues in the test example. Higher capital costs must be spent on high-quality products as opposed to products of a lower quality as high-quality, products require more capital. As adjusting the capacity can influence the storage and capital costs, the significance of the products during the dimensioning should be considered. Although there is no existing obligatory relationship between the significance and the quality of the products; the combination of problem-specific criteria of the significance of products and the demanded delivery reliability and flexibility have been omitted for reasons of simplification of the test design. Thus, the development of a design is distinguished into 3 product groups:

- High-quality products with high quality standards,
- Medium-quality products with medium quality standards,
- Low-quality products with low quality standards.

All ten products are parameterized every test series in a standardized way according to one of the three product groups.

For depicting different system load curves, different fluctuation types and amplitudes and frequencies of the time course of the changes are differentiated. The system load describes the production program to be finalized within the modeled production system in an objective way and according to deadlines. The system load data is subdivided into product and order data. For depicting different fluctuation types:

- increasing,
- decreasing and
- repeatedly fluctuating

system load curves are used. In terms of repeatedly fluctuating system load curves and regarding the fluctuation frequency there is a differentiation between fluctuations with:

- a high frequency (12 fluctuation cycles per reference period) and
- a low frequency (2 fluctuation cycles per reference period) and regarding fluctuation amplitude between fluctuations with:
- a low amplitude (+25% of the minimal load) and
- a high amplitude (+100% of the minimal load).

The system load curve is described according to concrete orders with defined order lot sizes and defined release dates. The production system continuously operates in 21 shifts[9] a week[10].

The technological processes and the qualitatively determined machines and plants[11] are the starting point for the dimensioning and have a significant impact on it.

This is why the work schedules and the processing time for the test planning is assumed to be invariant. Different types of resources must be considered when using the concept because of the high problem-specific variability of the resources. In addition to other numerous criteria, the resources' fixed costs are a big part of the quantitative shift of the resources. For example, a simple operating and stationary brick oven has low fixed costs due to considerably lower investment vs. a modern machine which has high fixed costs due to considerably higher investment. In the context of the examinations, the following resources have to be differentiated:

- low fixed costs
- medium fixed costs and
- high fixed costs

The expense of powering on and off resources influences their operating strategies. For instance, firing the oven costs more than turning on the modern machine. For this reason, there is a differentiation between:

- high cost rates for the activation and de-activation,
- medium cost rates for the activation and de-activation and
- low cost rates for the activation and de-activation

and between resources with:

- long activation and de-activation times,
- medium activation and de-activation times and
- short activation and de-activation times.

Another factor influencing regarding the selection of appropriate strategies for activating and de-activating capacities is the expense of maintaining the operational readiness between the activation and de-activation of the resources. Therefore, in the context of the examinations, it can be distinguished between:

- high costs for the operational readiness,
- medium costs for the operational readiness and
- low costs for the operational readiness.

As all the products of the production program have to be processed, and as there are no technological alternatives for the single production steps, the expenses for

---

[9] 8 hours per shift.

[10] Primarily for reasons of simplification of the problem it has been determined that there is a continuously operating production system.

[11] Resources.

processing the products and their derived variable costs do not influence the capacity shift and the determination of appropriate operating strategies. That is why a standardized variable cost rate has been determined for all the resources and all the tests.

The different problems for examining the use of the concept have been derived from the variability of the system load curves, the products and the resources. In order to keep the test design manageable, the derived problems are limited to selected parameter combinations combining the maximum and minimum but also the medium parameter specifications.

16.3.1.4 Execution and Evaluation

During the examinations the three described possibilities are used for the nine possible resource states, as shown in table 1 on page 9, per concept, so that there are 19683 possible decision rule combinations.

The value range of the parameters has been defined as follows:

Interval length t_{sz}:

value range : 1-10 days
step range: 1 day

Backlog of the shift t_{sb}:

value range : 6hrs -240hrs
step range: 6 hrs

Amount of resources at the beginning of the planning periods n_{rapb}

value range : 1-6 resources per type of resource
step range: 1 resource

Activated amount of resources at the beginning of the planning periods n_{raa}

value range : 1-6 resources per type of resource
step range: 1 resource

Thus, 7.085.880 variants result from the variant genesis. The variant selection, the dimensioning and the evaluation of the variant takes place within a conceptual system in the framework of the test execution.

A simulation of a variant takes 2-4 minutes, so the solution space cannot be calculated in total. Therefore, the optimizer helps to find appropriate solutions in a justifiable time. In the context of the examinations a generation size of 20 individuals12 and an amount of 20 generations is used. Thereby 390 variants can be examined. Figure 16.6 depicts a typical optimization process. It becomes clear, that the expense considerably increases.

[12] 10 paternal individuals and 10 maternal individuals.

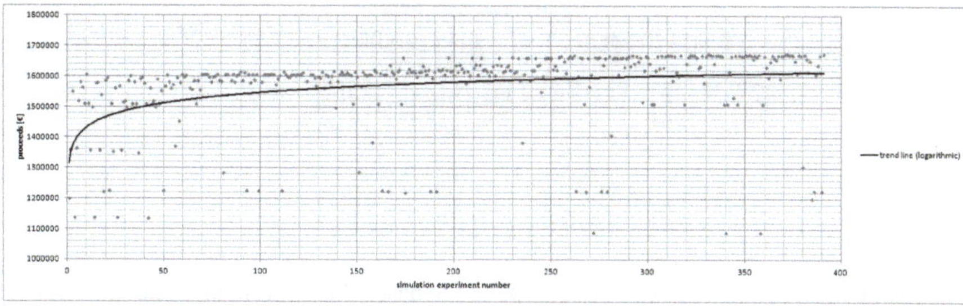

Fig. 16.6 Typical optimization process.

The simulation tests with the optimizer have shown that very different solutions can be found for the different problems. The differences are the calculated amount of resources and the activation and de-activation frequency of the resources. This is clarified by four different problems[13]. The four problems have a repeatedly low-frequent varying system load curve with high amplitude as shown in figure 16.7. The statistical calculation of the medium resource requirements according to formula (1) for this system load results in approx. two resources per resource type.

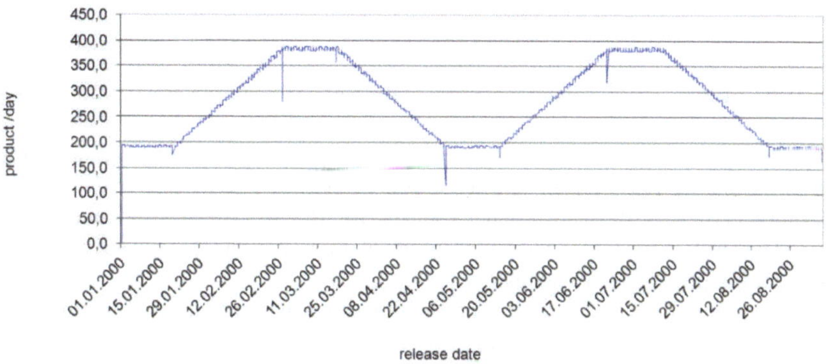

Fig. 16.7 repeatedly low-frequent varying system load curve with a high amplitude.

Problem 1 is characterized by low production costs, a short activation time, low activation costs, low fixed costs and high capacity costs. Figure 16.8 shows a circuit profile of one resource type of the best solution.

In the best solution from problem 1, each of three resources of a resource type are installed, whereas two resources are often activated and de-activated. The reasons for this frequent activation and de-activation of the resources are due to the low activation costs and the high capacity costs.

[13] The low, medium and high parameterizations respectively differ by the factor 10.

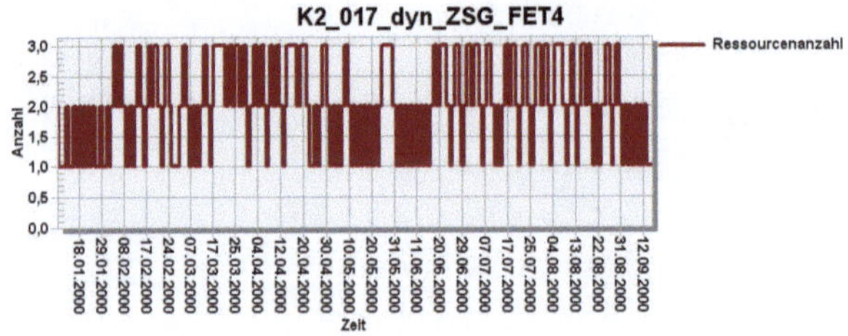

Fig. 16.8 Circuit profile of one resource type of the best solution from problem 1.

Problem 2 is characterized by low production costs, short activation times, low activation costs, high fixed costs and high capacity costs. Figure 16.9 shows a circuit profile of one resource type of the best solution.

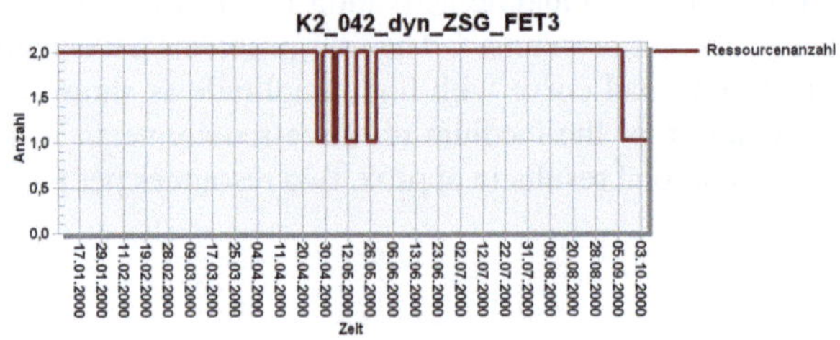

Fig. 16.9 Circuit profile of one resource type of the best solution from problem 2.

In contrast to problem 2, only two resources per resource type are scheduled due to the higher fixed costs. The reduction of the fixed costs turn out to be considerably higher as the additional expenses for the storage, capital and default costs. The shift of resources is so high in problem 2, that hardly any activation and de-activation processes are necessary.

Problem 3 is based on medium production cost, medium activation times, medium activation costs, medium fixed costs and medium capacity costs. Figure 16.10 shows a circuit profile of one resource type of the best solution.

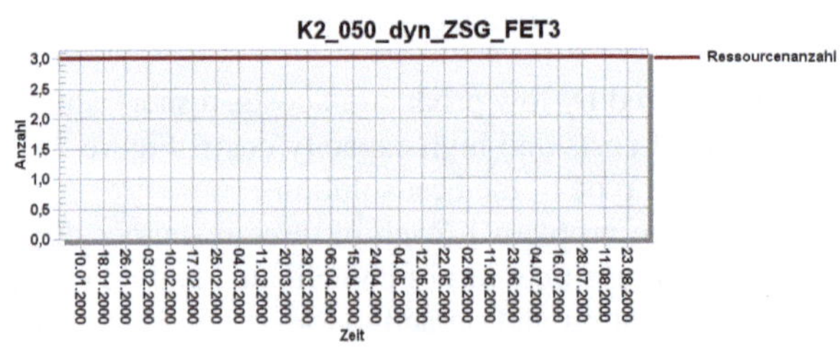

Fig. 16.10 Circuit profile of one resource type of the best solution from problem 3.

A continuous process of three resources per resource type is the best compromise for problem 3 between storage and capital costs, activation costs, fixed costs and capacity costs.

Problem 4 is characterised by high production costs, long activation times, high activation costs, low fixed costs and low capacity costs. Figure 16.11 shows a circuit profile of one resource type of the best solution.

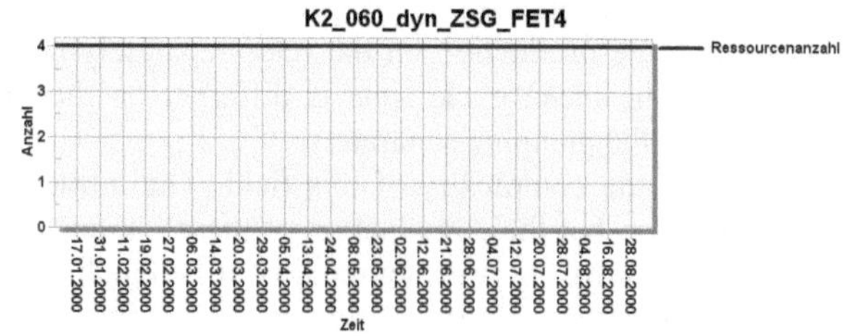

Fig. 16.11 Circuit profile of one resource type of the best solution from problem 4.

In contrast to problem 3, four resources per resource type are continuously operating in problem 4. Therefore, compared to the static calculated solution based on two resources per resource type, the system is strongly over-dimensioned. On closer examination of the variants with a small amount of resource (problem 3) it can be noticed that the storage and capital costs are considerably higher than the cost savings of the fixed costs. For problem 4 it is more effective to provide a higher amount of resources in order to minimise backlogs and processing times.

Besides the four depicted problems, 102 different problems in total have been examined. Therefore it is shown that the presented concept could somewhat give better solutions than the static solution method.

16.3.2 Case Study 2: Order Controlling in Engine Assembly with the Aid of Optimisers (by János Jósvai)

16.3.2.1 Production System and Its Complexity

Today the production tasks have got a very complex planning process. This is caused by the high amount of variants of one product. We can speak here about a vehicle or engine production. Most of the production structures are established as lines and have the task to produce several product types and several variants of the products. This means a very difficult planning and execution of production. The establishment of the production program is complicated, the times of work tasks are different, and the material delivery on the line and the inventory has to be taken into consideration, too.

The production planning has several goals, some of them are:

- the scheduling of the tasks to ensure delivery accuracy,
- to determine the lot size of product batches,
- to ensure smoothed workloads at the workplaces,

- to determine the buffer sizes in the production line,
- to handle the lead times – depending on the complexities of the products,
- to determine and handle the bottlenecks – can change with the system dynamic behaviour, etc.

Mostly the production system is not configured as a whole integrated line. To plan a system, which is separated by buffers between two or perhaps three main lines, has a lot of influential parameters. The main question is either to plan these part lines together, or to plan the production on the lines separately because of some reasons. For example if the mean cycle time is different on the lines then this could be a reason to make the planning separately.

These properties show the complexity of this field. The influence parameters are not only a large number, but the combination of these parameters causes a lot of option and problems to solve. In practice there is not enough time to fulfil the mathematical analysis manually, even if the right behaviour functions are ready to use.

There is another possible method which is useful to plan such complex systems. The modelling and dynamic simulation is able to answer most of the questions, and show the time dependent behaviour of the concerned production system. This modelling technique is the time discrete event controlled simulation.

The following sections would like to show and describe the modelling steps of a complex production system with a lot of products and three different line parts, which are connected with buffers.

16.3.2.2 Problem Definition

The considered production system was an engine production line with three separated line parts. These were connected by buffers. The simulation model and study had to investigate, how the line output, usage statistics changes with the different production sequences.

The product mix changes time to time, this had many influences and plus tasks while the planning of the model. We will see how it works when a product has to be changed in the model. This could mean for instance the end of production of one product type, or new type has to be launched on the line. This data handling procedure and the amount of handled data causes a great model size.

The modelling had to consider, that a lot of flexible parameters were needed to ensure enough planning roam. Lot size determination had to be fixed, that the actual pre-planned production program could be changed and set on new levels by the simulation.

Another main goal was to determine the computational achievable "right" production sequence. The hand-made production program should be optimized by the simulation. A genetic evolution algorithm was used to solve this difficult problem with a large search area.

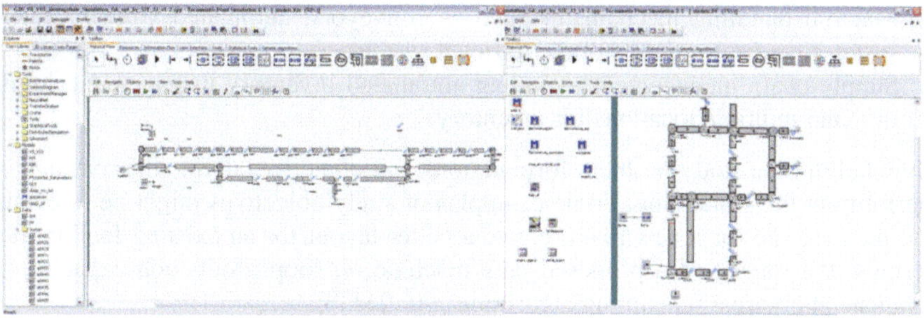

Fig. 16.12 Simulation models of the line parts

For planning the line balancing there was needed an option, to ensure handling functionality, when workload change has to be planned. The mounting tasks can be assigned to various places in the line. This means that the variation of workloads at the stations in the line has a large number. The line balancing has the goal to put the tasks in the right order after each other and approximately hold the average cycle time at one station. In case of production changes - product type, produced volume, technological, and production base time – there was a need to precalculate the changed line behaviour. There are different changes in the task load of the stations, we make such influences which determine the throughput, working portion of the stations and gives different optimal sequence combination of products.

16.3.2.3 Simulation and Scheduling

There are similarities and differences as well between general research- and simulation case studies. Simulation case studies are typically focused on finding answers to questions through simulation-based experiments. In the social science area, experimentation is considered to be a distinct research method separate from the case study. Social science case study researchers use observation, data collection, and analysis to try to develop theories that explain social phenomena and behaviours. Simulation analysts use observation and data collection to develop "as-is" models of manufacturing systems, facilities, and organizations. The analysts test their theories and modifications to those models through simulation experiments using collected data as inputs. Data sets may be used to exercise both "as-is" and "to-be" simulation models. Data sets may also be fabricated to represent possible future "to-be" conditions, e.g., forecast workloads for a factory. [16.21]

In [16.29], teaching simulation through the use of manufacturing case studies is discussed. He organizes case studies into four modules:

- Basic manufacturing systems organizations, such as work stations, production lines, and job shops.
- System operating strategies including pull (just-in-time) versus push operations, flexible manufacturing, cellular manufacturing, and complete automation.

- Material handling mechanisms such as conveyors, automated guided vehicle systems, and automated storage/retrieval systems.
- Supply chain management including automated inventory management, logistics, and multiple locations for inventory.

Simulation case study problem formulations and objectives define the reasons for performing the simulation. Some examples of study objectives might be to evaluate the best site for a new plant, create a better layout for an existing facility, determine the impact of a proposed new machine on shop production capacity, or evaluate alternative scheduling algorithms. [16.21]

Simulation textbooks typically recommend that a ten to twelve step process be followed in the development of simulation case studies. The recommended approach usually involves the following steps (Fig. 16.13): 1. problem formulation, 2. setting of objectives and overall project plan, 3. model conceptualization, 4. data collection, 5. model translation into computerized format, 6. code verification, 7. model validation, 8. design of experiments to be run, 9. production runs and analysis, 10. documentation and reporting, and 11. implementation [16.1].

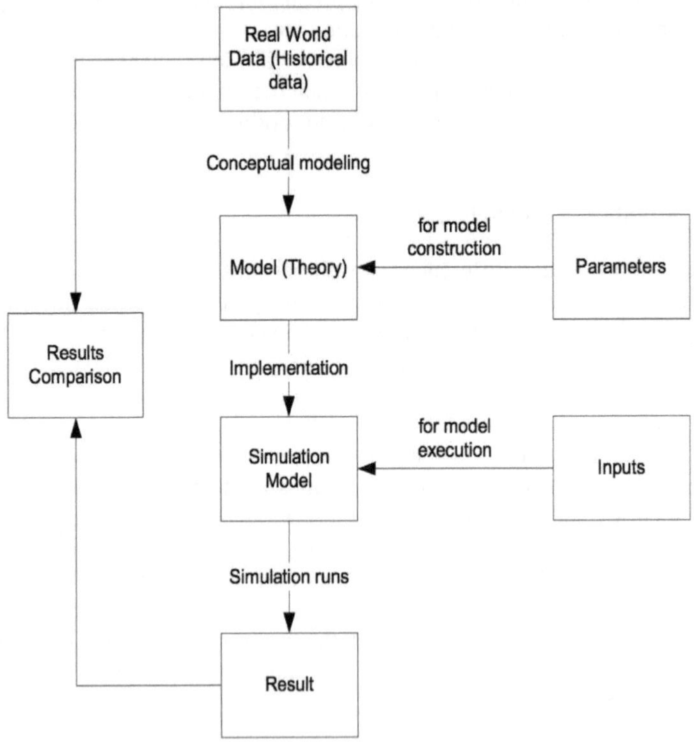

Fig. 16.13 Simulation modelling and executing steps [16.28].

What Is Manufacturing Simulation?

"...the imitation of the operation of a real-world process or system over time. Simulation involves the generation of an artificial history of the system and the observation of that artificial history to draw inferences concerning the operational characteristics of the real system that is represented. Simulation is an indispensable problem-solving methodology for the solution of many real-world problems. Simulation is used to describe and analyze the behaviour of a system, ask what-if questions about the real system, and aid in the design of real systems. Both existing and conceptual systems can be modelled with simulation." [16.1]

Manufacturing simulation focuses on modelling the behaviour of manufacturing organizations, processes, and systems. Organizations, processes and systems include supply chains, as well as people, machines, tools, and information systems. For example, manufacturing simulation can be used to:

- Model "as-is" and "to-be" manufacturing and support operations from the supply chain level down to the shop floor
- Evaluate the manufacturability of new product designs
- Support the development and validation of process data for new products
- Assist in the engineering of new production systems and processes
- Evaluate their impact on overall business performance
- Evaluate resource allocation and scheduling alternatives
- Analyze layouts and flow of materials within production areas, lines, and workstations
- Perform capacity planning analyses
- Determine production and material handling resource requirements
- Train production and support staff on systems and processes
- Develop metrics to allow the comparison of predicted performance against "best in class" benchmarks to support continuous improvement of manufacturing operations [16.20]

Genetic Algorithms

An implementation of a genetic algorithm begins with a population of (typically random) chromosomes. One then evaluates these structures and allocates reproductive opportunities in such a way that those chromosomes which represent a better solution to the target problem are given more chances to reproduce than those chromosomes which are poorer solutions.

The goodness of a solution is typically defined with respect to the current population. This particular description of a genetic algorithm is intentionally abstract because in some sense, the term genetic algorithm has two meanings. In a strict interpretation, the genetic algorithm refers to a model introduced and investigated by John Holland [16.10] and by students of Holland (e.g., DeJong [16.2]). It is still the case that most of the existing theory for genetic algorithms applies either solely or primarily to the model introduced by Holland, as well as variations on what will be referred to in this paper as the canonical genetic algorithm. Recent theoretical advances in modelling genetic algorithms also apply primarily to the canonical genetic algorithm [16.34].

In a broader usage of the term, a genetic algorithm is any population-based model that uses selection and recombination operators to generate new sample points in a search space. Many genetic algorithm models have been introduced by researchers largely working from an experimental perspective. Many of these researchers are application oriented and are typically interested in genetic algorithms as optimization tools. [16.18]

The use of genetic algorithms requires five components:

1. A way of encoding solutions to the problem - fixed length string of symbols.
2. An evaluation function that returns a rating for each solution.
3. A way of initializing the population of solutions.
4. Operators that may be applied to parents when they reproduce to alter their genetic composition such as crossover (i.e. exchanging a randomly selected segment between parents), mutation (i.e. gene modification), and other domain specific operators.
5. Parameter setting for the algorithm, the operators, and so forth. [16.13]

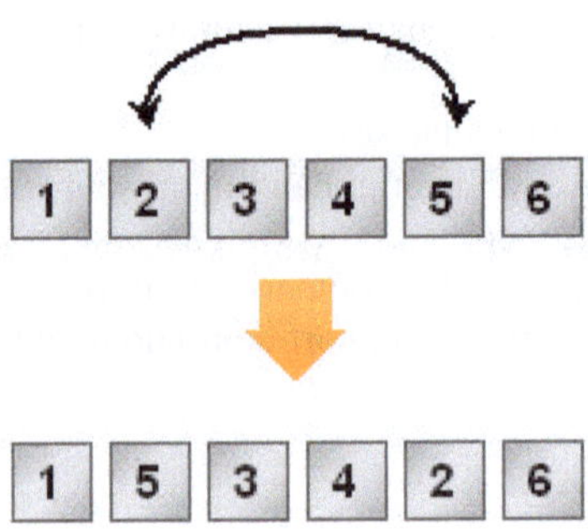

Fig.16.14 Mutation for a sequential task [16.30]

The simulation model uses the genetic algorithm for a sequential task. The logic to produce a new population is shown on Figure 16.14. Several test runs were made in order to identify the right settings of the algorithm. The statistical operators were configured after real life data test runs, to make the algorithm converge faster. The runs showed at last, that the population size has to be set to 10 and the simulated generations' numbers were 20. This was a main question among others, because the simulation running time was limited up to one and half an hour.

Scheduling

Scheduling has been defined as the art of assigning resources to tasks in order to insure the termination of these tasks in a reasonable amount of time. The general problem is to find a sequence, in which the jobs (e.g., a basic task) pass between the resources (e.g., machines), which is a feasible schedule, and optimal with respect to some performance criterion. A functional classification scheme categorizes problems using the following dimensions:

1. Requirement generation,
2. Processing complexity,
3. Scheduling criteria,
4. Parameter variability,
5. Scheduling environment.

Based on requirements generation, a manufacturing shop can be classified as an open shop or a closed shop. An open shop is "build to order", and no inventory is stocked. In a closed shop the orders are filled from existing inventory.

Processing complexity refers to the number of processing steps and workstations associated with the production process. This dimension can be decomposed further as follows:

1. One stage, one processor
2. One stage, multiple processors,
3. Multistage, flow shop,
4. Multistage, job shop.

The one stage, one processor and one stage, multiple processors problems require one processing step that must be performed on a single resource or multiple resources respectively.

In the multistage, flow shop problem each job consists of several tasks, which require processing by distinct resources; but there is a common route for all jobs. Finally, in the multistage, job shop situation, alternative resource sets and routes can be chosen, possibly for the same job, allowing the production of different part types.

The third dimension, scheduling criteria, states the desired objectives to be met. They are numerous, complex, and often conflicting. Some commonly used scheduling criteria include the following:

1. Minimize total tardiness,
2. Minimize the number of late jobs,
3. Maximize system/resource utilization,
4. Minimize in-process inventory,
5. Balance resource usage,
6. Maximize production rate.

The fourth dimension, parameters variability, indicates the degree of uncertainty of the various parameters of the scheduling problem. If the degree of uncertainty is insignificant, the scheduling problem could be called deterministic. For example, the expected processing time is six hours, and the variance is one minute. Otherwise, the scheduling problem could be called stochastic.

The last dimension, scheduling environment, defined the scheduling problem as static or dynamic. Scheduling problems in which the number of jobs to be considered and their ready times are available are called static. On the other hand, scheduling problems in which the number of jobs and related characteristics change over time are called dynamic. [16.14]

According to the previous classification the modelled system can be classified as:

- Open shop
- Multistage, flow shop
- The processing times are treated as deterministic
- Job characteristic is dynamic

16.3.2.4 Modeling and Simulation Runs

This model is a planning tool which is able to answer several questions of the complex production planning. The creation of the model followed the physical parameters of the real system. The iteration process of the modelling was difficult because it had to handle the product mounting time. The mounting times were gained from the real production system, but the collection and filtering was made inside the simulation model, to prepare the data ready for production inside the simulation.

Model Building

Plant Simulation provides a number of predefined objects for simulating the material flow and logic in a manufacturing environment. There are five types of main object groups from Plant Simulation:

- Material flow objects: Objects used to represent stationary processes and resources that process moving objects.
- Moving objects: Objects used to represent mobile material, people and vehicles in the simulation model and that are processed by material flow objects. Moving objects are more commonly referred to as MUs.
- Information flow objects: Objects used to record information and distribute information among objects in the model.
- Control objects: Objects inherently necessary for controlling the logic and functionality of the simulation model.
- Display and User interface objects: Objects used to display and communicate information to the user and to prompt the user to provide inputs at any time during a simulation run.

SimTalk is the programming language of Plant Simulation; it was specifically developed for application in Plant Simulation models. The Method objects are used to dynamically control and manipulate models. SimTalk programs are written inside method objects and executed every time the method is called during a simulation run.

The logical structure of the model was created on basis of Plant Simulation provided level structure. So it was a “simple” planning step to divide the model into specified functional levels. Different folders and frames are used in order to implement the line structure, the data handling for manufacturing programs and the basic data for the manufactured products. However, the scheduling of the production program has its own separate level.

The data input and output of the model work with the Excel Interface of Plant Simulation. Users can manipulate the parameter settings and see the results of the simulation runs on this easy way independently from Plant Simulation – no special simulation knowledge is asked.

User interface has been implemented for the model in order to handle the simulation model and the several built-in functions, which are to test the simulated line behaviour. This handling tool, helps the manufacturing engineer to plan tasks and solve rescheduling problems on the line.

Model Validation and Verification

Validation and verification of the model is formulated as follows: Model validation: process of demonstrating that a model and its behaviour are suitable representations of the real system and its behaviour w.r.t. intended purpose of model application.

Model verification: process of demonstrating that a model is correctly represented and was transformed correctly from one representation form into another, w.r.t. transformation and representation rules, requirements, and constraints. [16.24]

There are many techniques to validate and verify the model. The physical environment has high influence on the method which is adaptable to verify and validate the model. In this particular case together with experts from the enterprise a structured walkthrough was possible to use for this system model. For special throughput data of the line it was possible to make historical data validation.

Simulation Runs and Results

The regular use of the simulation was secured with the several setting function, among them the line speed, the different value setting of the palettes on the separated lines, lot size limitations, and daily production program definition function. The simulation test runs with manufacturing data brought the following most important results:

- The simulation model is capable for everyday usage.
- To bring more efficiency 2-3 days are to be handled with the rescheduling algorithm.
- It is able to reduce lead time with 1-10%, this depends on product mixtures.

The simulation model building and the test runs at the enterprise show that the simulation technique is suitable for the manufacturing planning. The model and the line connection mean in this case that the real data application could be made much better. This depends on both sides; the model structure has to be modified if the physical system is able to give over real time data. In this matter the rescheduling and the simulation tool could be not only the planning tool, but also it would be the production control tool.

Authors Biography, Contact

Since 2002, **Prof. Dr.-Ing. Egon Müller** is head of the Department of Factory Planning and Factory Management, Chemnitz University of Technology. Among others, he is actively involved in HAB (Scientific society of Industrial Management), GfSE (Society for Systems Engineering), VDI (Association of German Engineers) - Technical Division "Factory Planning", VDI (Association of German Engineers) - District Executive Board Chemnitz, Candidate Fellow of AIM (European Academy on Industrial Management), SoCol net member, Reviewer of the journal "Production Planning and Control" and Reviewer of ICPR.

Andreas Krauss studied industrial engineering at the Chemnitz University of Technology. Since 2005 he is working in the Department of Factory Planning and factory Management at Chemnitz University of Technology. He is a PhD candidate and specialized in production planning, simulation und virtual reality.

Contact

Andreas Krauß
Professur für Fabrikplanung und Fabrikbetrieb
Technische Universität Chemnitz
D-09107 Chemnitz
Germany

János Jósvai is working at the Széchenyi István University, Győr, Hungary. He is a PhD candidate, his field is manufacturing planning and simulation methods. He has several year experiences in material flow simulation of manufacturing systems and in production process planning. In the field of research and development he spent significant time aboard with international cooperation in matter of digital factory.

References

[16.1] Banks, J. (ed.): Handbook of simulation, Principles, Methodology, Advances, Application and Practice. JohnWiley & Sons Inc., Atlanta (1998)

[16.2] De Jong, K.: An Analysis of the Behavior of a Class of Genetic Adaptive Systems. PhD Dissertation. Dept. of Computer and Communication Sciences. Univ. of Michigan, Ann Arbor (1975)

[16.3] Dombrowski, U., Herrmann, C., Lacker, L., Sonnentag, S.: Modernisierung kleiner und mittlerer Unternehmen - Ein ganzheitliches Konzept. Springer, Heidelberg (2009)

[16.4] Domschke, W.: Modelle und Verfahren zur Bestimmung betrieblicher und innerbetrieblicher Standorte - ein Überblick. Zeitschrift für Operation Research Heft 19, S13–S41 (1975)

[16.5] Fisher, H., Thompson, G.L.: Probabilistic Learning Combinations of Local Job-Shop Scheduling Rules. In: Muth, J.F., Thompson, G.L. (eds.) Industrial Scheduling, pp. 225–251. Prentice-Hall, Englewood Cliffs (1963)
[16.6] Grundig, C.-G.: Fabrikplanung - Planungssystematik - Methoden - Anwendungen. Carl Hanser Verlag, München (2009)
[16.7] Gudehus, T.: Logistik Grundlagen Strategien Anwendungen. Springer, Berlin (1999)
[16.8] Günther, H.-O., Tempelmeier, H.: Produktion und Logistik. Springer, Heidelberg (2005)
[16.9] Hader, S.: Ein hybrider Ansatz zur Optimierung technischer Systeme. Dissertation, Technische Universität Chemnitz, Chemnitz (2001)
[16.10] Holland, J.: Adaptation in Natural and Artifical Systems. University of Michigan Press (1975)
[16.11] Hopp, W.J., Spearman, M.L.: Factory Physics. McGraw-Hill, Boston (2008)
[16.12] Horbach, S.: Modulares Planungskonzept für Logistikstrukturen und Produktionsstätten kompetenzzellenbasierter Netze. Wissenschaftliche Schriftenreihe des IBF, Heft 70, Chemnitz (2008)
[16.13] Jones, A., Riddick, F., Rabelo, L.: Development of a Predictive-Reactive Scheduler Using Genetic Algorithms and Simulation-based Scheduling Software, National Institute of Standards and Technology, Ohio University (1996), `http://www.nist.gov` (accessed May 18, 1996)
[16.14] Jones, A., Rabelo, L.: Survey of Job Shop Scheduling Techniques, National Institute of Standards and Technology, California Polytechnic State University (1998), `http://www.nist.gov` (accessed May 18, 2009)
[16.15] Käschel, J., Teich, T.: Produktionswirtschaft - Band 1: Grundlagen, Produktionsplanung und -steuerung. Verlag der Gesellschaft für Unternehmensrechnung und Controlling m.b.H., Chemnitz (2007)
[16.16] Kobylka, A.: Simulationsbasierte Dimensionierung von Produktionssystemen mit definiertem Potential an Leistungsflexibilität. Wissenschaftliche Schriftenreihe des IBF, Heft 24, Chemnitz (2000)
[16.17] Kuhn, A., Tempelmeier, H., Arnold, D., Isermann, H.: Handbuch Logistik. Springer, Berlin (2002)
[16.18] Kühn, W.: Digitale Fabrik - Fabriksimulation für Produktionsplaner. Wien, Hanser (2006)
[16.19] März, L., Krug, W., Rose, O., Weigert, G.: Simulation und Optimierung in Produktion und Logistik - Praxisorientierter Leitfaden mit Fallbeispielen. Springer, Heidelberg (2011)
[16.20] McLean, C., Leong, S.: The Role of Simulation in Strategic Manufacturing, Manufacturing Simulation and Modeling Group National Institute of Standards and Technology (2002), `http://www.nist.gov` (accessed May 18, 2009)
[16.21] McLean, C., Shao, G.: Generic Case Studies for Manufacturing Simulation Applications, National Institute of Standards and Technology (2003), `http://www.nist.gov` (accessed May, 18 2009)
[16.22] Nyhuis, P., Reinhart, G., Abele, E.: Wandlungsfähige Produktionssysteme - Heute die Industrie von morgen gestalten. Impressum Verlag, Hamburg (2008)
[16.23] Pfeiffer, A.: Novel Methods for Decision Support in Production Planning and Control. Thesis (PhD), Budapest University of Technology and Economics (2007)
[16.24] Rabe, M., Spieckermann, S., Wenzel, S.: Verifikation und Validierung für die Simulation in Produktion und Logistik. Springer, Berlin (2008)

[16.25] Schenk, M., Wirth, S.: Fabrikplanung und Fabrikbetrieb. Methoden für die wandlungsfähige und vernetzte Fabrik. Springer, Berlin (2004)

[16.26] Schmigalla, H.: Fabrikplanung - Begriffe und Zusammenhänge. Hanser-Verlag, München (1995)

[16.27] Schönsleben, P.: Integrales Logistikmanagement, Operations and Supply Chain Management in umfassenden Wertschöpfungsnetzwerken. Springer, Berlin (2007)

[16.28] Shao, G., McLean, C., Brodsky, A., Amman, P.: Parameter Validation Using Constraint Optimization for Modeling and Simulation, Manufacturing Simulation and Modeling Group, National Institute of Standards and Technology (2008), `http://www.nist.gov` (accessed May 18, 2009)

[16.29] Standridge, C.: Teaching Simulation Using Case Studies. In: Proceedings of the 32nd on Winter Simulation Conference, Orlando, Florida, USA, December 10-13, pp. 1630–1634 (2000)

[16.30] Tecnomatix Technologies Ltd, Tecnomatix Plant Simulation Help (2006)

[16.31] VDI 3633: VDI-Richtlinie Simulation von Logistik-, Materialfluss und Produktionssystemen - Grundlagen. Verein Deutscher Ingenieure. Blatt 1. Beuth-Verlag, Berlin (2010)

[16.32] VDI 3633: VDI-Richtlinie Simulation von Logistik-, Materialfluss und Produktionssystemen - Grundlagen. Verein Deutscher Ingenieure. Blatt 7. Beuth-Verlag, Berlin (2001)

[16.33] Vollmann, T.E., Berry, W.L., Whybark, D.C., Jacobs, F.R.: Manufacturing Planning and Control Systems for Supply Chain Management. McGraw-Hill, New York (2005)

[16.34] Vose, M.: Modeling Simple Genetic Algorithms. In: Whitley, D. (ed.) Foundations of Genetic Algorithms, vol. 2, pp. 63–73. Morgan Kaufmann (1993)

[16.35] Westkämper, E., Zahn, E.: Wandlungsfähige Produktionsunternehmen - Das Stuttgarter Unternehmensmodell. Springer, Heidelberg (2009)

[16.36] Whitley, D.: A Genetic Algorithm Tutorial. Statistics and Computing 4, 65–85 (1995)

[16.37] Wunderlich, J.: Kostensimulation - Simulationsbasierte Wirtschaftlichkeitsregelung komplexer Produktionssysteme. Dissertation, Universität Erlangen-Nürnberg, Erlangen (2002)

[16.38] Zäpfel, G.: Strategisches Produktions-Management. Wien, Oldenbourg (2000)

Author Index

Subject Index

D

E

F

S

T

U

V

W

X

MIX
Papier aus verantwortungsvollen Quellen
Paper from responsible sources
FSC® C105338

If you have any concerns about our products,
you can contact us on
ProductSafety@springernature.com

In case Publisher is established outside the EU,
the EU authorized representative is:
Springer Nature Customer Service Center GmbH
Europaplatz 3, 69115 Heidelberg, Germany

Printed by Libri Plureos GmbH
in Hamburg, Germany